# Methods in Enzymology

Volume 297
PHOTOSYNTHESIS: MOLECULAR BIOLOGY
OF ENERGY CAPTURE

# METHODS IN ENZYMOLOGY

EDITORS-IN-CHIEF

## John N. Abelson    Melvin I. Simon

DIVISION OF BIOLOGY
CALIFORNIA INSTITUTE OF TECHNOLOGY
PASADENA, CALIFORNIA

FOUNDING EDITORS

## Sidney P. Colowick and Nathan O. Kaplan

*Methods in Enzymology*

*Volume 297*

# Photosynthesis: Molecular Biology of Energy Capture

EDITED BY

## Lee McIntosh

DEPARTMENT OF ENERGY
MICHIGAN STATE UNIVERSITY
EAST LANSING, MICHIGAN

ACADEMIC PRESS

San Diego   London   Boston   New York   Sydney   Tokyo   Toronto

Academic Press
*a division of Harcourt Brace & Company*
525 B Street, Suite 1900, San Diego, California 92101-4495, USA
http://www.academicpress.com

Academic Press Limited
24-28 Oval Road, London NW1 7DX, UK
http://www.hbuk.co.uk/ap/

International Standard Book Number: 0-12-182198-6

PRINTED IN THE UNITED STATES OF AMERICA
98  99  00  01  02  03  MM  9  8  7  6  5  4  3  2  1

# Table of Contents

## Section I. Genetic Approaches to Dissect Complex Functions

## Section II. Photosynthetic Complexes: Function/Structure

## Section III. Gene Expression of Photosynthetic Components

## Section IV. Biogenesis and Adaptation of Photosynthetic Components

## Section V. Photosynthetic Mutants: Construction and Biological/Biochemical/Biophysical Analyses

# Contributors to Volume 297

Article numbers are in parentheses following the names of contributors.
Affiliations listed are current.

KLAUS APEL (16), *Institut für Pflanzenwissenschaften, CH 8092 Zürich, Switzerland*

GREGORY ARMSTRONG (16), *Institut für Pflanzenwissenschaften, CH 8092 Zürich, Switzerland*

ALICE BARKAN (4), *Institute of Molecular Biology, University of Oregon, Eugene, Oregon 97403*

SCOTT E. BINGHAM (21), *Department of Plant Biology, Arizona State University, Tempe, Arizona 85287-1601*

TERRY M. BRICKER (22), *Department of Biological Sciences, Louisiana State University, Baton Rouge, Louisiana 70803*

RICHARD K. BRUICK (13), *Department of Cell Biology, The Skaggs Institute of Chemical Biology, The Scripps Research Institute, La Jolla, California 92037*

YUPING A. CAI (1), *Department of Molecular and Cell Biology, University of California, Berkeley, California 94720-3206*

PARAG R. CHITNIS (8), *Department of Biochemistry and Biophysics, Iowa State University, Ames, Iowa 50011*

VAISHALI P. CHITNIS (8), *Department of Biochemistry and Biophysics, Iowa State University, Ames, Iowa 50011*

AMYBETH COHEN (13), *Department of Cell Biology, The Skaggs Institute of Chemical Biology, The Scripps Research Institute, La Jolla, California 92037*

MICHAEL F. COHEN (1), *Section of Microbiology, Division of Biological Sciences, University of California, Davis, California 95616*

JEFFREY A. CRUZ (9), *Department of Microbiology, Louisiana State University, Baton Rouge, Louisiana 70803*

FEVZI DALDAL (6), *Department of Biology, University of Pennsylvania, Philadelphia, Pennsylvania 19104-6018*

PAUL P. DIJKWEL (12), *Department of Molecular Genetics, John Innes Centre, Colney, Norwich NR4 7UH, United Kingdom*

BRUCE A. DINER (23), *Central Research and Development Department, Experimental Station, E. I. du Pont de Nemours and Company, Wilmington, Delaware 19880-0173*

DION G. DURNFORD (15), *Department of Biology, University of New Brunswick, Fredericton, New Brunswick, Canada E3B 6E1*

JESUS M. ERASO (10), *Department of Microbiology and Molecular Genetics, The University of Texas Medical School at Houston, Houston, Texas 77030*

JAN ERIKSSON (11), *Department of Biochemistry, The Arrhenius Laboratories, Stockholm University, S-10691 Stockholm, Sweden*

JEAN-MICHEL ESCOUBAS (15), *Défense et Résistance chez les Invertébrés Marins (DRIM), Université de Montpellier II, 34095 Montpellier, France*

PAUL G. FALKOWSKI (15), *Environmental Biophysics and Molecular Biology Program, Institute of Marine and Coastal Sciences, Rutgers University, New Brunswick, New Jersey 08901-8521*

GÜNTER FRITZSCH (5), *Max-Planck-Institut für Biophysik, D-60528 Frankfurt a.M., Germany*

HAILE GHEBRAMEDHIN (11), *Department of Biochemistry, The Arrhenius Laboratories, Stockholm University, S-10691 Stockholm, Sweden*

ix

JOHN H. GOLBECK (7), *Department of Bio-chemistry and Molecular Biology, The Pennsylvania State University, University Park, Pennsylvania 16802*

MARK GOMELSKY (10), *Department of Microbiology and Molecular Genetics, The University of Texas Medical School at Houston, Houston, Texas 77030*

ARTHUR R. GROSSMAN (19), *Department of Plant Biology, Carnegie Institution of Washington, Stanford, California 94305*

MARK HARKER (17), *Department of Genetics, Institute of Life Sciences, The Hebrew University of Jerusalem, Jerusalem, 91904 Israel*

JOSEPH HIRSCHBERG (17), *Department of Genetics, Institute of Life Sciences, The Hebrew University of Jerusalem, Jerusalem, 91904 Israel*

CHRISTER JANSSON (11), *Department of Biochemistry, The Arrhenius Laboratories, Stockholm University, S-10691 Stockholm, Sweden*

PING JIN (8), *Department of Agronomy, Iowa State University, Ames, Iowa 50011*

SAMUEL KAPLAN (10), *Department of Microbiology and Molecular Genetics, The University of Texas Medical School at Houston, Houston, Texas 77030*

AN KE (8), *Division of Biology, Kansas State University, Manhattan, Kansas 66506*

DAVID M. KEHOE (19), *Department of Plant Biology, Carnegie Institution of Washington, Stanford, California 94305*

KAREN L. KINDLE (3), *Plant Science Center, Cornell University, Ithaca, New York 14853*

HANS-GEORG KOCH (6), *Department of Biology, University of Pennsylvania, Philadelphia, Pennsylvania 19104-6018*

HYEONMOO LEE (21), *Department of Plant Biology, Arizona State University, Tempe, Arizona 85287-1601*

STEPHEN P. MAYFIELD (13), *Department of Cell Biology, The Skaggs Institute of Chemical Biology, The Scripps Research Institute, La Jolla, California 92037*

RICHARD E. MCCARTY (9), *Department of Biology, Johns Hopkins University, Baltimore, Maryland 21218*

LEE MCINTOSH (2), *MSU-DOE Plant Research Laboratory and Department of Biochemistry, Michigan State University, East Lansing, Michigan 48824*

JOHN C. MEEKS (1), *Section of Microbiology, Division of Biological Sciences, University of California, Davis, California 95616*

SABEEHA MERCHANT (18), *Department of Chemistry and Biochemistry, University of California, Los Angeles, Los Angeles, California 90095*

HANNU MYLLYKALLIO (6), *Department of Biology, University of Pennsylvania, Philadelphia, Pennsylvania 19104-6018*

ONDREJ PRASIL (15), *Laboratory of Photosynthesis, Institute of Microbiology, Academy of Sciences of Czech Republic, Trebon CZ 379 81, Czech Republic*

CINDY PUTNAM-EVANS (22), *Department of Biology, East Carolina University, Greenville, North Carolina 27858*

JEANETTE M. QUINN (18), *Department of Chemistry and Biochemistry, University of California, Los Angeles, Los Angeles, California 90095*

STEVEN RODERMEL (14), *Department of Botany, Iowa State University, Ames, Iowa 50011*

FRED ROOK (12), *Department of Molecular Genetics, John Innes Centre, Colney, Norwich NR4 7UH, United Kingdom*

GAZA SALIH (11), *Department of Biochemistry, The Arrhenius Laboratories, Stockholm University, S-10691 Stockholm, Sweden*

SJEF C. M. SMEEKENS (12), *Department of Molecular Cell Biology, University of Utrecht, 3584 CH Utrecht, The Netherlands*

JUN SUN (8), *Department of Biochemistry and Biophysics, Iowa State University, Ames, Iowa 50011*

WIM F. J. VERMAAS (20), *Department of Plant Biology, and Center for the Study of Early Events in Photosynthesis, Arizona State University, Tempe, Arizona 85287-1601*

ANDREW N. WEBBER (21), *Department of Plant Biology, Arizona State University, Tempe, Arizona 85287-1601*

RONNEY WIKLUND (11), *Department of Biochemistry, The Arrhenius Laboratories, Stockholm University, S-10691 Stockholm, Sweden*

C. PETER WOLK (1), *MSU-DOE Plant Research Laboratory, Michigan State University, East Lansing, Michigan 48824-1312*

JITUO WU (22), *Department of Plant Pathology, Louisiana State University, Baton Rouge, Louisiana 70803*

ALEXEI A. YELISEEV (10), *Department of Microbiology and Molecular Genetics, The University of Texas Medical School at Houston, Houston, Texas 77030*

CHRISTOPHER B. YOHN (13), *Developmental Genetics Program, Skirball Institute of Biomolecular Medicine, New York, New York 10016*

JIANPING YU (2), *MSU-DOE Plant Research Laboratory, Michigan State University, East Lansing, Michigan 48824*

JILL H. ZEILSTRA-RYALLS (10), *Department of Microbiology and Molecular Genetics, The University of Texas Medical School at Houston, Houston, Texas 77030*

# Preface

The sequence of a chloroplast gene encoding a protein was first published in 1980, not so very long ago. It was not a coincidence that the integration of molecular approaches in plants initially took place, to a large degree, in the area of photosynthesis. The depth and quality of the physiological, biochemical, and biophysical advances in photosynthesis made it one of the best targets for the "new" molecular approaches. We have now moved from studying "photosynthetic" genes to their modification and reinsertion into bacteria, cyanobacteria, algae, and higher plants. Integration of structure and function studies in photosynthetic research is the modern paradigm, no longer a new phenomenon. This *Methods in Enzymology* volume demonstrates how far we have come and makes it obvious that future work has many challenges but far fewer technical limitations.

My gratitude is extended to all the authors contributing to this volume and others not represented, but without whom it would not have been possible. I also extend my thanks to those at Academic Press for all their help and patience.

Lee McIntosh

# METHODS IN ENZYMOLOGY

VOLUME 73. Immunochemical Techniques (Part B)
*Edited by* JOHN J. LANGONE AND HELEN VAN VUNAKIS

VOLUME 74. Immunochemical Techniques (Part C)
*Edited by* JOHN J. LANGONE AND HELEN VAN VUNAKIS

VOLUME 75. Cumulative Subject Index Volumes XXXI, XXXII, XXXIV–LX
*Edited by* EDWARD A. DENNIS AND MARTHA G. DENNIS

VOLUME 76. Hemoglobins
*Edited by* ERALDO ANTONINI, LUIGI ROSSI-BERNARDI, AND EMILIA CHIANCONE

VOLUME 77. Detoxication and Drug Metabolism
*Edited by* WILLIAM B. JAKOBY

VOLUME 78. Interferons (Part A)
*Edited by* SIDNEY PESTKA

VOLUME 79. Interferons (Part B)
*Edited by* SIDNEY PESTKA

VOLUME 80. Proteolytic Enzymes (Part C)
*Edited by* LASZLO LORAND

VOLUME 81. Biomembranes (Part H: Visual Pigments and Purple Membranes, I)
*Edited by* LESTER PACKER

VOLUME 82. Structural and Contractile Proteins (Part A: Extracellular Matrix)
*Edited by* LEON W. CUNNINGHAM AND DIXIE W. FREDERIKSEN

VOLUME 83. Complex Carbohydrates (Part D)
*Edited by* VICTOR GINSBURG

VOLUME 84. Immunochemical Techniques (Part D: Selected Immunoassays)
*Edited by* JOHN J. LANGONE AND HELEN VAN VUNAKIS

VOLUME 85. Structural and Contractile Proteins (Part B: The Contractile Apparatus and the Cytoskeleton)
*Edited by* DIXIE W. FREDERIKSEN AND LEON W. CUNNINGHAM

VOLUME 86. Prostaglandins and Arachidonate Metabolites
*Edited by* WILLIAM E. M. LANDS AND WILLIAM L. SMITH

VOLUME 87. Enzyme Kinetics and Mechanism (Part C: Intermediates, Stereochemistry, and Rate Studies)
*Edited by* DANIEL L. PURICH

VOLUME 88. Biomembranes (Part I: Visual Pigments and Purple Membranes, II)
*Edited by* LESTER PACKER

VOLUME 89. Carbohydrate Metabolism (Part D)
*Edited by* WILLIS A. WOOD

VOLUME 90. Carbohydrate Metabolism (Part E)
*Edited by* WILLIS A. WOOD

# Section I

# Genetic Approaches to Dissect Complex Functions

# [1] Transposon Mutagenesis of Heterocyst-Forming Filamentous Cyanobacteria

By MICHAEL F. COHEN, JOHN C. MEEKS, YUPING A. CAI, AND C. PETER WOLK

## Introduction

The vegetative cells of certain filamentous cyanobacteria can differentiate morphologically and functionally into one or more of the following: heterocysts, the sites of nitrogen fixation under aerobic conditions; akinetes, also known as spores; and hormogonia, gliding filaments comprised exclusively of small vegetative cells. Photosynthetic oxygen evolution and carbon dioxide fixation are reduced in hormogonia compared to their rates in vegetative cells,[1] and are suppressed completely in heterocysts.[2] Facultatively heterotrophic heterocyst-forming cyanobacteria such as *Nostoc punctiforme* strain ATCC 29133 (PCC 73102; hereinafter referred to as *N. punctiforme*) provide an experimental system for study of the regulation of expression of genes that encode components of the photosynthetic apparatus.

Unicellular and filamentous cyanobacteria establish symbiotic associations with a variety of eukaryotic protists, metazoans, and plants. *N. punctiforme*, for example, was isolated from a symbiotic association with the cycad *Macrozamia* sp.[3] When the eukaryotic partner is heterotrophic, the symbiotic cyanobacterium functions to supply reduced carbon; but when *Nostoc* spp. are in symbiotic association with the bryophyte *Anthoceros punctatus* and other plants, cyanobacterial photosynthetic $CO_2$ fixation is suppressed and activity of cyanobacterial ribulose-bisphosphate carboxylase/oxygenase is inhibited,[4] presumably facilitating the establishment and stable maintenance of the symbiotic association. Because *N. punctiforme* can grow heterotrophically, mutants of it that are affected in photosynthesis should be isolable. In particular, placement of reporter operons within photosynthetically active genes that are regulated during symbiosis could help to elucidate key features of the symbiotic interaction.

[1] E. L. Campbell and J. C. Meeks, *Appl. Environ. Microbiol.* **55,** 125 (1989).
[2] C. P. Wolk, A. Ernst, and J. Elhai, in "The Molecular Biology of Cyanobacteria" (D. A. Bryant, ed.), p. 769. Kluwer Academic Publishers, Dordrecht, 1994.
[3] R. Rippka and M. Herdman, "Pasteur Collection of Cyanobacterial Strains in Axenic Culture, Vol. I: Catalogue of Strains." Institut Pasteur, Paris, 1992.
[4] N. A. Steinberg and J. C. Meeks, *J. Bacteriol.* **171,** 6227 (1989).

As is true also of other diazotrophic cyanobacteria, *N. punctiforme* supplies reduced nitrogen to its symbiotic partner.

Both *Anabaena* (*Nostoc*) sp. strain PCC 7120 (hereinafter referred to as PCC 7120) and *N. punctiforme* grow rapidly and homogeneously in liquid medium, and form visible colonies on plates within 4–7 days after plating. Whereas both strains form heterocysts, only *N. punctiforme* has been found to form akinetes and hormogonial filaments, to grow heterotrophically,[5] and to establish a symbiotic association[6]; however, unlike in PCC 7120,[7,8] mutations of *N. punctiforme* have not been complemented with a cosmid library.[9]

Adjacent cells of filamentous cyanobacteria share an end wall, and therefore are strongly linked, and (at least, on average) each of the cells has multiple chromosomal equivalents. As a consequence, mutant copies of the genome segregate only with difficulty. Selection for antibiotic resistance that results from the presence of a transposon enhances segregation, and antibiotic-resistant colonies that are to be screened for phenotypes of interest are necessarily mutant in some way. Therefore, transposition, relative to other means of mutagenesis, facilitates the isolation of null mutants, and can mark the sites of mutation by the antibiotic resistance gene(s) of the transposon. The conjugal transfer of transposons from *Escherichia coli* has recently been used to generate a variety of mutants that affect heterocyst development[10–14] and symbiotic competence,[6] principally in the cyanobacteria PCC 7120 and *N. punctiforme*. In the two strains mentioned, there is a high probability that the marker and the mutation are coupled. Thus, transposons provide an effective means for identifying genes that correspond to mutant phenotypes. In addition, transposons can stabilize mutations, facilitate mapping[10,15] and—in combination with reporter genes—can directly aid analysis of the transcription of the genes that have been interrupted.

[5] M. L. Summers, J. G. Wallis, E. L. Campbell, and J. C. Meeks, *J. Bacteriol.* **177,** 6184 (1995).
[6] M. F. Cohen, J. G. Wallis, E. L. Campbell, and J. C. Meeks, *Microbiol.* **140,** 3233 (1994).
[7] C. P. Wolk, Y. Cai, L. Cardemil, E. Flores, B. Hohn, M. Murry, G. Schmetterer, B. Schrautemeier, and R. Wilson, *J. Bacteriol.* **170,** 1239 (1988).
[8] W. J. Buikema and R. Haselkorn, *J. Bacteriol.* **173,** 1879 (1991).
[9] J. G. Wallis, Ph.D. Thesis, University of California, Davis (1993).
[10] C. P. Wolk, Y. Cai, and J.-M. Panoff, *Proc. Natl. Acad. Sci. U.S.A.* **88,** 5355 (1991).
[11] A. Ernst, T. Black, Y. Cai, J.-M. Panoff, D. N. Tiwari, and C. P. Wolk, *J. Bacteriol.* **174,** 6025 (1992).
[12] D. Borthakur and R. Haselkorn, *J. Bacteriol.* **171,** 5759 (1989).
[13] T. A. Black, Y. Cai, and C. P. Wolk, *Mol. Microbiol.* **9,** 77 (1993).
[14] E. L. Campbell, K. D. Hagen, M. F. Cohen, M. S. Summers, and J. C. Meeks, *J. Bacteriol.* **178,** 2037 (1996).
[15] T. Kuritz, A. Ernst, T. A. Black, and C. P. Wolk, *Mol. Microbiol.* **8,** 101 (1993).

Choice of a Transposon

Heterocyst-forming cyanobacteria contain insertion sequences (ISs); at least five such sequences have been characterized at the molecular level, including several families of ISs in PCC 7120.[16-19] These IS elements can move within the host genome, perhaps as a consequence of stress; as they transpose, they induce mutations.[18,20] On spontaneous transposition, endogenous IS elements of *Fremyella diplosiphon* (syn. *Calothrix* sp. strain PCC 7601) have insertionally inactivated genes that are involved in synthesis of components of phycobilisomes; electroporation increases their rate of transposition.[21] However, the regulation of the transposition of these IS elements has not been studied and there is no known selection for mutants other than loss of function. No useful, selectable transposon engineered from a cyanobacterial IS has yet been described. We focus here on the use of transposons derived from heterotrophic bacteria.

Transposon Tn901 (which confers resistance to ampicillin) was transferred by transformation to the unicellular cyanobacterium *Synechococcus* sp. strain PCC 7942 (syn. *Anacystis nidulans* R2),[22] where it transposed spontaneously to an endogenous plasmid. A second transposition produced a methionine auxotroph.[23] The frequency of transposition, and thus of mutagenesis, was very low.[24] Analysis of sites at which the transposon Tn5 and derivatives of it insert in PCC 7120[12,15] and *N. punctiforme*[6] shows no apparent site or sequence preference. This property makes Tn5 an appropriate choice as a mutagen.[25] However, transposition of Tn5 from the same *Synechococcus* 7942 plasmid was even less frequent than transposition of Tn901.[24] In contrast, relatively high frequencies of mutation result from transposition of derivatives of wild-type Tn5, when delivered by conjugation from *E. coli* to PCC 7120[10] and to *N. punctiforme*[6] in suicide (nonreplicating) plasmids. (Tn5, delivered by conjugal transfer of a nonreplicating

[16] I. Bancroft and C. P. Wolk, *J. Bacteriol.* **171,** 5949 (1989).
[17] Y. Cai, *J. Bacteriol.* **173,** 5771 (1991).
[18] J. Alam, J. M. Vrba, Y. Cai, J. A. Martin, L. J. Weislo, and S. E. Curtis, *J. Bacteriol.* **173,** 5778 (1991).
[19] D. Mazel, C. Bernard, R. Schwarz, A. M. Castets, J. Houmard, and N. Tandeau de Marsac, *Mol. Microbiol.* **5,** 2165 (1991).
[20] Y. Cai and C. P. Wolk, *J. Bacteriol.* **172,** 3138 (1990).
[21] B. U. Bruns, W. R. Briggs, and A. R. Grossman, *J. Bacteriol.* **171,** 901 (1989).
[22] C. A. M. J. J. van den Hondel, S. Verbeek, A. van der Ende, P. J. Weisbeek, W. E. Borrias, and G. A. van Arkel, *Proc. Natl. Acad. Sci. U.S.A.* **77,** 1570 (1980).
[23] N. Tandeau de Marsac, W. E. Borrias, C. J. Kuhlemeier, A. M. Castets, G. A. van Arkel, and C. A. M. J. J. van den Hondel, *Gene* **20,** 111 (1982).
[24] C. J. Kuhlemeier and G. A. van Arkel, *Methods Enzymol.* **153,** 199 (1987).
[25] D. E. Berg, *in* "Mobile DNA" (D. E. Berg and M. M. Howe, eds.), p. 185. American Society for Microbiology, Washington, D.C., 1989.

TABLE I

TRANSPOSONS DERIVED FROM Tn5 CITED IN THIS CHAPTER[a]

| Transposon | Size (bp) | Resistance markers | Promoter for antibiotic operon | Factor-independent terminator | Reporter gene | oriV | Reference |
|---|---|---|---|---|---|---|---|
| Tn5-764 | 4145 | Km/Nm | psbA | Present | None | p15A | 29, this chapter |
| Tn5-765 | 6572 | Km/Nm | psbA | Present | V.f. luxAB | p15A | 29, this chapter |
| Tn5-770 | 4948 | CmEm | Native | Present | None | p15A | This chapter |
| Tn5-771 | 7375 | CmEm | Native | Present | V.f. luxAB | p15A | This chapter |
| Tn5-800 | 5000 | CmEm | Native | None | None | pMB1 | This chapter |
| Tn5-801 | 7427 | CmEm | Native | None | V.f. luxAB | pMB1 | This chapter |
| Tn5-1058 | 5395 | Km/NmBmSm | psbA | Present | None | p15A | 10 |
| Tn5-1062 | 7832 | Km/NmBmSm | psbA | Present | V.h. luxAB | p15A | This chapter |
| Tn5-1063 | 7834 | Km/NmBmSm | psbA | Present | V.f. luxAB | p15A | 10 |
| Tn5-1065 | 8117 | Km/NmBmSm | psbA | Present | T7 gene 1 | p15A | 30 |
| Tn5-1087b | 5331 | CmEm | Native | None | None | pMB1 | 11 |
| Tn5-1088a,b | 7251 | CmEmSmSp | Native | None | None | pMB1 | 11 |
| Tn5-1140 | 8910 | CmEm | Native | None | lacZ | pMB1 | This chapter |

[a] For reasons presented in Ref. 31, we do not recommend, for routine reporting, further derivatives described therein that contain entire *lux* operons as a reporter. The geneology of the transposons listed above is as follows: Tn5-1058→Tn5-764[→(Tn5-765, Tn5-770→Tn5-771)], Tn5-1062, Tn5-1063, Tn5-1065, Tn5-1087b[→Tn5-800→(Tn5-801, Tn5-1140)], Tn5-1088a,b.

plasmid, has also been reported to have integrated into the genomes of a marine filamentous *Pseudanabaena* sp. and two unicellular cyanobacteria[26]; unfortunately, it was not verified in those cases that Tn5 had actually transposed; see below.) As in *E. coli* and other bacteria examined,[25] a 9-bp direct repeat of target sequences is generated on insertion, although 8-bp repeats are occasionally generated.[27,28]

We recommend the use of Tn5-based transposon Tn5-1058 or three groups of derivatives of it (Table I). These transposons have several advantages relative to wild-type Tn5. First, an *oriV* from the medium-copy-number plasmid p15A (the same *oriV* that functions in pACYC177[32]) is engineered into many of these derivatives to facilitate the recovery in *E. coli* of the transposon together with flanking genomic sequences from

[26] K. Sode, M. Tatara, H. Takeyama, J. G. Burgess, and T. Matsunaga, *Appl. Microbiol. Biotechnol.* **37,** 369 (1992).
[27] F. Fernández-Piñas, F. Leganés, and C. P. Wolk, *J. Bacteriol.* **176,** 5277 (1994).
[28] Y. Cai and A. N. Glazer (1996).
[29] Y. Cai and C. P. Wolk, *J. Bacteriol.* **179,** 258 (1997).
[30] C. P. Wolk, J. Elhai, T. Kuritz, and D. Holland, *Mol. Microbiol.* **7,** 441 (1993).
[31] F. Fernández-Piñas and C. P. Wolk, *Gene* **150,** 169 (1994).
[32] A. C. Y. Chang and S. N. Cohen, *J. Bacteriol.* **134,** 1141 (1978).

resultant mutants.[10] Second, relative to wild-type Tn5, these derivatives have an enhanced frequency of transposition. Furthermore, in those that retain the Tn5 determinants for resistance to the antibiotics neomycin (Nm) and kanamycin (Km), bleomycin (Bm), and streptomycin (Sm), the resistances have been enhanced by replacement of the wild-type promoter with the psbA promoter from the chloroplast genome of the higher plant, Amaranthus hybridus. The sequence of this promoter approaches the consensus sequence of the eubacterial $\sigma^{70}$-family of promoters that yield a high level of transcription in vegetative cells of unicellular and filamentous cyanobacteria.[33] In most of those same derivatives, a rho-independent terminator was inserted downstream from the highly expressed antibiotic resistance operon to minimize antisense interference with the expression of the transposase gene. Streptomycin resistance in cells of E. coli[34] and PCC 7120[10] was further enhanced by introduction of a 6-bp deletion in the 3' portion of the Sm[r] gene (in Tn5-1058, Tn5-1063, and certain others).

In one group of derivatives of Tn5-1058 (e.g., Tn5-770), only the antibiotic resistance genes have been replaced. Use of these transposons may be advantageous, compared with transposons that confer resistance to Nm, when the cyanobacterium to be mutagenized shows high natural resistance to Nm. (Our results also suggest that strains that bear Tn5-1058 may derive metabolic nitrogen when exposed to high levels of Nm.[35]) Plasmid pRL764SX[29] (Fig. 1), bearing Tn5-764, facilitates the generation of derivatives of Tn5 to fit specific needs. In this plasmid, unique restriction sites flank the oriV (origin of replication in E. coli), the oriT (origin of conjugal transfer), and the antibiotic resistance segments, allowing facile replacement of these components.

Those in a second group of derivatives of Tn5-1058 (see Table I) bear a promoterless reporter gene or operon (e.g., luxAB from Vibrio fischeri (Tn5-765, Tn5-771, Tn5-1063], or Vibrio harveyi [Tn5-1062], lacZ [Tn5-1140], T7 gene 1 [Tn5-1065]) very close to the L end of the transposon, enabling formation of transcriptional (not translational) fusions on transposition into the genome. Mutant strains of PCC 7120[2,10,29] bearing the luxAB-containing transposons can show environmentally responsive production of light. In Tn5-1065 the promoterless gene 1 of coliphage T7 gene 1, encoding RNA polymerase, functions as an amplifier in a reporting cascade.[30] The gene 1 product, induced from a transcriptional fusion, activates transcription of a T7-specific promoter controlling, for example, luxAB. Expression of the reporter gene can be enhanced relative to direct promoter

---

[33] S. E. Curtis and J. A. Martin, this chapter, Ref. 2, p. 613.
[34] P. Mazodier, O. Genilloud, E. Giraud, and F. Gasser, Mol. Gen. Genet. **204,** 404 (1986).
[35] Y. Cai (1996).

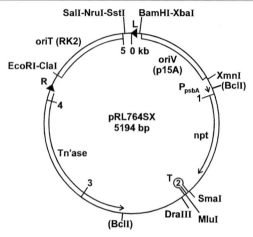

FIG. 1. Simplified schematic of pRL764SX bearing transposon Tn5-764. *oriV*(p15A), origin of replication (from plasmid p15A); $P_{psbA}$, promoter of the *Amaranthus hybridus* gene that encodes the 32-kDa protein of photosystem II; *npt*, gene that encodes neomycin phosphotransferase, which confers resistance to neomycin and kanamycin; T, *rho*-independent terminator of *lpp* gene; Tn'ase, gene (with modified promoter incorporated[36]) that encodes the transposase of transposon Tn5; *oriT*(RK2), transfer origin of broad-host-range plasmid RK2; R and L: right and left ends, respectively, of Tn5. Unique enzyme sites are shown that permit modular replacement of the *oriV* and the antibiotic-resistance determinant. (The *Xba*I site shown is not unique, but is the only *Xba*I site that can be cut when the plasmid is taken from a *dam*⁺ strain of *E. coli*.)

reporter fusions.[30] However, high activity of T7 gene 1 appears toxic to PCC 7120 (perhaps its product recognizes some sequence in the genome), so that this system is effective only for genes that are weakly transcribed during growth.

The Tn5 derivatives (Table I) bear an RK2 *oriT* region on their respective plasmids to allow for conjugal transfer from *E. coli*.[10] They can be introduced into PCC 7120 and *N. punctiforme* by conjugation with *E. coli* and by electroporation, but not by simple transformation. In a final group of derivatives of Tn5-1058, a group that includes Tn5-800, Tn5-801 and Tn5-1087b, only the transposase-encoding IS*50*R and the *oriT* are retained as a *Bam*HI-*Bcl*I cassette, easily combined with many other plasmids that provide an *oriV* and a selectable marker.

An advantage of the series of Tn5 derivatives mentioned that is apparently not shared by wild-type Tn5,[37,38] at least in PCC 7120, is the stabilization of mutations. This difference may be attributable to the greater promo-

[36] J. C. P. Yin, M. P. Krebs, and W. S. Reznikoff, *J. Mol. Biol.* **199,** 35 (1988).
[37] J. Liang, L. Scappino, and R. Haselkorn, *Proc. Natl. Acad. Sci. U.S.A.* **89,** 5655 (1992).
[38] J. Liang, L. Scappino, and R. Haselkorn, *J. Bacteriol.* **175,** 1697 (1993).

tion of the antibiotic-resistance operon in the derivative transposons, but may also be attributable in part to greater excision of the wild-type transposon. Excision is stimulated by the presence of extensive inverted repeats at the termini of a transposon.[25] In the derivative transposons of Table I, the lengthy (ca. 1.5-kb) repeat represented by IS50L in wild-type Tn5 has been reduced to less than 0.1 kb. In none of the numerous PCC 7120 mutants studied so far has the Tn5-derived transposon been shown to have moved from its original point of insertion.[28]

Further Considerations

The cyanobacterium to be transposon-mutagenized must be in some way transformable, so that the transposon can enter it, the transposon DNA must escape restriction and degradation in the recipient, and the transposase gene and its products must function in the recipient. Preliminary results indicate that *F. diplosiphon,* which is the preferred organism for studies of complementary chromatic adaptation, is susceptible to mutagenesis with certain derivatives of Tn5-1058 that lack sites for the restriction endonuclease *Sph*I.[39] All of ca. 70 exconjugants of *Anabaena variabilis* strain ATCC 29413 that had been mated with *E. coli* bearing Tn5-1088a and Tn5-1088b were both spectinomycin[r] and Em[r], suggesting that the strain may be susceptible to transposon mutagenesis.[40] Two of seven analyzed derivatives of *Nostoc ellipsosporum* that had received pRL1063a (bearing Tn5-1063) were products of transposition.[41] In the other five, the following series of events appears to have taken place. Native insertion sequence IS891N (whose sequence differs by one bp from the corrected sequence of IS891 [GenBank accession no. M30792]) transposed into the DNA of incoming plasmid pRL1063a. The resulting plasmid then integrated into the genome via homologous recombination of its copy of IS891N with a corresponding element in the genome of *N. ellipsosporum.*[41] Exconjugants that result from such recombinational events (or from the very unlikely possibility that the transposon-bearing plasmid can replicate in the cyanobacterium) rather than from transposition can be identified by the continued presence of the *oriT* region of the plasmid.

PCC 7120 contains at least three type II restriction endonucleases, isoschizomers of *Ava*I, *Ava*II and *Ava*III,[42] that initiate degradation of DNA that has unmethylated sites for those nucleases. Chloramphenicol[r]

[39] J. Cobley, personal communication (1996).
[40] C. P. Wolk (1996).
[41] F. Leganés, F. Fernández-Piñas, and C. P. Wolk, *Mol. Microbiol.* **12,** 679 (1994).
[42] T. Thiel, this chapter, Ref. 2, p. 581.

(Cm$^r$) helper plasmids pRL528 and pRL623, derivatives of pDS4101,[43] carry genes that code for AvaI and Eco47II methylases that modify sequences recognized by AvaI and AvaII, respectively; pRL623 also carries the gene that encodes the EcoT22I methylase, which confers resistance to AvaIII. Use of pRL528 or pRL623 in E. coli to premethylate the plasmid carrying the transposon is essential for efficient transfer to, and transposition in, PCC 7120.[10,44,45] Plasmids pRL1045 and pRL1124, used with Cm$^r$ transposons such as Tn5-1087b, are Km$^r$ derivatives of pACYC177[32] that bear the same methylase genes as pRL528 and pRL623, respectively.[45] No type II restriction endonuclease activity has been detected in N. punctiforme, and no beneficial result was observed on premodification of the transposon-bearing plasmid with pRL528. Although we routinely transfer replicating plasmids to N. punctiforme by electroporation,[5,14] attempted transfer of Tn5-1063 by electroporation of pRL1063a has yielded no transposition events.[46]

Protocol for Mutagenesis

*Conjugation*

The triparental mating procedures used to transposon-mutagenize N. punctiforme[6] differ from those devised for PCC 7120,[10,12,44,47] illustrating the possible need to optimize conditions when developing a transposition protocol for yet another strain. The triparental system consists of a (cyanobacterial) recipient strain, an E. coli donor conjugal strain carrying RP4 (a broad host range conjugal plasmid) or a closely related plasmid,[44] and an E. coli donor cargo strain carrying the plasmid to be transferred (in this case the plasmid bearing the transposon) and (if necessary for mobilization and/or modification) a helper plasmid. Unlike the oriT site on many pMB1-derived cargo plasmids (pMB1 is a progenitor of pBR322), the RK2 oriT in the plasmids carrying the Tn5-1058 derivatives requires only the transacting factors supplied by the broad host range conjugal plasmid; thus, helper plasmids such as pDS4101 and pRL528 are not needed to mobilize those transposons. However, as mentioned earlier, experiments using PCC 7120 supplement the cargo plasmid in E. coli with a methylating plasmid

[43] F. Finnegan and D. Sherratt, *Mol. Gen. Genet.* **185,** 344 (1982).
[44] J. Elhai and C. P. Wolk, *Methods Enzymol.* **167,** 747 (1988).
[45] J. Elhai, A. Vepritskiy, A. M. Muro-Pastor, E. Flores, and C. P. Wolk, *J. Bacteriol.* **179,** 1998 (1997).
[46] M. F. Cohen, Ph.D. Thesis, University of California, Davis (1996).
[47] T. Thiel and C. P. Wolk, *Methods Enzymol.* **153,** 232 (1987).

to methylate *Ava*I, *Ava*II, and *Ava*III sites on the DNA of the transposon-bearing cargo plasmid prior to transfer. Transfer of RK2 *oriT*-containing plasmids to *N. punctiforme*[6] is enhanced by use of conjugal plasmid pRK2013,[48] whose copy number is greater than that of RK2.

The *E. coli* donor strains are grown in Luria–Bertani (LB) liquid medium with selecting antibiotics(s) (generally 25 $\mu$g/ml Km for the conjugal strain carrying pRK2013; 50 $\mu$g/ml of Ap and/or Km for the strain carrying RP4; and 50 $\mu$g/ml Km plus 25 $\mu$g/ml Cm for the strain carrying a Km$^r$ cargo plasmid such as pRL1058 [bearing Tn5-1058] and a Cm$^r$ helper such as pRL528, or a Cm$^r$ cargo plasmid such as pRL1087b and a Km$^r$ helper such as pRL1045). The cells are harvested by centrifugation at 2000$g$ for 10 min, washed in LB medium lacking antibiotics, and resuspended to the desired cell density (see below).

The recipient cyanobacterium is prepared similarly for both mating procedures. To reduce the chance of spontaneous mutation to a Fox$^-$ phenotype (inability to fix $N_2$ in the presence of oxygen),[11] cultures of both strains are maintained on $N_2$, and then subcultured 1 : 200 (v/v) to medium that contains nitrate (PCC 7120) or ammonium (*N. punctiforme;* 2.5 m$M$, buffered with 5 m$M$ 3-[*N*-morpholino]propanesulfonic acid [MOPS]) for growth of the culture to be transposon mutagenized. Liquid cultures of less than 4 $\mu$g/ml of chlorophyll *a* (Chl *a*) of both cyanobacteria (such relatively dilute cultures are strongly recommended) are subjected to mild cavitation in a sonic bath[49] to break up clumps of filaments (*N. punctiforme*) to facilitate spread-plating and, by fragmenting the filaments to a length of 3 to 10 cells, to help ensure that there will be only one transposition event per colony-forming unit (cfu; because a viable filament is a cfu, the fewer cells per filament the lower the probability of multiple transpositions per cfu). The fragmented filaments are washed once by centrifugation at 1000 $g$ for 5 min, resupended in growth medium (for *N. punctiforme*, supplemented with MOPS-buffered $NH_4Cl$) and incubated under photoautotrophic growth conditions for 6–10 hr. It may be possible to increase the frequency of transposition into certain genes by inducing their transcription shortly before mating,[25,35] e.g., by removal of combined nitrogen for heterocyst-formation and nitrogen-fixation genes, or by a shift to light of different quality and/or quantity for light harvesting-related genes.

Conjugal contact is established by spreading a mixture of cells of the conjugal and the cargo/helper *E. coli* donor strains and the recipient cyanobacterium on detergent-free membrane filters (Millipore HATF or Nucle-

---

[48] D. H. Figurski and D. R. Helinski, *Proc. Natl. Acad. Sci. U.S.A.* **76,** 1648 (1979).
[49] C. P. Wolk and E. Wojciuch, *Arch. Mikrobiol.* **91,** 91 (1973).

pore REC-85) that lie atop cyanobacterial minimal medium (BG11[50] or AA[51] with or without combined nitrogen) solidified with agar (1% Difco Bacto agar purified by the method of Braun and Wood[52]) and supplemented with 5% (PCC 7120) or 0.5% (*N. punctiforme*) (v/v) separately autoclaved LB medium. The density of the parental strains spread on the filters is an important difference between the two mating procedures. For a single plate mating with PCC 7120, the conjugal and donor strains of *E. coli* are grown overnight, diluted 1 : 40, and 2.5–3 hr later, cells from a 10-ml portion of each culture are harvested by centrifugation, separately washed, concentrated, combined with the other *E. coli* strain, and mated with PCC 7120 cells containing a total of 15 $\mu$g of Chl *a* per filter (1 $\mu$g Chl *a* corresponds to approximately $3 \times 10^6$ cells of PCC 7120[44]). The cells are spread by placing the mixture of cells atop five 1-mm glass beads (Sigma Chemical Co., St. Louis, MO; acid-washed, thoroughly rinsed with distilled water, and autoclaved) resting on the mating filter, then shaking vigorously in the plane of the filter.

Such a high density of *E. coli* cells is detrimental to the viability of *N. punctiforme*[6]; therefore, approximately 20-fold fewer *E. coli* cells are used. Roughly equal amounts of the conjugal and cargo donor cells are harvested, washed, and combined, and the mixture is resuspended to an $OD_{600}$ of 9 to 10. A 0.5-ml aliquot of this mixture of *E. coli* donor cells is then mixed in a microcentrifuge tube with a 0.5-ml suspension of *N. punctiforme* cells containing 50 $\mu$g Chl *a* (1 $\mu$g of Chl *a* corresponds to about $5.7 \times 10^6$ cells of *N. punctiforme*[6]). The mixture is then centrifuged at 4000g for 30 sec, all but about 150 $\mu$l decanted, and the remainder gently mixed using a wide-bore pipette tip (the mixture should be somewhat pasty) and spread on the filters. A 30-ml culture of each of the *E. coli* donor strains at $OD_{600}$ of about 1 is sufficient for 12 plate matings of *N. punctiforme*. The higher numbers of cyanobacterial cells used in the *N. punctiforme* mating procedure compensates partially for the lower number of *E. coli* donor cells, but the yield of transpositions is nonetheless about five-fold lower than from a typical PCC 7120 mating.

To allow time for plasmid transfer and for expression of transposon-encoded antibiotic resistance, the mating filters are incubated for at least 48 hr (for *N. punctiforme*, not more than 4 days) in the light at 30° (PCC 7120) or 28° (*N. punctiforme*) prior to exposure to antibiotics. During this period, noticeable growth should occur. *N. punctiforme* has a higher mating

---

[50] R. Rippka, J. Deruelles, J. B. Waterbury, M. Herdman, and R. Y. Stanier, *J. Gen. Microbiol.* **111**, 1 (1979).

[51] M. B. Allen and D. I. Arnon, *Plant Physiol.* **30**, 366 (1955).

[52] A. C. Braun and H. N. Wood, *Proc. Natl. Acad. Sci. U.S.A.* **48**, 1776 (1962).

efficiency and grows more rapidly on plates when air is supplemented with 1% (v/v) $CO_2$. $CO_2$ supplementation has not been tried for plate growth of PCC 7120.

*Selection for Colonies in which Transposition Has Occurred*

The filters are then transferred to medium supplemented with one or more antibiotics. Cyanobacteria vary in their natural sensitivity to different antibiotics. Neomycin is often more selective than kanamycin in filamentous strains. For PCC 7120, selection after mating is carried out with either Nm at 400 $\mu$g/ml, Sm at 5 $\mu$g/ml, or a combination of Nm at 25 $\mu$g/ml plus Sm at 1 $\mu$g/ml when Tn5-1063 or similar $Nm^rSm^r$ transposons are used[10]; for *N. punctiforme* a combination of Nm at 10 $\mu$g/ml plus Sm at 1 $\mu$g/ml is used.[6] At low levels of antibiotics exconjugant strains in which the antibiotic resistance genes are driven by the very strong $P_{psbA}$ promoter from *A. hybridus* become antibiotic resistant due to the presence of transferred, albeit nonreplicating (but not degraded) copies of the transposon-bearing plasmid. One of us (YC) recommends transfer of the filters to fresh selective medium when biliproteins from lysed cells have diffused into, and colored, the agar medium beneath the filters. Transfer of the mating filters to fresh selective medium tends to accelerate the clearing of the background and the growth of exconjugant colonies. Exconjugants appear as dark green colonies atop a white or pale yellow background of dead cells about 1–3 weeks after the start of antibiotic selection. At this point the colonies on the filters can be screened for mutant phenotypes. For example, to screen for mutants defective in nitrogen fixation, filters are transferred to a medium without combined nitrogen (subsequent transfer to the same medium may be desirable) and colonies that develop yellowing edges within 7 days are further characterized. To prevent growth of wild-type cells, the filters should be transferred to fresh Nm-containing agar medium at regular intervals or the Nm concentration increased to, e.g., 100–400 $\mu$g/ml.

*Purifying Transposition-Derived Colonies*

A colony formed on a mating filter can be streaked on selective agar medium, or can be inoculated directly into an antibiotic-supplemented minimal liquid medium. Culturing and subculturing is an effective means of ridding the exconjugants of contaminating *E. coli* cells.[44] Cultures are tested for retention of viable *E. coli* by putting a drop of culture on an LB agar plate and incubating overnight at 37°. An axenic cyanobacterial culture will show no growth on the LB plate. Once an axenic culture is obtained, Southern analysis is used to indicate whether the transposon has inserted into the chromosome of the exconjugant by transposition.

*Segregation of the Mutant and Wild-Type Chromosomes*

Many cyanobacteria, including PCC 7120 and *N. punctiforme*, are polyploid; e.g., on average a cell of PCC 7120 has been calculated to contain more than 10 copies of the chromosome (see Ref. 20). Complete segregation of a mutant genotype is normally desirable, although for certain mutations it may not be attainable. Continuous subculturing in liquid medium with sufficiently strong antibiotic selection, coupled with occasional mild cavitation in a sonic bath to fragment the filaments[49] facilitates the process, but plating following fragmentation is essential to ensure segregation from all wild-type copies of the genome. Southern analysis of genomic DNA with an appropriate probe is used to establish whether the wild-type chromosome is present or absent (see below).

Difficulty in obtaining completely segregated mutants could indicate inadequate antibiotic selection or that the mutated gene is important to cellular survival under given growth conditions. While the former difficulty is easily addressed by increasing the concentration of antibiotic used, the latter may require procedures beyond the scope of this article. Change of growth conditions and nutrient supply may be tried. Several transposon-generated mutants of PCC 7120 have been obtained in which the insertions are within genes essential for growth, such as the *rpl*15 gene coding for the 50S ribosomal protein L32. A common characteristic of these mutants is their inconsistency in growth. When cells of such a mutant are inoculated from a selective agar medium to a selective liquid medium, survival and growth of the inoculum in the liquid medium is unpredictable; once a liquid culture is obtained, streaking on a selective agar medium often gives spotty, instead of homogeneous, cell growth in areas of predicted high cell density.[28]

Reconstructing the Mutation

In two mutants of PCC 7120 that were tagged with wild-type Tn5,[37,38] the mutations were not linked to the site of insertion of the transposon. In *N. ellipsosporum*[41] and *Anabaena* sp. strain M-131[16] and perhaps in PCC 7120,[53] conjugation may stimulate mutagenic transposition of IS elements. Although there is no evidence for such a response in *N. punctiforme*, it is essential to reconstruct the mutation with the transposon or with an interposon in wild-type cells to determine whether the mutant phenotype is a direct result of the specific insertion and not the consequence of a secondary mutation. While regeneration of the mutation can be done with a simple

[53] C. C. Bauer, personal communication cited in Ref. 37.

interposon such as an antibiotic-resistance cassette, such a construction may generate a novel phenotype.[54]

To reconstruct the mutation, the transposon with flanking cyanobacterial genomic DNA must first be recovered from the mutant. The phenol-chloroform glass-bead method for isolation of genomic DNA,[20] applicable to virtually all bacteria including those cyanobacteria that we have tried, is suitable for this purpose. If one wants to avoid handling phenol and obtain genomic DNA of high molecular mass, multistep lysis–nucleic acid precipitation protocols are also available.[6,55]

Genomic DNA prepared from the mutant is digested with a restriction endonuclease that does not cut within the transposon, but cuts the cyanobacterial genomic DNA frequently[10]; the digested DNA is diluted and then ligated to circularize the genomic fragments. For Tn5-1058 or Tn5-1063, candidate enzymes include *Cla*I, *Eco*RI, *Eco*RV, *Pvu*I, *Sca*I, and *Spe*I. Successful transformation of *E. coli* should occur only with those circularized fragments that contain the p15A *oriV* carried by the Tn5 derivatives. We recommend transformation or electroporation using *E. coli* strains DH10B or DH5α-MCR (which lack host-specific and methylation-dependent restriction systems[56]). The presence of methylation-dependent restriction systems in numerous other strains of *E. coli*, such as the popular DH5α, can greatly reduce the transformation frequency of isolated DNA because many cyanobacteria have variously methylated DNA.[57] The Tn5 derivatives mentioned earlier that have an *oriV* from p15A provide relatively stable maintenance of the transposon-derived plasmids in *E. coli*. If the cloned DNA appears at all unstable, the host *E. coli* should be grown at 30° and promptly stored at −80° rather than repeatedly subcultured.

Provided that the length of DNA flanking each side of the transposon suffices for homologous recombination (if not, the transposon should be excised from the cyanobacterial chromosome with a different restriction endonuclease), the recovered plasmid is prepared for interposon mutagenesis of the wild-type strain, usually using *sacB*-mediated positive selection for double recombinants.[6,13,20] The minimum length of homologous flanking DNA with which we have tested and observed a recombination event in *N. punctiforme* is 700 bp, whereas a 200-bp length suffices in PCC 7120 for a second crossover. First, a frameshift mutation is introduced into the transposase gene by filling in the *Not*I site (if unique in the plasmid).[6] This

[54] T. A. Black and C. P. Wolk, *J. Bacteriol.* **176**, 2282 (1994).
[55] C. Franche and T. Damerval, *Methods Enzymol.* **167**, 803 (1988).
[56] S. G. Grant, J. Jesse, F. R. Bloom, and D. Hanahan, *Proc. Natl. Acad. Sci. U.S.A.* **87**, 4645 (1990).
[57] R. N. Padhy, F. G. Hottat, M. M. Coene, and P. P. Hoet, *J. Bacteriol.* **170**, 1934 (1988).

step can often be omitted in PCC 7120 reconstructions because homologous recombination often occurs at a much higher frequency than transposition. Second, a positive selection cassette, such as those that contain *sacB*, is placed between the two outer ends of the flanking DNA. The *sacB* gene product confers sensitivity to sucrose in many organisms (references cited in Ref. [20]). A *sacB*-containing vector, pRL1075 (Cm[r]Em[r]), has been especially designed for ligation to the recovered transposon-derived plasmids,[13] and Sm[r]Sp[r] variants (pRL1130a[29] [Fig. 2] and pRL1130b) made for use with Cm[r]Em[r] transposons such as Tn5-1087b. In addition to the *sacB* gene from *Bacillus subtilis* and an RK2 *oriT*, pRL1075 and pRL1130a and b contain symmetrical polylinkers that flank a pMB1 *oriV*; this *oriV* is replaced by the linearized excised chromosomal fragment which supplies replication functions from its p15A *oriV*.

We routinely verify sucrose sensitivity (Suc[s]) conferred by the resulting plasmid in *E. coli* prior to transfer of the plasmid into the wild-type cyanobacterium. The plasmid is mobilized from an *E. coli* strain to wild-type *N. punctiforme* by pRK2013 or to PCC 7120 by RP4, by a standard triparental

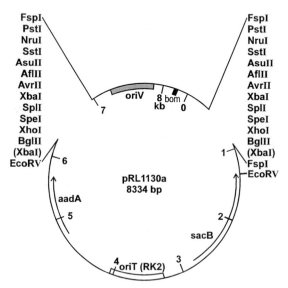

FIG. 2. Simplified schematic of *sacB*-containing positive selection vector pRL1130a. *oriV* and *oriT*(RK2) as in Fig. 1; *bom*, basis of mobilization (for pRL1130a, not normally used); *sacB*, gene that confers sensitivity to the host strain to sucrose; *aadA*, gene that confers resistance to streptomycin and spectinomycin. For reconstructions of transposon-induced mutations, transposon- (and thereby *oriV*-)containing plasmids excised from the chromosome are inserted between the inverted polylinkers of pRL1130a, replacing the *oriV*- and *bom*-containing segment of that plasmid.

mating with (PCC 7120) or without (*N. punctiforme*) a methylation-helper plasmid. Plasmids generated from pRL1075 that bear Tn5-1063 (or a derivative of it in which the transposase has been inactivated) are Bm$^r$Cm$^r$Em$^r$Nm$^r$Sm$^r$ and Suc$^s$. Single recombinants are typically maintained on solid medium with Em at 15 to 20 μg/ml (*N. punctiforme*) or 10 μg/ml (PCC 7120) or grown in liquid medium with 5 μg/ml Em and 10 μg/ml Nm (*N. punctiforme*) or with Em and Nm both at 2 μg/ml (PCC 7120). To accumulate double recombinants, a single-recombinant colony is inoculated into liquid medium with Nm (10 or 50 μg/ml for *N. punctiforme* or PCC 7120, respectively), but without Em, and incubated under growth conditions for several weeks.[6,20] Cells from such a culture are then used for positive selection for double recombinants (Suc$^r$). Filaments are first subjected to brief, mild cavitation, washed, and plated, usually in several cell densities, onto cyanobacterial growth medium supplemented with 5% (w/v) sucrose and Nm at 10 μg/ml (*N. punctiforme*)[6] or 200 μg/ml (PCC 7120).[20] (When the *Eco*RV site immediately 3' from *sacB* of pRL1075 has been used for cloning, 10% [w/v] sucrose can be more effective than 5%.) Use of higher light intensity tends to accelerate the killing of single-recombinant background cells that bear a functional copy of *sacB*, and the formation of sucrose-resistant colonies. One should not assume that all sucrose-resistant colonies are double recombinants, because inactivation of *sacB* in single recombinants by point mutation or by insertion of cyanobacterial IS elements is not uncommon.[20,35] These "pseudo-double recombinants" can be easily identified in this case by simply checking for their ability to grow on Em-containing medium, unless they have accumulated a rare second mutation in the Em$^r$ gene. True double recombinants have lost the Em$^r$-, *sacB*-containing fragment from pRL1075, and so are Em$^s$. Nonetheless, Southern analysis to confirm the reconstruction is highly recommended. Further studies of the mutated genomic DNA are warranted if the double recombinant strain bearing the reconstruction exhibits the same phenotype as the original transposon-insertion mutant. If the original mutant phenotype is interesting but evidently not caused by the transposon, it may be cloned by complementation from cosmid libraries (PCC 7120).

## Acknowledgments

This work was supported by the U.S. National Science Foundation (grants IBN 92-06139 and IBN 95-14787 to JCM), DOE grant DE-FG01-90ER20021 (CPW), and a Floyd and Mary Schwall Fellowship in Medical Research to MFC. We thank Elsie Campbell and Tom Hanson for advice on experimental details and critical review of the manuscript.

## [2] Isolation and Genetic Characterization of Pseudorevertants from Site-Directed PSI Mutants in Synechocystis 6803

By Jianping Yu and Lee McIntosh

Introduction

Reversion of mutation results from a change in DNA that either exactly reverses the original alteration (a true reversion) or compensates for it (a suppressor mutation). Suppressor mutations yield valuable information about other sites and/or other polypeptides that interact with the site altered by the original (primary) mutation. An organism with a suppressor mutation can be called a *pseudorevertant*. Selection of pseudorevertants can yield an array of suppressor mutations.

Suppressor selection has been applied to *Synechocystis* sp. PCC 6803 for the molecular analysis of the structure, function, and assembly of photosystem I (PSI). This methodology involves the isolation of spontaneous pseudorevertants from site-directed PSI mutants. Suppressor mutations can then be located to a specific gene or operon by phenotypic complementation and/or identified by DNA sequencing. In this way, specific amino acids and electron carriers may be assigned to a PSI structural model, and insights may be gained into the roles of specific amino acids on the assembly/ stability of the complex and on modulating electron transport. Isolation and characterization of pseudorevertants has recently been applied in the study of PSI in *Chlamydomonas reinhardtii*.[1]

*Synechocystis* 6803 is a model organism for the study of oxygenic photosynthesis. Its genome (3.57 Mb) has recently been sequenced,[2] and it is naturally competent and can be easily transformed. It also has an active homologous recombination mechanism.[3] Furthermore, it can grow photoheterotrophically on glucose, which has allowed propagation of PSII mutants that are incapable of photosynthesis,[3] and it can grow under light-activated heterotrophic growth (LAHG) conditions, permitting isolation

[1] H. Lee, S. E. Bingham, and A. N. Webber, *Photochem. Photobiol.* **64**, 46 (1996).
[2] T. Kaneko, S. Sato, H. Kotani, A. Tanaka, E. Asamizu, Y. Nakamura, N. Miyajima, M. Hirosawa, M. Sugiura, S. Sasamoto, T. Kimura, T. Hosouch, A. Matsuno, A. Muraki, N. Nakazaki, K. Naruo, S. Okumura, S. Shimpo, C. Takeuchi, T. Wada, A. Watanabe, M. Yamada, M. Tasuda, and S. Tabata, *DNA Res.* **3**, 109 (1996).
[3] J. G. K. Williams, *Methods Enzymol.* **167**, 766 (1988).

of PSI mutants in the absence of PSI function.[4] Site-directed mutagenesis has been successfully used in this organism to study structure–function relationships both in PSII[3] and in PSI.[5-7]

PSI consists of at least 11 different proteins in cyanobacteria.[8] All PSI proteins are believed to be present as one copy per P700 reaction center. They vary considerably in their molecular weights, hydrophobicities, and locations with respect to the lipid bilayer. In addition to proteins, the PSI complex contains approximately 100 Chl $a$ molecules, two $\beta$-carotenes, two phylloquinone molecules, and three [4Fe-4S] clusters. The cofactors of PSI are bound to the PsaA, PsaB, and PsaC proteins. Trimers of PSI have been observed by electron microscopy in the photosynthetic membranes of cyanobacteria and are considered to be *in vivo* functional units.[9] On the basis of the electron density maps, a structural model for trimeric PSI from *Synechococcus elongatus* was initially proposed at a resolution of 6 Å.[10] From new data, many aspects of this model have been later refined to resolutions of 4.5 and 4.0 Å.[11-13] Resolution of PSI structure at the atomic level will require further X-ray diffraction analyses of better crystals of PSI. These fine structure determinations are being complemented by topological explorations, electron microscopy, and genetic analysis.

Genetic inactivation of the *psaA, psaB,* or *psaC* genes in *Synechocystis* 6803 produced mutants that lack PSI function and depend on glucose for growth.[14,15] Photosynthesis-deficient PSI mutants have also been created by site-directed mutagenesis of these genes.[5-7] An additional phenotype, inhibition of mixotrophic growth by white light of moderate intensities, has

[4] S. L. Anderson and L. McIntosh, *J. Bacteriol.* **173,** 2761 (1991).

[5] L. B. Smart, P. V. Warren, J. H. Golbeck, and L. McIntosh, *Proc. Natl. Acad. Sci. U.S.A.* **90,** 1132 (1993).

[6] R. Schulz, L. B. Smart, J. Yu, and L. McIntosh, *in* "Photosynthesis: from Light to Biosphere" (P. Mathis, ed.), Vol. II, pp. 119–122. 1995.

[7] J. Yu, Y.-S. Jung, I. Vassiliev, J. H. Golbeck, and L. McIntosh, *J. Biol. Chem.* **272,** 8032 (1977).

[8] P. R. Chitnis, *Plant Physiol.* **111,** 661 (1996).

[9] E. J. Boekema, A. F. Boonstra, J. P. Dekker, and M. Rogner, *J. Bioenerg. Biomembr.* **26,** 17 (1994).

[10] N. Krauss, W. Hinrichs, I. Witt, P. Fromme, W. Pritzkow, Z. Dauter, C. Betzel, K. S. Wilson, H. T. Witt, and W. Saenger, *Nature* **361,** (1993).

[11] W. D. Schubert, O. Klukas, N. Krauß, W. Saenger, P. Fromme, and H. T. Witt, *in* "Photosynthesis: From Light to Biosphere" (P. Mathis, ed.), Vol. II, pp. 3–10. Kluwer Academic Publishers, Dordrecht/Boston/London, 1995.

[12] P. Fromme, H. T. Witt, W. D. Schubert, O. Klukas, W. Saenger, and N. Krauss, *Biochim. Biophys. Acta* **1275,** 76 (1996).

[13] N. Krauss, W.-D. Schubert, O. Klukas, P. Fromme, H. T. Witt, and W. Saenger, *Nature Struct. Biol.* **3,** 965 (1996).

[14] L. B. Smart, S. L. Anderson, and L. McIntosh, *EMBO J.* **10,** 3289 (1991).

[15] J. Yu, L. B. Smart, Y.-S. Jung, J. Golbeck, and L. McIntosh, *Plant Mol. Biol.* **29,** 331 (1995).

also been observed for some PSI mutants.[5,7] Characterization of these site-directed mutations has yielded valuable information on biogenesis of PSI and on functions of electron carriers. One limitation of the site-directed mutagenesis approach is that it relies on proposed models or predictions. Studies using pseudorevertants will add additional power to the mutagenesis approach and may potentially reveal information about PSI biogenesis and function that may not have been predicted. This paper describes a procedure for isolating spontaneous pseudorevertants from PSI mutants of *Synecho-cystis* 6803, localization and identification of the suppressor mutations, and some of the properties of the genetic transformation system in this cyano-bacterium.

## Isolation and Genetic Characterization of Pseudorevertants from PSI Mutants

### Isolation of Pseudorevertants from PsaC Mutants

The PsaC subunit, encoded by *psaC*, provides the ligands for two [4Fe-4S] clusters, $F_A$ and $F_B$. The proposed cysteine ligands have been studied by a combination of site-directed mutagenesis and *in vitro* reconstitution. Mutant PsaC derived from overexpression in *Escherichia coli* was reconstituted with a biochemical preparation of PSI cores. These experiments showed that Cys-51 in PsaC is a ligand to the $F_A$ cluster, whereas Cys-14 is a ligand to the $F_B$ cluster.[16] PsaC mutants C51D, C51S, C51A, C14D, C14S, and C14A were introduced into *Synechocystis* 6803, in which a cysteine ligand to $F_A$ or $F_B$, respectively, is changed to aspartate, serine, or alanine by site-directed mutagenesis. Physiologic characterization has shown that these mutations have decreased PSI capacity and efficiency, making PSI limiting for whole-chain photosynthesis. As a result, these mutants cannot grow photoautotrophically under white light and their mixotrophic growth (in the presence of glucose) is inhibited by white light (22–60 $\mu$mol m$^{-2}$ s$^{-1}$).[7] To isolate pseudorevertants, $10^8$ cells from a mid-log phase culture grown under LAHG conditions[4] were spread onto 90-cm plates with solid BG-11 medium supplemented with gentamicin (glucose is added for mixo-trophic growth). The plates were placed at 30° under 10 $\mu$mol m$^{-2}$ s$^{-1}$ white light for 2 days then transferred to 22 $\mu$mol m$^{-2}$ s$^{-1}$ white light for up to 40 days. Colonies have been found to appear spontaneously under these selective conditions at frequencies between $10^{-6}$–$10^{-8}$ in several tests. Hundreds of the isolates were serially streaked three to four times, with about half of them able to grow on the fourth and successive plates. Why other isolates could appear on the first selective plates but could not grow on

[16] J. Zhao, N. Li, P. V. Warren, J. H. Golbeck, and D. A. Bryant, *Biochemistry* **31**, 5093 (1992).

the successive plates is not clear. Following the fourth serial plating, selected isolates were characterized physiologically and genetically, as discussed later.

Photoautotrophic and mixotrophic growth under white light (22 $\mu$mol m$^{-2}$ s$^{-1}$) was used to isolate and characterize pseudorevertants. Isolates obtained from one selective condition were tested for their ability to grow in other conditions by streaking on plates. Three classes of pseudorevertants have been found: those capable of growth under (1) photoautotrophic conditions, but not mixotrophic conditions; (2) mixotrophic conditions, but not photoautotrophic conditions; and (3) both conditions. Pseudorevertants generally grow more slowly than wild-type under these conditions.

*Genetic Confirmation for Suppressor Mutations*

Because there are multiple copies of the genome in each *Synechocystis* cell,[3] revertant colonies are serially streaked at least four times on selective medium in order to achieve genetic homogeneity. The presence of the primary mutations was checked using a rapid genetic screening based on polymerase chain reaction (PCR). Target fragments containing *psaC* were directly amplified from cyanobacterial cells using the following primers: (1) ATATTATTTTTTTCGACTTTA and (2) GATCAAAAATTGGAA TAATG. Purified PCR products were used for restriction analysis and direct sequencing (see Materials and Methods section for more details). Although true reversions have been found in C14S revertants (codon TCT for serine was reverted to TGT for cysteine), most of the revertants of PsaC mutants contained the primary mutations, and no second-site mutations were found in *psaC*. These results suggest that suppressor mutations in other genes may be responsible for growth under the selective conditions. However, the metabolism of cyanobacteria is very flexible. For example, flavodoxin can substitute for ferredoxin under conditions of iron limitation.[17] It is important therefore to demonstrate that the revertants have come from genetic mutations rather than from physiologic acclimation. DNA applied directly to the surface of a lawn of *Synechocystis* 6803 embedded in agar can result in the transformation of these cells.[18] This observation provides a simple procedure named *dot transformation* for the rapid assay of large numbers of DNA samples that may phenotypically complement a primary mutation. Dot transformation was used in characterization of revertants in this study. Total DNA was isolated from revertant colonies and was used to transform the primary PSI mutants. An example of such

---

[17] K. K. Ho and D. W. Krogmann, *in* "The Biology of Cyanobacteria" (N. G. Carr and B. A. Whitton, eds.), pp. 191–214. Los Angeles: University of California Press, 1982.
[18] V. A. Dzelzkalns and L. Bogorad, *EMBO J.* **7**, 333 (1988).

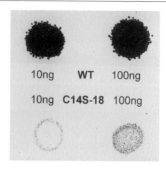

FIG. 1. *In situ* dot transformation demonstrates suppressor mutation. Log-phase culture 0.1 ml (OD$_{730}$ = 0.3) of a primary PSI mutant C14S was added to 3 ml melted 0.7% Bacto-agar at 50°, then the mixture was poured onto a 90-cm petri dish containing 50 ml BG-11 solid medium supplemented with gentamycin. Total DNA (10 and 100 ng) isolated from wild-type and a pseudorevertant C14S-18 was dropped in 5 $\mu$l on the lawn at marked areas. The plates were placed under 10 $\mu$mol m$^{-2}$ s$^{-1}$ white light for 4 days before transfer to 22 $\mu$mol m$^{-2}$ s$^{-1}$ white light. The plate was photographed 40 days after transformation. In a control experiment total DNA isolated from C14S was dropped on a lawn of C14S cells, and no cluster of colonies appeared at the marked areas in 40 days. Note that some colonies appeared spontaneously outside of transformed areas. The protocol was modified from V. A. Dzelskalns and L. Bogorad, *EMBO J.* **7**, 333 (1988).

transformation assay is shown in Fig. 1. All the isolates tested so far demonstrated specific DNA modifications.

## Locate Suppressor Mutations by Phenotypic Complementation

Mutations in PsaC such as C51D result in unstable assembly of the complex and lower electron transfer efficiency in assembled complexes. Conceivably the defects could be compensated by suppressor mutations in PsaA/PsaB dimer, which may restore complex stability and/or electron transfer efficiency. Dot transformation was used to locate suppressor mutations to a particular DNA fragment by phenotypically complementing primary mutant cells using specific fragments amplified from the revertants. Suppressor mutations in 10 pseudorevertant isolates from C51D have been localized to PsaA/PsaB in this way. Primers used to amplify psaA–psaB operon are (1) CCAGTTCCCGATGAAGGATTTC and (2) CCGGTCAACCTGCGACAGAA. The 10 pseudorevertants show consistently different photoautotrophic growth rates on plate, indicating that the mutant phenotype that results from C51D can be suppressed by several different mutations on PsaA/PsaB. In contrast, no suppressor mutation for C14S has been localized to any tested genes for PSI subunits (psaA–psaB, psaC, psaD, and psaE).

## Secondary Mutation at the Same Site as the Primary Mutation

The primary photochemical reactions catalyzed by the PSI reaction center occur on a core heterodimer composed of the subunits PsaA and PsaB, and a small subunit PsaC. The interaction between the PsaC subunit and the core heterodimer has been studied by molecular modeling and it was proposed that the two conserved regions between the cysteine ligands of the $F_X$ cluster on PsaA and PsaB (CDGPRGGTC) are two flexible loops that form a surface-exposed cavity to accommodate the PsaC subunit.[19] To test the proposal, an arginine residue $R561_{PsaB}$, which may interact with a negatively charged residue on the PsaC subunit in *Synechocystis* 6803, was changed to glutamate by site-directed mutagenesis. The mutant R561E cannot grow photoautotrophically due to the reduced level of PSI.[6] To further understand the role of $R561_{PsaB}$ in assembly/stability of PSI complex, spontaneous revertants have been isolated from R561E at a rate of $4 \times 10^{-7}$ under photoautotrophic growth conditions. The spontaneous mutations have been localized to the psaA–psaB operon in many revertants by phenotypic complementation. Sequencing of PCR fragments amplified from one of the revertants revealed a secondary mutation at $561_{PsaB}$, which converts codon GAA for glutamate to AAA for lysine. The reversion restores photoautotrophic growth to a rate comparable to wild-type cells. This result strongly suggests a role for a positively charged residue at this site in stable assembly of the PSI complex.

## Genetic Transformation in *Synechocystis* 6803

Some characteristics of genetic transformation in *Synechocystis* 6803 have been studied previously.[3] When dot transformation is used in the characterization of pseudorevertants, some colonies grow under selective conditions as a result of transformation, and other colonies appear on the plates spontaneously (see Fig. 1 for an example). Therefore, success is dependent on a higher transformation frequency over the proportion of cells containing spontaneous mutations. Spontaneous mutations have been found to accumulate in cultures of the primary mutants, making aged cultures unusable for suppression testing. To avoid such a problem, a fresh liquid culture should be used for suppression tests. To achieve a higher transformation frequency, several parameters have been studied: growth conditions of the primary mutants, methods for preparation of transforming DNA, competency of primary PSI mutants, and transformation techniques.

---

[19] S. M. Roddy, R. Schulz, L. McIntosh, and J. Biggins, *Photosynth. Res.* **42,** 185 (1994).

*Comparison of Growth Conditions for Primary PSI Mutants*

Isolation of *Synechocystis* 6803 mutants deficient in PSI function was initially made possible by growing cells under LAHG conditions.[4] The PsaC mutants, including C51D, can grow under LAHG or mixotrophic conditions with dim white light (5 $\mu$mol m$^{-2}$ s$^{-1}$).[7] The frequencies of transformation of C51D cells grown under these conditions were tested using plasmid pCG, which contains a 3.5-kb insert including a wild-type *psaC* gene.[7] As shown in Table I, mixotrophically grown cells were transformed with frequencies about three times that of LAHG cells. Although the reason for this differential is not known, the LAHG grown cells may have to go through an additional physiologic acclimation to continuous light. Note that the transformation frequency for both LAHG and mixotrophically grown PSI mutant cells varies considerably with different transforming DNA molecules, in agreement with a previous report with autotrophically grown wild-type cells.[3]

*Comparison of Methods for Preparation of Transforming DNA*

To study whether transformation is sensitive to impurities in transforming DNA, wild-type cells were transformed to kanamycin resistance using plasmid pKW1194[3] prepared in four different ways: (1) direct use of a crude preparation made by ethanol precipitation of *E. coli* cell lysate

TABLE I
COMPARISON OF TRANSFORMATION FREQUENCIES FOR LAHG AND
MIXOTROPHICALLY GROWN C51D CELLS[a]

| Growth conditions | pCG amount (ng) | Transformation frequency ($10^{-5}$ cell) |
| --- | --- | --- |
| LAHG | 2 | 6 |
| Mixotrophic | 2 | 20 |
| LAHG | 20 | 15 |
| Mixotrophic | 20 | 40 |

[a] C51D cells ($10^7$) from LAHG culture (OD$_{730}$ = 0.33) and mixotrophic culture (OD$_{730}$ = 0.50) were harvested (OD$_{730}$ = 0.25 corresponds to 1 × $10^8$ cells ml$^{-1}$) by centrifugation at 4500$g$ for 6 min at room temperature. The cell pellet was suspended in 20 $\mu$l BG-11 medium with 22 or 200 ng of plasmid pCG. The mixture of cells and DNA was incubated for 2 hr in a sterile test tube under 30°, 12 $\mu$mol m$^{-2}$ s$^{-1}$. Fresh medium was added to final volume of 500 $\mu$l, then 50 $\mu$l of diluted mixture (containing $10^6$ cells, 2 or 20 ng pCG) was spread onto BG-11 agar plates supplemented with 1 $\mu$g ml$^{-1}$ gentamicin. The plates were placed under 30°, 12 $\mu$mol m$^{-2}$ s$^{-1}$, for 3 days before transfer to 22 $\mu$mol m$^{-2}$ s$^{-1}$. Colonies were counted 10 days after transformation. No colonies appeared in 10 days in a control experiment from which pCG DNA was omitted.

(Wizard miniprep system, Promega, Madison, WI); or further treatment of the crude preparation by the following means: (2) purification using Wizard miniprep system (Promega); (3) purification using QiaExII system (Qiagen, Chatsworth, CA); (4) purification by phenol-chloroform extraction. WT-Gm is a strain carrying a Gm resistance gene downstream of the *psaC2* but otherwise wild type.[7] Cells (3 ml) of this strain grown in BG-11 liquid medium supplemented with 1 $\mu$g ml$^{-1}$ gentamycin were pelleted when $OD_{730}$ was 0.41. The cells were resuspended in 3 ml of fresh BG-11. A 300-$\mu$l portion was transferred to a 15-ml Corning tube, and was mixed with 30 or 300 ng of plasmid. The mixture was then incubated for 20 hr at 30° under 22 $\mu$mol m$^{-2}$ s$^{-1}$ white light and agitated several times to keep the cells suspended. This mixture (100 $\mu$l/plate) was when spread onto 9-cm BG-11 plates supplemented with 5 $\mu$g ml$^{-1}$ kanamycin and incubated as before. Colonies were counted 9 days after transformation. Transformation frequencies between $3 \times 10^{-3}$ and $5 \times 10^{-3}$ were observed for all four DNA preparations when 10 ng plasmid was mixed with $10^6$ cells, and transformation frequencies between $5 \times 10^{-3}$ and $1 \times 10^{-2}$ were observed for all four DNA preparations when 100 ng plasmid was mixed with $10^6$ cells. No colonies appeared in a control experiment from which the plasmid was omitted.

### Competency of the PSI Mutant Cultures at Various Stages of Log Phase

Mixotrophically grown C51D cells in early log phase ($OD_{730} = 0.15$), mid-log phase ($OD_{730} = 0.30$), and late log phase ($OD_{730} = 0.65$) were plated onto BG11 medium. Total DNA (10 ng in 5 $\mu$l) isolated from the wild-type strain was dropped on cell lawns at marked areas, and the plates were put under 5 $\mu$mol m$^{-2}$ s$^{-1}$ light for 4 days before transfer to 22 $\mu$mol m$^{-2}$ s$^{-1}$ white light. Some clusters of colonies were observed at the areas on all the plates 10 days following transformation. This shows that the PSI mutant cells were competent throughout the log phase. In addition, total DNA (10 ng in 5 $\mu$l) isolated from the wild-type strain was dropped on plates containing the mid-log-phase cells 2 days and 5 hours, respectively, before the plates were transferred to 22 $\mu$mol m$^{-2}$ s$^{-1}$ white light. Transformation was observed in 10 days, showing the PSI mutant cells stayed competent for at least 4 days under dim light on plates.

## Materials and Methods

### Strains and Growth Conditions

Experiments were performed using a glucose-tolerant strain of *Synechocystis* sp. PCC 6803, which was acclimated for growth on solid medium in the dark. PSI mutants were maintained under LAHG conditions.[4] Cultures were grown in BG-11 medium prepared according to Williams[3] except laboratory-purified Bacto-agar was used for LAHG.[4] Antibiotics were added in the following concentrations: kanamycin (Km), 5 mg/liter; gentamicin (Gm), 1 mg/liter. Photoautotrophic and mixotrophic growth were on solid media with or without supplemental glucose in a chamber providing continuous light. Cool white fluorescent bulbs made by General Electric were used. The light intensity was varied by covering plates with layers of cheese cloth and was monitored using a L1-185A photometer (LICOR, Lincoln, NE). Fully grown plates containing glucose may be wrapped with parafilm and stored in the dark at 4° for up to 6 months. Photoautotrophically grown cells should be stored in BG11 with 15% glycerol at −80°.

*DNA Manipulations*

Polymerase chain reaction (PCR) was performed using a PTC 200 thermal cycler (MJ Research Inc., Watertown, MA) and *Taq* polymerase (Boehringer Mannheim, Indianapolis, IN). Cells picked from a medium-size colony or an equivalent amount of cells collected from liquid culture were washed once with water and used as template. For amplification of long fragments (≥3 kb), total DNA was used as template, and *Pwo* polymerase (Boehringer Mannheim) was used at one-fifth total polymerase activity in addition to *Taq* polymerase. Amplification products were checked in agarose gel and may be directly used for dot transformation. For restriction analysis or sequencing, PCR products were purified using a Wizard PCR purification kit (Promega) or Ultrafree-MC 30,000 NMWL filter unit (Millipore Corp., Bedford, MA), which is recommended for long fragments. DNA sequencing was performed using dye-terminator chemistry for fluorescence sequencing at the Michigan State University DNA Sequencing Facility. The procedure for preparation of cyanobacterial DNA was adapted from Ohad and Hirschberg[20] with two "loopsful" of cells scraped from plates or cells from 10 ml liquid culture being used to extract DNA. RNA was removed from the preparation by incubation with RNase A, and DNA was recovered by ethanol precipitation.

Acknowledgment

We thank Yong Wang for her contribution in characterization of R561E-derived revertants.

[20] N. Ohad and J. Hirschberg, *Plant Cell* **4,** 272 (1992).

## [3] High-Frequency Nuclear Transformation of *Chlamydomonas reinhardtii*

By KAREN L. KINDLE

### Introduction

Because both the chloroplast and nuclear genomes can be transformed easily, quickly, and with high efficiency, *Chlamydomonas reinhardtii* has become a valuable eukaryotic model for addressing a variety of questions in photosynthesis and chloroplast biology. Although similar methods are used for introducing DNA into the chloroplast and nucleus, the nature of the recombination events and the availability of selectable markers and reporter genes are different, the latter reflecting the essential prokaryotic or eukaryotic nature of gene expression in the two compartments. Whereas all integration and replacement events in the chloroplast genome appear to occur via homologous recombination events that require short regions of identity,[1] the overwhelming majority of integration events in the nucleus occur by nonhomologous recombination at ectopic genomic locations.[2,3] Nuclear integration is sometimes associated with deletions and/or re-arrangements of the introduced DNA and/or DNA at the genomic integration site. The rarity of homologous recombination between the transforming plasmid and chromosomal DNA has so far limited the utility of nuclear transformation for gene-targeted disruption. However, because integration is fairly random, insertions sometimes disrupt the function of actively expressed genes, resulting in a null mutant phenotype. The inserted DNA becomes a molecular tag, which allows the DNA flanking the insertion to be cloned via plasmid rescue or hybridization strategies.[3]

Nuclear transformation can be used for several purposes. First, to confirm or test gene function, a wild-type gene, or one that has been altered *in vitro*, can be introduced into a mutant defective at the corresponding nuclear locus. The most efficient selectable markers now available for *Chlamydomonas* nuclear transformation complement auxotrophic or non-photosynthetic mutations. Even if the gene under study has no selectable phenotype, it can be introduced by cotransformation with a selectable marker. In our hands, the frequency of transformants that express nonselected genes varies between 25 and 90%. A second application, mentioned

[1] J. E. Boynton and N. W. Gillham, *Methods Enzymol.* **217,** 510 (1993).
[2] K. L. Kindle, R. A. Schnell, E. Fernández, and P. A. Lefebvre, *J. Cell Biol.* **109,** 2589 (1989).
[3] L.-W. Tam and P. A. Lefebvre, *Genetics* **135,** 375 (1993).

earlier, is to use the insertion events that occur during nuclear transformation as a mutagen to generate tagged null alleles. Because *Chlamydomonas* vegetative cells are haploid, any phenotype is immediately apparent. Although many potentially interesting mutations would be lethal as knockouts, mutants that are completely defective in photosynthesis are viable in *C. reinhardtii*, since the organism can use acetate as a reduced carbon and energy source. Therefore, mutants defective in the nuclear-encoded structural or regulatory genes required to assemble the various protein complexes that carry out photosynthetic electron transport can be isolated as nonphotosynthetic transformants that exhibit high chlorophyll fluorescence. Bennoun and Béal[4] have developed instrumentation for measuring fluorescence induction kinetics directly on an agar plate. The software permits simultaneous screening of several hundred colonies for abnormal photosynthetic electron transport by measuring fluorescence yields at four time points during the fluorescence decay. A single plate can be completely screened in about a minute. Finally, nuclear transformation can potentially be used to generate or phenocopy mutants defective in the genomic copy of a cloned gene, by gene-targeted disruption or antisense approaches, respectively. Although these latter approaches are not yet well developed in *Chlamydomonas,* gene-targeting does occur at a low frequency[5,6] and this approach has been used to generate mutations in the *nit8* gene[7] and possibly in the *psaF* gene.[8] With a predictable phenotype and/or a sufficiently powerful screen, the efficiency of random insertional mutagenesis may allow mutations in specific nuclear genes to be isolated, making it unnecessary to develop gene-targeted disruption techniques.[8-10]

Selectable Markers

The most efficient selectable markers for nuclear transformation are *Chlamydomonas* genes that complement mutations in the corresponding structural genes. The most widely used markers have been *NIT1,* which encodes nitrate reductase,[2,11] and *ARG7,* which encodes argininosuccinate lyase,[12] which are described in the following protocols. The *NIC7* gene, which complements a nicotinamide-requiring mutation,[13] also appears to

[4] P. Bennoun and D. Béal, *Photosyn. Res.* **51,** 161 (1997).
[5] O. A. Sodeinde and K. L. Kindle, *Proc. Natl. Acad. Sci. U.S.A.* **90,** 9199 (1993).
[6] N. Gumpel, J.-D. Rochaix, and S. Purton, *Curr. Genet.* **26,** 438 (1994).
[7] J. A. E. Nelson and P. A. Lefebvre, *Mol. Cell Biol.* **15,** 5762 (1995).
[8] J. Farah, F. Rappaport, Y. Choquet, P. Joliot, and J.-D. Rochaix, *EMBO J.* **14,** 4976 (1995).
[9] E. J. Smart and B. R. Selman, *Mol. Cell Biol.* **11,** 5053 (1991).
[10] C. G. Wilkerson, S. M. King, A. Koutoulis, G. J. Pazour, and G. B. Witman, *J. Cell Biol.* **129,** 169 (1995).
[11] E. Fernández, R. Schnell, L. P. W. Ranum, S. C. Hussey, C. D. Silflow, and P. A. Lefebvre, *Proc. Natl. Acad. Sci. U.S.A.* **86,** 6449 (1989).
[12] R. Debuchy, S. Purton, and J.-D. Rochaix, *EMBO J.* **8,** 2803 (1989).
[13] P. J. Ferris, *Genetics* **141,** 543 (1995).

be a good transformation marker, since reversion of the *nic7* mutation is rare and background growth of mutant cells can be reduced by growing them in the presence of 3-acetylpyridine. Although genes that are required for photosynthesis can be used for selection, they are not particularly useful for studies of photosynthesis or chloroplast development and are not considered here.

Drug resistance markers can be used in any genetic background, which circumvents the need to construct appropriately marked strains as transformation recipients. A *Chlamydomonas* gene (*CRY1*) that confers resistance to the cytosolic translation inhibitors cryptopleurine and emetine carries a mutation in the coding region of cytosolic ribosomal protein S14. Although the *CRY1* gene confers cryptopleurine resistance at a high frequency, it is semidominant, so existing ribosomes must be depleted by a period of nitrogen starvation and regenerated before the resistance can be expressed.[14] Although foreign genes are not readily expressed in *Chlamydomonas*, low-frequency transformation (~1% of the rate with *Chlamydomonas* genes) has been reported with bacterial genes conferring resistance to phleomycin (*ble*, which encodes a bleomycin-binding protein)[15] and spectinomycin/ streptomycin (*aadA*, which encodes aminoglycoside adenine transferase).[16] The difficulty in expressing foreign genes in *Chlamydomonas* has been attributed to the high frequency of nuclear codons with G or C in the third position and to gene silencing; lack of gene expression has been correlated with hypermethylation.[16,17] Silencing of both foreign genes and reintroduced *Chlamydomonas* genes is less of a problem when they confer a selective advantage (our unpublished observations and Ref. 16).

Reporters

Although genes that are good reporters in vascular plants, such as *uidA* (encodes beta glucuronidase; GUS), have not been expressed in the nuclear genome of *Chlamydomonas* to date,[17] two *Chlamydomonas* genes confer easily discernible phenotypes. The first is the *PC1* gene, which encodes NADPH:protochlorophyllide oxidoreductase (POR), the enzyme responsible for light-dependent protochlorophyll(ide) reduction. Although wild-type strains of *Chlamydomonas* can make chlorophyll and are green whether they are grown in the dark or the light, strains carrying mutations in both *pc1* and in the light-independent protochlorophyllide reduction pathway (*y* mutants) are yellow and very light sensitive. After transformation with the *PC1* gene, they form green colonies in the light.[18] The gene

[14] J. A. E. Nelson, P. B. Savereide, and P. A. Lefebvre, *Mol. Cell. Biol.* **14**, 4011 (1994).
[15] D. R. Stevens, J.-D. Rochaix, and S. Purton, *Mol. Gen. Genet.* **251**, 23 (1996).
[16] H. Cerutti, A. M. Johnson, N. W. Gillham, and J. E. Boynton, *Genetics* **145**, 97 (1997).
[17] J. E. Blankenship and K. L. Kindle, *Mol. Cell. Biol.* **12**, 5268 (1992).
[18] J. Li and M. P. Timko, *Plant Mol. Biol.* **30**, 15 (1996).

is thus both a selectable marker and a screenable one, since the colors of the transformant colonies vary in shade, which appears to be related to the level of expression (our published observations). A second, probably more useful reporter gene is arylsulfatase,[19] which can be detected easily in colonies by spraying plates with the chromogenic substrate 5-bromo-4-chloro-3-indolyl sulfate ($XSO_4$) or assayed in solution with $p$-nitrophenyl sulfate or $\alpha$-naphthyl sulfate.[20,20a] Because arylsulfatase is normally induced in response to sulfur starvation, it can be used as a reporter of chimeric gene expression in normal sulfur-sufficient conditions in any genetic background. Arylsulfatase protein turns over very slowly, however, and is therefore inappropriate for some applications.[21]

Promoters

A number of nuclear genes are regulated in response to growth conditions and represent possible sources of inducible promoters. These include genes encoding nitrate reductase, which is induced when cells are switched from ammonium- to nitrate-containing medium,[11,17] arylsulfatase and cytochrome $c_6$, which are induced when cells are starved for sulfur or copper, respectively,[22,23] and carbonic anhydrase, which is induced in response to low $[CO_2]$.[24]

When *Chlamydomonas* genes are reintroduced into the nuclear genome, their expression is sometimes significantly lower than that of the endogenous gene. For example, though the *cabII-1* gene is highly expressed in *Chlamydomonas*, molecularly tagged or chimeric *Chlamydomonas* genes containing this promoter were expressed at a low level when transformed into the nuclear genome.[2,17] A promising promoter for efficient transcription of chimeric genes is the *rbcS2* promoter, which permitted high-level expression of an altered *Chlamydomonas* tubulin gene.[25]

Methods for Introducing DNA into Cells

Several methods have been developed for introducing DNA into the nuclear genome of *Chlamydomonas*, including particle bombardment,[2,12] vortexing with glass beads[26] or silicon carbide whiskers,[27] and electropora-

[19] J. P. Davies, D. P. Weeks, and A. R. Grossman, *Nucl. Acids Res.* **20**, 2959 (1992).
[20] E. L. de Hostos, R. K. Togasaki, and A. Grossman, *J. Cell Biol.* **106**, 29 (1988).
[20a] M. Ohresser, R. F. Matagne, and R. L. Loppes, *Curr. Genet.* **31**, 264 (1997).
[21] S. Jacobshagen, K. L. Kindle, and C. H. Johnson, *Plant Mol. Biol.* **31**, 1173 (1996).
[22] J. P. Davies, F. Yildiz, and A. R. Grossman, *Plant Cell* **6**, 53 (1994).
[23] J. M. Quinn and S. Merchant, *Plant J.* **7**, 623 (1995).
[24] P. Villand, M. Eriksson, and G. Samuelsson, *Biochem. J.* **327**, 51 (1997).
[25] K. G. Kozminski, D. R. Diener, and J. L. Rosenbaum, *Cell Motil. Cytoskeleton* **25**, 158 (1993).
[26] K. L. Kindle, *Proc. Natl. Acad. Sci. U.S.A.* **87**, 1228 (1990).
[27] T. G. Dunahay, *Biotechniques* **15**, 452 (1993).

tion.[28] The vortexing methods require no specialized equipment, and are very easy and efficient; we prefer glass beads because of the biohazard associated with silicon whiskers. Glass-bead transformation can be scaled up without loss of efficiency, but it does require a cell wall-deficient transformation recipient, either a strain carrying a *cw* mutation, or cells that have been rendered cell wall deficient by incubation in gamate lytic enzyme (GLE). In contrast, particle bombardment is more efficient with walled cells, and the frequency of homologous recombination events appears to be higher with this method.[5] However, the overall efficiency of transformation is somewhat lower than with glass-bead transformation, and the number of DNA copies that are integrated into the genome is higher,[2,26] which is a disadvantage for recovering tagged alleles from insertional mutants. Electroporation results in the integration of a small number of gene copies, and it is unnecessary to remove the cell wall for this procedure. However, with early protocols overall transformation efficiencies are lower than with glass-bead transformation.[28] Recently, a very high efficiency electroporation protocol has been reported for cell wall-deficient *Chlamydomonas* strains.[28a]

## Nuclear Transformation of *Chlamydomonas* using *NIT1* or *ARG7*

### *Strains*

The Chlamydomonas Genetics Center at Duke University (http://www.botany.duke.edu/DCMB/chlamy.htm) maintains and distributes a collection of *Chlamydomonas* strains and plasmids. Detailed information on culturing and maintaining *Chlamydomonas* strains can be found in *The Chlamydomonas Sourcebook*.[29] For transformation with *NIT1*, a *nit1NIT2* strain is required, preferably in the *cw15* genetic background. *NIT1* encodes the structural gene for nitrate reductase; *NIT2* is a regulatory gene that is required for expression of *NIT1*. Many of the strains in the Chlamydomonas Genetics Center collection are derived from a "wild-type" background that actually carries *nit1* and *nit2* mutations and hence are inappropriate for transformation with *NIT1*.[29] For future genetic crosses, the mating type should be considered; chloroplast DNA is inherited uniparentally from the mt⁺ strain, mitochondrial DNA from the mt⁻ strain. Furthermore, it is worthwhile to establish that the transformation recipient mates well before generating transformants. Appropriate strains for transformation with *NIT1* are CC-2454 or CC-2677 (*nit1-305cw15mt⁻*). For transformation with

---

[28] L. E. Brown, L. Sprecher, and L. R. Keller, *Mol. Cell. Biol.* **11,** 2328 (1991).

[28a] K. Shimogawara, S. Fujiwara, A. Grossman, and H. Usuda, *Genetics* **148,** 1821 (1997).

[29] E. H. Harris, *in* "The Chlamydomonas Sourcebook: A Comprehensive Guide to Biology and Laboratory Use." Academic Press, San Diego, 1989.

*ARG7*, both *arg2* and *arg7* mutants lack argininosuccinate lyase and can be complemented by *ARG7*; *arg2* shows intragenic complementation with some *arg7* alleles. CC-1618 is an appropriate strain for transformation (*arg7cw15mt⁻*). Double mutant strains CC-2986 (*arg2nit1mt⁺*), CC-3379 (*arg2nit1mt⁻*), CC-3396 (*arg7nit1mt⁻*) or CC-3397, CC-3398, or CC-3399 (all *arg7nit1mt⁺*) are useful for sequential transformation by *ARG7* and *NIT1*. Note that arginine interferes with the selection for nitrate reductase, so *ARG7* transformation must be carried out first.

## DNA

pMN24[11] (Chlamydomonas Genetics Center P-387) contains the *Chlamydomonas NIT1* gene cloned into pUC119, whereas pArg7.8[12] (Chlamydomonas Genetics Center P-389) contains the *ARG7* gene cloned into pUC19. Plasmid DNA should be purified by CsCl centrifugation or an alternative method to give DNA of comparable purity. We find a slight enhancement in transformation efficiency by using DNA linearized at a site in the polylinker (two- to three-fold[26]), and linearization is critical for insertional mutagenesis, because it greatly increases the chance that the plasmid DNA will integrate intact and hence be clonable by plasmid rescue or "bookshelf" library hybridization approaches.[3] Linearized DNA should be phenol extracted and ethanol precipitated prior to use in transformation.

The amount of DNA that should be used depends on the nature of the experiment. For cotransformation experiments, especially ones where screening for the unselected marker is labor intensive, a two- to five-fold molar ratio of unselected to selected DNA is suggested. We have shown previously that addition of more than ~2 $\mu$g plasmid DNA for $5 \times 10^7$ cells in a volume of 0.3 ml results in no further increase in the number of transformants recovered, but increases the number of inserted plasmid copies.[26] For cotransformation experimetns 1 $\mu$g of plasmid DNA containing the selectable marker is therefore suggested with correspondingly more of the unselected DNA. For insertional mutagenesis, the amount of DNA should be kept as low as possible without severely affecting the recovery of transformants, because isolation of the genomic DNA sequences that flank the plasmid insertion is greatly facilitated in strains that contain only a single insertion event. In this case, preliminary experiments with DNA ranging in amount from 0.1 to 5 $\mu$g are suggested to determine recovery of transformants.

## Glass Bead Transformation

The number of selective agar plates that will be required depends on the application. Using the following protocol, 100–300 transformant colonies can be obtained per plate. For cotransformation, 1 or 2 plates per construct is usually ample. For mutagenesis, the number of plates is determined by the expected frequency of mutations of interest, and the number of colonies that can reasonably be screened. Forty milliliters of cultured cells are required for each plate.

*Materials*

20% Polyethylene glycol (PEG)-8000 (Sigma P-2139, St. Louis, MO), autoclaved. PEG solutions are somewhat unstable, so they should be prepared fresh every 1–2 months. There is some controversy as to whether PEG is always necessary or helpful; we have found that with *cw15* mutants and cells treated with GLE, PEG substantially increases the number of transformants.

Sterile, disposable, conical, 15-ml polypropylene tubes.

Glass beads, 0.4–0.5 mm. To wash, soak in sulfuric acid for several hours, then rinse extensively with deionized water. Dry and bake at 250° for 3 hr. It is convenient to aliquot glass beads in the amounts that will be used: e.g., 0.3, 0.6, or 3 g, depending on the number of cells to be transformed.

Selective liquid medium: SGII+KNO$_3$[26] for *NIT1* transformation, HSA or TAP[29] for *ARG7* transformation.

Selective agar plates: above medium solidified with 1.5% agar. For *NIT1* transformations, the agar should be rinsed extensively with deionized water to wash out contaminating ammonium.

GLE, ~20 ml per liter of original culture, prepared as described later.

*Protocol*

All manipulations should be carried out in a sterile hood using sterile glass and plasticware.

1. Grow the appropriate *Chlamydomonas* strain with shaking in nonselective liquid medium, 500 ml/2 liter flask until cells reach a density of ~2 × 10$^6$/ml. We use SGII+NH$_4$NO$_3$[26] for *NIT1* transformations and HSA or TAP[29] + 40 $\mu$g/ml arginine for *ARG7* transformations.

2. Spin down cells at moderate speed (4000*g*) for 5 min and pour off the growth medium.

3. For *cw15* strains, resuspend in 7.5 ml per liter of original culture volume in liquid selective medium: SGII+KNO$_3$[26] for *NIT1* transformations or HSA or TAP for *ARG7* transformations. For *NIT1*, we have found that transformation frequency can be enhanced significantly by allowing the cells to shake at this high density in SGII+KNO$_3$ liquid for 4 hr in the light. Whether this is due to induction of *NIT1* gene expression or to gamete formation as the cells are starved for nitrogen is not known, but *CW*$^+$ cells

treated this way are competent for transformation and do not need further treatment with GLE.[30]

4. Efficient transformation of $CW^+$ cells by $ARG7$ requires that they be treated with GLE. Although this enzyme is not commercially available, it can be prepared very easily, as described later. Resuspend cells in GLE, using 1/25–1/100 of the original culture volume, and incubate for 45–90 min at 30°. The effectiveness of cell wall removal can be tested by determining the cells' sensitivity to detergent. Count in a hemacytometer duplicate samples treated or not with 0.05% Nonidet P-40. After incubation with GLE, harvest the cells by centrifugation and wash them twice with selective medium. Carry out the transformation immediately, because the cells begin to regenerate cell walls as soon as they are washed out of GLE.

5. Work with 3–4 tubes at a time. Add to a sterile, conical 15-ml polypropylene tube in order 300 $\mu$l cells, DNA, 100 $\mu$l 20% PEG, and 300 mg sterile glass beads. Vortex for 15–30 sec at top speed on a Fisher (Springfield, NJ) Vortex Genie 2 mixer. Use a pipette tip to remove the cells from the glass beads and spread them immediately on selective agar plates. (Note that for drug resistance markers, it may be necessary to allow the cells to express the gene in liquid culture for some time before spreading them on a selective plate). Because PEG is toxic, cells should not be left in 5% PEG longer than necessary. If desired, the cells may be washed by diluting them in selective liquid medium and harvesting them by gentle centrifugation before plating. For insertion mutagenesis it is convenient to scale up the vortex procedure 10-fold and to plate 350 $\mu$l vortexed cells per plate. Allow the liquid to soak into the plates, leaving them open in a sterile hood if necessary. Seal the plates with Parafilm and incubate them in the light; transformant colonies should be visible in 5–6 days.

### Preparation of Gamete Lytic Enzyme

GLE is synthesized during gametogenesis and released into the medium when gametes of opposite mating types are mixed; it degrades both gamete and vegetative cell walls. The yield of lytic activity depends on the efficiency of mating, so high-efficiency mating strains (e.g., CC-620 and CC-621) are recommended. Sterile media and plasticware are used throughout.

1. Grow cultures of opposite mating types photoautotrophically in HS liquid medium[29] on a 14-hr light, 10-hr dark–light cycle, bubbled or shaken with 5% $CO_2$, until they reach a density of ~$10^6$/ml. Alternatively, cells

---

[30] J. A. E. Nelson and P. A. Lefebvre, *in* "Methods in Cell Biology," Vol. 47 Cilia and Flagella (W. Dentler and G. Witman, eds), pp. 513. Academic Press, New York, 1995.

can be grown mixotrophically in HSA medium until they reach a concentration $\sim 3 \times 10^6$ cells/ml.

2. Harvest cells by centrifugation at 2500$g$ for 5 min, resuspend them at $2 \times 10^6$/ml in nitrogen-free HS medium, and incubate overnight in the light, with shaking.

3. Harvest cells as above and resuspend them at $5 \times 10^7$/ml in N-free HS medium, then add the two mating types together. Incubate in the light for 1 hr without shaking.

4. Harvest cells by centrifuging at 4000$g$ for 10 min, then spin the supernatant at 30,000$g$ for 30 min.

5. Filter the culture supernatant through a 0.45-$\mu$m filter to remove any remaining cells, then aliquot into sterile 15-ml tubes and freeze at $-20°$ for short-term storage, $-80°$ for longer term storage. This filtration is optional as *Chlamydomonas* cells do not survive freezing in aqueous media.

## Particle Bombardment

The glass-bead transformation method described earlier is adequate for most applications. For the *NIT1* gene, it appeared that the ratio of homologous recombination to ectopic integration events was higher with particle bombardment than with glass-bead transformation,[5] so particle bombardment may be the method of choice for attempting to generate mutants by targeted gene disruption. Furthermore, because walled cells can be transformed efficiently by particle bombardment, there is no need to use GLE. For particle bombardment, it is important that the total amount of DNA be kept low ($\leq 2.5$ $\mu$g in the protocol below) to prevent the tungsten particles from clumping. Furthermore, the number of cells plated should not be too high ($10^6$–$10^7$) since colonies form very slowly if the lawn is too dense. The efficiency of the method has improved since early reports,[2] largely because of the use of high-pressure helium in the newer model PDS1000/He particle delivery system from Bio-Rad (Hercules, CA).

### Materials

Tungsten M10 particles (available from Sylvania or Bio-Rad), 60 mg/ml. Sterilize by vortexing for 1–2 min in 1 ml 100% ethanol. Collect particles by brief centrifugation, remove ethanol, wash three times, and resuspend in 1 ml of sterile distilled water. Because tungsten particles oxidize, use them fresh or store them frozen. Gold particles can also be used and are somewhat less toxic, but tungsten is cheaper and gives reproducible results in our hands.

Flying disks (macrocarriers), flying disk holders, stopping screens, and rupture disks, all available from Bio-Rad, are soaked in 70% ethanol for 15 min and dried in a sterile hood.

2.5 $M$ CaCl$_2$.

0.1 $M$ spermidine (Sigma-S2626). This should be prepared monthly because it is unstable. It is important to use the free base form.

*Protocol*

1. Grow, harvest, and resuspend cells in 1/10 to 1/100 of the original culture volume, as described earlier. Spread 0.4 ml cell suspension in the center of a 90-mm culture dish containing the appropriate selective agar and allow the plates to dry in a sterile hood. Alternatively, cells can be resuspended in soft agar and poured onto an agar plate for bombardment.[1]

2. Precipitate DNA. Add 25 $\mu$l of thoroughly suspended M10 tungsten particles (60 mg/ml) to a sterile microcentrifuge tube. Add in this order, vortexing after each addition, 2.5 $\mu$g DNA (in a volume of 2.5–10 $\mu$l), 25 $\mu$l 2.5 $M$ CaCl$_2$, 10 $\mu$l 0.1 $M$ spermidine. Allow the mixture to sit on ice for 10 min, and then collect particles by spinning briefly in a microcentrifuge. Wash the particles with 250 $\mu$l 95% ethanol, mix, spin, and remove the supernatant. Resuspend the particles in 15 $\mu$l 95% ethanol.

3. Bombardment. Aseptically transfer a flying disk to a flying disk holder. Vigorously vortex the DNA-coated particles to be sure they are uniformly suspended and pipette 5 $\mu$l of the particles onto the center of the flying disk. If the particles are clumpy, they may be dispersed by incubating the microcentrifuge tube for 5–10 sec in a sonifier bath. Allow the ethanol to evaporate (1–2 min). Transfer the flying disk to the Bio-Rad PDS-1000/He, and carry out bombardment according to the manufacturer's instructions, using rupture disks that burst between 650 and 900 psi. In our standard setup, there is 1 cm between the rupture disk and the flying disk, and 1 cm between the flying disk and the stopping screen. The plate to be bombarded is placed on the bottom shelf of the chamber, which is evacuated to 27 in. Hg for the bombardment. Following bombardment, seal the plates with parafilm. Transformant colonies are visible within 7–10 days, and can vary in number from 10 to >100 transformants per plate.

Considerations for Choosing a Transformation Strategy

Both *NIT1* and *ARG7* are very effective nuclear transformation markers. More consistent transformation rates seem to be obtained with *ARG7*. *NIT1* transformation occasionally fails or works poorly for reasons that are unclear. Some people have had a very difficult time establishing nuclear transformation in their labs with *NIT1*. Furthermore, background growth of untransformed *nit1*$^-$ cells can be substantial if they are not washed before plating or if the agar is not adequately washed. This can be a problem for some screens, such as those for high chlorophyll fluorescence. On the other hand, anecdotal evidence suggests that *NIT1* insertion mutants are more frequently tagged by the inserted DNA, and insertion events are less frequently associated with deletions at the insertion site. Furthermore, we have found that *ARG7* transformation sometimes fails if the cells are incubated in low light or darkness following the procedure. Because many photosynthetic mutants are light sensitive, *NIT1* would appear to be the superior marker for generating insertional mutants that are unable to carry out photosynthesis. On the other hand, for cotransformation experiments or for experiments in which a particular mutant is sought, but there is no need to clone the disrupted DNA, *ARG7* may be the marker of choice. Although the *NIC7* marker has not been used extensively, it appears promising.

It is likely that the dominant drug resistance markers will gain popularity, both because they can be used in any strain and because of their smaller size. In our hands, the *ble* gene[15] is a significantly more efficient transformation marker than the *aadA* gene[16]; however, phleomycin is expensive (our unpublished observations). In large-scale mutant hunts, the lower transformation rates may be problematic, and the fact that phleomycin is itself mutagenic may mean that the frequency of tagged alleles will be lower.

Although nuclear transformation is easier with cell wall-deficient strains, subequent genetic crosses may be much more difficult because *cw* strains often do not mate efficiently and may have a high degree of meiotic product lethality after zygospore germination. Given that preparation of GLE is simple, and apparently dispensable for *NIT1* transformations, serious consideration should be given to using *CW*+ strains for insertional mutagenesis. On the other hand, it is difficult to isolate intact chloroplasts from *CW*+ strains because intact chloroplasts and walled cells fractionate together on Percoll gradients.[31]

*Chlamydomonas* nuclear transformation provides a powerful tool for studying photosynthesis and chloroplast biogenesis in a model eukaryote with excellent classical genetics. The capability of generating insertional mutations that eliminate nonvital chloroplast functions and cloning wild-type versions of the mutated genes should greatly increase our understanding of proteins that regulate chloroplast structure and function. With the cloned genes and corresponding mutants in hand, detailed structure–function analyses can be carried out by mutating the gene *in vitro*, and determining its phenotype *in vivo* in the appropriate transformed strains. These approaches should provide an unparalleled opportunity to under-

[31] K. L. Kindle and S. D. Lawrence, *Plant physiol.* **116,** 1179 (1998).

stand the intracellular signaling mechanisms that govern the biogenesis and function of chloroplasts in this unicellular photosynthetic eukaryote.

## Acknowledgments

Work in the author's lab has been supported by grants from the National Science Foundation, currently grants MCB9406540 from the Cell Biology Program and MCB9406550 from the Biochemical Genetics Program. I am grateful to Clare Simpson and Elizabeth Harris for comments on the manuscript.

# [4] Approaches to Investigating Nuclear Genes that Function in Chloroplast Biogenesis in Land Plants

*By* ALICE BARKAN

## Introduction

Chloroplasts belong to an organelle family called the *plastids*, members of which share the same genome but assume different forms in the different cell types of multicellular plants. All plastid forms arise from progenitor organelles called *proplastids*. The most dramatic morphologic change accompanying the differentiation of a proplastid into a chloroplast is the elaboration of the thylakoid membrane, an event that requires a massive increase in the expression of chloroplast and nuclear genes encoding thylakoid proteins and the precise targeting and assembly of these proteins within the membrane. Biochemical approaches have been used for many years to describe the changes in chloroplast gene expression that accompany chloroplast differentiation and to probe mechanisms by which the thylakoid membrane is assembled. Complementary approaches are required, however, to establish the mechanisms underlying chloroplast gene regulation and protein targeting *in vivo* since only a subset of factors will be identifiable in biochemical assays and it can be difficult to deduce their spectrum of roles *in vivo*. Recently, genetic approaches have begun to yield insights into mechanisms of chloroplast biogenesis. Nonetheless, the genetic analysis of chloroplast biogenesis is in its infancy and much remains to be learned from this approach.

## Choice of Organism

The three most genetically tractable photosynthetic eukaryotes are the single-celled algae *Chlamydomonas reinhardtii*, the dicot *Arabidopsis thali-*

*ana*, and the monocot *Zea mays. Chlamydomonas* is unique among these in that both the chloroplast and nuclear genomes can be transformed and nonphotosynthetic individuals yield progeny. In maize and *Arabidopsis*, mutations that result in photosynthetic defects must be propagated in the heterozygous state. Therefore, it is difficult in these organisms to identify second-site suppressors of photosynthetic defects, an approach that has proven useful with *Chlamydomonas*.[1-3] Although *Chlamydomonas* is in many ways an ideal experimental organism, studies of *Chlamydomonas* will not necessarily elucidate regulatory mechanisms in plants. Thus far, in fact, the variety of mutant phenotypes obtained in plants and *Chlamydomonas* differ substantially, suggesting that regulatory mechanisms in plants and algae may have diverged considerably.

Among land plants, maize offers a unique set of properties that facilitates the informative analysis of mutant phenotypes and the molecular cloning of mutant genes. Cloning genes by transposon tagging has become straightforward in maize. Biochemical studies are facilitated by the large seed, which supports the growth of nonphotosynthetic seedlings to the four-leaf stage, providing adequate leaf material for biochemical analysis quickly and conveniently. Developmental studies are aided by the fact that proplastids, maturing and mature chloroplasts are spatially separated in a gradient along the length of the seedling leaf.[4] *Arabidopsis* offers the advantages of a small genome and short generation time, but the small seed and plant are not well suited to the biochemical analysis of chloroplast defects. Nonetheless, several nuclear mutations that disrupt chloroplast gene expression have been identified in *Arabidopsis*[5,6] and studies in *Arabidopsis* will provide an important complement to analogous studies in large seeded plants. Described later are the experimental rationale and methodologies used in my laboratory to identify nuclear genes in maize that modulate chloroplast gene expression and that facilitate the targeting of proteins to the thylakoid membrane.

[1] T. A. Smith and B. D. Kohorn, *J. Cell Biol.* **126**, 365 (1994).
[2] J.-D. Rochaix, M. Kuchka, S. Mayfield, M. Schirmer-Rahire, J. Girard-Bascou, and P. Bennoun, *EMBO J.* **8**, 1013 (1989).
[3] K. L. Kindle, H. Suzuki, and D. B. Stern, *Plant Cell* **6**, 187 (1994).
[4] R. Leech, M. Rumsby, and W. Thompson, *Plant Physiol.* **52**, 240 (1973).
[5] J. Meurer, K. Meierhoff, and P. Westhoff, *Planta* **198**, 385 (1996).
[6] R. D. Dinkins, H. Bandaranayake, B. R. Green, and A. J. F. Griffiths, *Curr. Genet.* **25**, 282 (1994).

## Identification of Mutations that Disrupt Thylakoid Membrane Biogenesis

The phenotype of "high chlorophyll fluorescence" (*hcf*) has been widely used to identify higher plant mutants with defects in photosynthetic electron transport.[5–8] The screen for *hcf* mutants involves illuminating mutagenized F2 families with an ultraviolet light in a darkened room and identifying those families that segregate seedlings with increased levels of chlorophyll fluorescence.[9] Because most *hcf* mutants are noticeably chlorophyll deficient, we find it more convenient to screen for pale green seedlings that die after the development of three to four leaves, when nonphotosynthetic maize seedlings exhaust endosperm stores. Most pale green, seedling lethal mutants are deficient for one or more of the photosynthetic enzyme complexes that contain chloroplast-encoded subunits: photosystem II (PSII), cytochrome $b_6f$ (cyt $b_6f$), photosystem I (PSI), the thylakoid adenosine triphosphate (ATP) synthase, or ribulose 1,5-bisphosphate carboxylase (Rubisco). Many mutants identified in this way do exhibit increased chlorophyll fluorescence, but in some cases the increase is subtle and would likely be missed in a large-scale screen for the *hcf* phenotype. The degree of pigment deficiency associated with lesions in photosynthetic electron transport, the thylakoid ATP synthase, and/or Rubisco is influenced by light conditions. For some mutants, a chlorophyll deficiency is obvious only when seedlings are grown in full sunlight. Therefore, if the initial mutant screen is performed in growth chambers or in a winter greenhouse, a screen for the high chlorophyll fluorescence phenotype should be used to supplement visual screening for pigment deficiencies.

Mutations that cause a severe and global disruption of chloroplast gene expression or of thylakoid membrane biogenesis result not in the pale green phenotype described earlier, but in an albino phenotype. Plants harboring such mutations exhibit a distinctive ivory pigmentation that is easily distinguished from the paper-white phenotype results from lesions in the carotenoid biosynthetic pathway.[10,11] Carotenoid biosynthetic mutants such as *w3* or *vp2* can be obtained from the Maize Genetics Cooperative (University of Illinois) for comparison. An ivory phenotype can result from lesions in a wide variety of functions that are fundamental to the control of chloroplast biogenesis. In maize, for example, an ivory phenotype is associated with

---

[7] D. Miles, *Maydica* **39**, 35 (1994).
[8] A. Barkan, D. Miles, and W. Taylor, *EMBO J.* **5**, 1421 (1986).
[9] D. Miles, *in* "Methods in Chloroplast Molecular Biology" (R. Hallick, M. Edelman, and N.-H. Chua, eds.), pp. 75–106. Elsevier, New York, 1982.
[10] C.-D. Han, W. Patrie, M. Polacco, and E. H. Coe, *Planta* **191**, 552 (1993).
[11] C.-D. Han, R. J. Derby, P. S. Schnable, and R. A. Martienssen, *Maydica* **40**, 13 (1995).

the crs2 mutation, which disrupts the splicing of many chloroplast mRNAs,[12] and with mutations in a nuclear gene encoding a chloroplast-localized SecY homolog, which plays a key role in assembling the thylakoid membrane.[13] In tobacco, mutations that prevent the synthesis of the chloroplast-encoded RNA polymerase result in an albino phenotype.[14] Although ivory phenotypes are likely to reflect lesions in many fundamental aspects of chloroplast biogenesis, it can be difficult to gain insight into the primary defect in such mutants by phenotypic analysis alone, since these mutations cause many pleiotropic effects. The challenge is to distinguish primary defects from secondary effects. Yellow and paper-white phenotypes are likely to result from mutations disrupting chlorophyll or carotenoid biosynthetic pathways, respectively, and so are less likely to be informative with regard to mechanisms of chloroplast biogenesis.

## Transposon Mutagenesis and Mutant Propagation

Transposon tagging is one of the most convenient methods for cloning nuclear genes identified by mutations and is at present the only practical method available in maize. The *Mutator* (*Mu*) transposable element family is particularly well suited for generating and cloning mutations that disrupt chloroplast biogenesis. The *Mu* family consists of several different sequence classes, each flanked by similar terminal inverted repeats.[15] These elements transpose only in "*Mu*-active" maize lines, which harbor one or more copies of the autonomous member of the *Mu* family. *Mu*-active lines exhibit a high forward mutation rate and new mutations are distributed throughout the genome. The high mutation rate and lack of positional bias are advantageous for a mutagenesis of functions related to chloroplast biogenesis.

To generate new mutants, *Mu*-active plants are simultaneously self-pollinated and outcrossed to a non-*Mu* inbred line. (For methods to assess *Mu* activity, see Ref. 16.) The outcross progeny of those plants that were not heterozygous for a photosynthetic mutation (as revealed by examination of their selfed progeny) are then self-pollinated to generate F2 ears. Thirty kernels of each F2 ear are planted and screened for the segregation of nonphotosynthetic phenotypes. Any mutation segregating on an F2 ear arose during the development of the outcrossed *Mu*-active grandparent. Approximately 5% of the F2 ears typically segregate pigment-deficient

[12] B. Jenkins, D. Kulhanek, and A. Barkan, *Plant Cell* **9**, 283 (1997).
[13] L. Roy and A. Barkan, *J. Cell Biol.* **141**, 385 (1998).
[14] L. A. Allison, L. D. Simon, and P. Maliga, *EMBO J.* **15**, 2802 (1996).
[15] V. L. Chandler and K. J. Hardeman, *Adv. Genet.* **30**, 77 (1992).
[16] P. S. Chomet, *in* "The Maize Handbook" (M. Freeling and V. Walbot, eds.), pp. 243–249. Springer-Verlag, New York, 1994.

mutants. Many of these exhibit the yellow or paper-white phenotypes suggestive of defects in the chlorophyll or carotenoid biosynthetic pathways and are therefore not of interest for studies of chloroplast biogenesis. Approximately 1% segregate ivory mutants and approximately 1% segregate pale green mutants lacking components of the thylakoid membrane. These mutants can then be analyzed in detail to define the basis for the chloroplast defect.

New pigment-deficient mutants should be inspected for the appearance of dark-green leaf sectors, since somatic instability provides strong evidence that the mutation is caused by a transposable element insertion. The heterozygous siblings of mutants of interest should be propagated by outcrossing to non-*Mu* inbred lines followed by self-pollination to recover homozygous mutants. Repeated self-pollination of *Mu*-active plants results in severe lack of vigor and should be avoided. Because there is currently no convenient method for "turning off" *Mu* transposition at will, new mutations are constantly arising in *Mu*-active lines and pigment-deficient mutants are common among these. Therefore, it is dangerous to assume that all pigment-deficient progeny of a mutant of interest actually carry the same mutation as the parent. It is imperative to confirm the genotype of every ear used for experiments either by scoring for a defining feature of the mutant phenotype or, if the gene is cloned, by DNA testing.

## Mutant Characterization: Using Mutant Phenotypes to Elucidate Gene Function

Mutations that disrupt chloroplast gene expression or protein targeting can result in a variety of lesions in photosynthesis and with the loss of thylakoid proteins. Initial mutant characterization should be designed to pinpoint the site of the photosynthetic lesion so that more in-depth studies can be focused appropriately. One approach toward this end has been to measure the activities of the photosynthetic electron transport chain.[5,9] Such activity measurements are especially useful if one is interested in studying biochemical and biophysical aspects of photosynthesis or if one is interested only in lesions affecting a specific activity (e.g., PSII). However, if one's goal is to identify mutations that disrupt chloroplast gene expression, the critical information is the abundance, rather than the activities, of chloroplast-gene products. Likewise, defects in protein targeting result in the loss of thylakoid proteins and in increased accumulation of precursor forms,[17,18] both of which require protein analysis for their detection.

---

[17] R. Voelker and A. Barkan, *EMBO J.* **14,** 3905 (1995).
[18] R. Voelker, J. Mendel-Hartvig, and A. Barkan, *Genetics* **145,** 467 (1997).

Western blots can be used to quickly provide a snapshot of the protein deficiencies in mutant chloroplasts. Total leaf proteins are fractionated by sodium dodecyl sulfate–polyacrylamide gel electrophoresis (SDS–PAGE), transferred to nitrocellulose membranes, and stained with Ponceau S. Rubisco deficiencies are revealed simply by inspecting the pattern of stained proteins, since the Rubisco band is prominent in extracts of normal leaves. The abundance of thylakoid complexes is then assayed by probing with an antiserum cocktail that will detect one subunit of the PSI core complex, one subunit of the PSII core complex, one subunit of the ATP synthase, and one subunit of cyt $b_6f$. Because all core subunits of these complexes becomes destabilized if any single subunit is unavailable for assembly,[19] it is generally true that the abundance of each single subunit reflects (within a few-fold) the abundance of all of the other closely associated subunits. A particularly useful cocktail contains antisera to the *psaD*, *petD*, *psbA*, and *atpA* gene products, which resolve well on gels. With the protocols described later, it is straightforward for a single individual to complete this analysis with 100 new mutants in a 1-week period.

Growth conditions can have a major effect on the outcome of these assays. Although intense lighting is ideal for the initial identification of mutants, high-intensity light can cause secondary protein deficiencies that confuse the interpretation of the mutant phenotype. This is especially true for mutations in genes that function in the biogenesis of the cyt $b_6f$ complex. Mutations that cause the specific loss of cyt $b_6f$ subunits when plants are grown in a growth chamber cause, in addition, a 10-fold loss of PSII core proteins when mutants are grown in sunlight.[20] Therefore, plants to be used for phenotypic studies should be grown in a growth chamber under light intensities of no more than 400 $\mu E/m^2/s$. Our standard conditions are 16-hr days (400 $\mu E/m^2/s$) at 28° and 8-hr nights at 25°.

Protein Extraction Procedure

Excise a 1-cm segment of the second or third leaf of a 10- to 14-day-old seedling (which should have two or three leaves). Tissue can be used immediately for protein extraction or can be stored frozen at −80° for extraction at a later date. For quantification purposes, extract protein from one or two wild-type sibling plants grown in parallel with mutants.

Grind leaf tissue in a small mortar and pestle. If using fresh tissue, grind the unfrozen tissue directly in 0.3 ml ice-cold homogenization buffer (see

[19] A. Barkan, R. Voelker, J. Mendel-Hartvig, D. Johnson, and M. Walker, *Physiol. Plant.* **93,** 163 (1995).
[20] R. Voelker and A. Barkan, *Mol. Gen. Genet.* **249,** 507 (1995).

below) for approximately 30 sec. If using frozen tissue, grind in liquid nitrogen to a fine powder. Add 0.3 ml ice-cold homogenization buffer and continue grinding until thawed. It is essential to grind forcefully to release the proteins in bundle sheath cells. Failure to do this will reduce the yield of Rubisco.

Transfer the ground slurry to a 1.5-ml microfuge tube. Rinse mortar and pestle with 100 $\mu$l homogenization buffer and add to the microfuge tube. Store this total protein extract at $-80°$. The fibers in this preparation do not interfere with the gel analysis.

Caution should be taken to prevent the thawing of frozen tissue prior to the addition of homogenization buffer. Prolonged incubation of samples on ice or repeated freeze–thaw cycles can lead to protein degradation. It is best to prepare no more than 12 samples at a time prior to transferring samples to $-80°$. Particularly precious samples should be frozen in aliquots.

The following two homogenization buffers are both suitable for the analysis of most proteins. However, the 16-kDa subunit of the oxygen evolving complex of PSII is significantly more stable in buffer B than in buffer A.

Homogenization buffer A: 40 m$M$ 2-mercaptoethanol, 10% sucrose, 100 m$M$ Tris-HCl, pH 7.2, 5 m$M$ ethylenediaminetetraacetic acid (EDTA), 5 m$M$ ethylene-bis(oxyethylenenitrilo)tetraacetic acid (EGTA), 10 $\mu M$ (or 500 KIU/ml or 2 $\mu$g/ml) aprotinin, 2 m$M$ phenylmethylsulfonyl fluoride (PMSF). The PMSF is added fresh shortly prior to use, from a 60 m$M$ stock in isoproponol.

Homogenization buffer B: 100 m$M$ Tris-HCl, pH 7.5, 1 m$M$ EDTA, 1 m$M$ EGTA, 2 m$M$ PMSF, 2 $\mu$g/ml leupeptin, 2 $\mu$g/ml pepstatin, 2.5 m$M$ dithiothreitol (DTT). The PMSF, leupeptin, pepstatin, and DTT should be added fresh prior to each use. For 10 ml of working solution, add 500 $\mu$l of 40 m$M$ PMSF, 2 $\mu$l of 10 mg/ml leupeptin, 20 $\mu$l of 1 mg/ml pepstatin, and 25 $\mu$l of 1 $M$ DTT.

Quantification of Protein Extracts Prior to Electrophoretic Analysis

Five to 15 $\mu$g of protein are suitable for small format gels such as the Mini-PROTEAN II (Bio-Rad, Richmond, CA). For the purpose of defining protein deficiencies in mutants, it is more important that different samples contain the same amount of protein than that a particular quantity of protein be applied to each lane. The following method provides a quick estimate of the relative protein concentrations in different samples.

Spot 1.5 $\mu$l of each sample onto Whatman (Clifton, NJ) 3MM filter paper. Immerse the filter in a solution of Coomassie Blue-R250 (325 ml water, 125 ml isopropanol, 50 ml acetic acid, 250 mg Coomassie Blue-R250)

for 5–10 sec. Rinse with water to remove background staining. Compare staining intensities and adjust sample volumes with homogenization buffer to equalize protein concentrations. To aid in assessing relative concentrations, it is helpful to spot several dilutions of one sample (e.g., 0.5× and 2×) to illustrate the degree to which spot intensity changes with protein concentration. This method gives a good first approximation of relative protein concentrations. After samples are gel fractionated for the first time, concentrations can be further fine-tuned for subsequent analyses, based on their staining intensities in the gel.

### SDS–PAGE and Immunoblot Analysis to Define Chloroplast Protein Deficiencies

Proteins are fractionated in SDS–polyacrylamide gels prepared with the buffer system of Laemlli[21] according to standard procedures.[22] Three to 10 $\mu$l of protein samples prepared as described earlier should contain an appropriate amount of protein for the minigel format. To avoid aggregation of hydrophobic thylakoid proteins, samples should not be boiled. Instead, they should be denatured for 5–10 min at 70°. The inclusion of 4 $M$ urea in the gel and sample buffer results in a sharpening of several hydrophoic protein bands (e.g., the D1 and D2 subunits of PSII and the $psbB$ gene product) but also causes significant changes in the mobilities of several other proteins (e.g., cytochrome $f$ and plastocyanin). The 13.5% gels (acrylamide : bis-acrylamide = 30 : 0.8) resolve proteins ranging in size from approximately 10 to 70 kDa, a useful range for the analysis of most of the abundant chloroplast proteins.

Electrophoretic transfer to nitrocellulose membranes can be performed according to standard procedures.[23] For 1-mm-thick 13% mini-protein gels, transfer of most proteins is complete after 1 hr at 100 V, in a transfer buffer consisting of 25 m$M$ Tris base, 190 m$M$ glycine, and 20% methanol. Visualize proteins that are bound to the membrane by staining with Ponceau S.[23] A permanent record of the staining pattern can be made by photocopying the filter through a sheet of yellow acetate to increase contrast. Blots can either be probed with antibody immediately or air dried and stored for probing at a later date.

[21] U. K. Laemmli, *Nature* **227,** 680 (1970).
[22] J. Sambrook, E. F. Fritsch, and T. Maniatis, *in* "Molecular Cloning." Cold Spring Harbor Press, Cold Spring Harbor, NY, 1989.
[23] E. Harlow and D. Lane, *in* "Antibodies: A Laboratory Manual." Cold Spring Harbor Press, Cold Spring Harbor, NY, 1988.

*Probing with Antisera*

Wet the filter with water. Place in small tray containing 10 ml of blocking solution [TBST (50 m$M$ Tris-HCl, pH 7.5, 150 m$M$ NaCl, 0.1% Tween-20) containing 4% nonfat dried milk]. Incubate with gentle rocking for 30 min at room temperature. Pour off blocking solution and replace with 10 ml of antibody diluted into TBST containing 1% dried milk and 0.02% sodium azide (as a preservative). Optimal antibody concentrations must be determined empirically, but are typically between 1 : 500 and 1 : 5000 when working with crude antisera. Incubate for 1–3 hr at room temperature, or overnight at 4° with gentle rocking. Pipet off the diluted antibody solution and store at 4° for reuse. Diluted antibody solutions can be reused many times with little loss of sensitivity. Rinse the filter with water and wash it three times for 5 min with 20–30 ml TBST. Pour off the final rinse and add 10 ml secondary antibody [commercially available goat-anti-rabbit I$_g$G coupled to horse radish peroxidase (HRP), at a dilution of 1 : 10 : 000 in TBST + 1% milk]. Incubate for 1–2 hr at room temperature with gentle rocking. Discard secondary antibody, rinse filter briefly with water and then wash it four times for 5 min in 30 ml TBST.

The HRP-conjugated secondary antibody can be detected with commercially available detection reagents (e.g., Amersham ECL) according to the manufacturer's instructions. Alternatively, the following "home-brew" chemiluminescent detection reagent provides increased sensitivity at reduced cost.[24]

*Stock Solutions*

(A) 90 m$M$ $p$-coumaric acid (Sigma C-9008) in dimethyl sulfoxide (DMSO); (B) 250 m$M$ Luminol (Fluka 09253) in DMSO (protect from light); (C) 2% H$_2$O$_2$ (protect from light); (D) 100 m$M$ Tris-HCl, pH 8.5. All four stocks can be stored at room temperature for several months.

*Detection*

Combine 10 ml Tris stock with 50 $\mu$l of each of solutions A, B, and C. Soak blot in this mixture for several minutes. Transfer blot to an acetate folder, keeping it moist with detection solution. Expose to X-ray film. Exposures vary between several seconds and several minutes. The signal remains strong for 10–15 mins. To boost the signal after it fades, supply more substrate by placing the blot briefly in the same batch of detection solution.

---

[24] R. Schneppenheim, U. Budde, N. Dahlmann, and P. Rautenberg, *Electrophoresis* **12,** 367 (1991).

Identifying the Molecular Basis for Chloroplast Protein Deficiencies

A convenient first step toward determining whether protein deficiencies result from defects in chloroplast gene expression is to assay the structure and abundance of chloroplast mRNAs encoding missing proteins. If changes in mRNA structure and abundance are not detected, translation of the same mRNAs can then be probed by examining their association with polysomes and in pulse-labeling assays.

Total leaf RNA is conveniently prepared with Trizol reagent (Bethesda Research Lab, Gaithersburg, MD) according to the manufacturer's instructions. Leaf tissue (100 mg) yields 50–100 $\mu$g of RNA. Northern blots should be hybridized with gene-specific probes corresponding to each of the chloroplast-encoded subunits of those photosynthetic complexes that fail to accumulate to normal levels.

Northern Blot Protocol

*Buffers*

10× MOPS: 200 m$M$ MOPS, 80 m$M$ sodium acetate, 10 m$M$ EDTA, adjust pH to 7.0 with NaOH. Store at 4°.

RNA sample buffer: 1.2 ml deionized formamide, 0.4 ml 37% formaldehyde solution, 0.24 ml 10× MOPS, a few crystals of bromphenol blue and xylene cyanol (sufficient to make solution a deep blue). Store in 0.5 ml aliquots at −20°.

Reservoir buffer: 100 ml 10× MOPS, 100 ml 37% formaldehyde solution, 800 ml doubly distilled $H_2O$.

*Gel Preparation and Electrophoresis*

For each 120 ml of gel, melt 1.3 g agarose in 90 ml doubly distilled $H_2O$ in a microwave oven. Replace water lost to evaporation. Add 12 ml 10× MOPS and let cool until solution is approximately 60°. Add 20 ml formaldehyde solution, mix, and pour gel immediately. Let gel harden until it comes to room temperature. Place gel in electrophoresis chamber and immerse in reservoir buffer to a depth of a few millimeters. While preparing samples, prerun gel at 4 V/cm for 15–30 min (no longer). Add 4 $\mu$g of total leaf RNA to 12 $\mu$l of RNA sample buffer. Minimize sample volume to increase resolution, but use at least 2.5 volumes of sample buffer per volume of RNA. Heat to 70° for 5 min to denature RNA and apply to gel. Electrophorese at 4–5 V/cm for 3–4 hr. Circulate the buffer during electrophoresis.

*Transfer*

Rinse gel briefly in 5× SSC (1× SSC is 0.15 *M* NaCl, 0.015 *M* sodium citrate, pH 7). Transfer by capillary blotting overnight in 5× SSC to an uncharged nylon hybridization membrane (e.g., MSI Magna Nylon) as described.[22] Cross-link RNA to the membrane with a Stratagene Stratalinker or equivalent UV cross-linker.

To visualize the rRNA bound to the membrane, the filter is stained with methylene blue. A photocopy of the stained membrane provides a permanent record of the abundance and mobility of the rRNA bands in the gel. Soak the blot for 30 sec in methylene blue stain (0.03 g methylene blue dissolved in 10 ml 3 *M* NaOAc, pH 5.2, and 90 ml of doubly distilled $H_2O$). The staining solution can be reused numerous times. Rinse with several changes of doubly distilled $H_2O$ until rRNA bands stand out clearly against the background. The stained blot can be air dried and stored for long periods at room temperature prior to probing.

*Hybridization*

Hybridizations can be performed in either a hybridization oven or in sealable plastic bags. Prehybridize blots for 30–60 min at 65° in hybridization solution (7% SDS, 0.5 *M* sodium phosphate, pH 7, 1 m*M* EDTA). Add hybridization probe (denatured, if probe is double-stranded DNA) and incubate overnight at 65°. Blots are washed five times for 20 min at 65°. For homologous DNA probes, washes are in 0.2× SSC, 0.5% SDS. For homologous RNA probes, washes are in 0.1× SSC, 0.5% SDS. Heterologous probes may require lower hybridization and washing stringencies.

## Assays of Chloroplast Protein Synthesis

If chloroplast mRNAs encoding missing proteins show no apparent defects on Northern blots, the mutation likely disrupts either chloroplast translation, protein targeting, or the assembly of photosynthetic complexes. Pulse-labeling experiments allow one to assess rates of protein synthesis[25] and to detect decreases in the rates of proteolytic processing of proteins that are processed during their translocation across chloroplast membranes.[17]

Two pulse-labeling methods are described, one involving intact seedlings (*in vivo* labeling) and the other involving isolated chloroplasts (*in organello* labeling). The *in vivo* radiolabeling method is simpler to perform and allows one to radiolabel both nuclear and chloroplast-encoded proteins. Since most of the proteins synthesized with cleavable targeting sequences are

[25] A. Barkan, M. Walker, M. Nolasco, and D. Johnson, *EMBO J.* **13,** 3170 (1994).

encoded by nuclear genes, this method must be used to detect defects in protein targeting. Disadvantages of the *in vivo* method are that an effective "chase" is not possible and the minimum effective pulse length is rather long at 20 min. Since chloroplast proteins that fail to assemble properly may be subject to extremely rapid degradation (i.e., half-life of several minutes), failure to detect synthesis in a 20-min labeling period does not prove that the defect is at the level of protein synthesis. Therefore, for chloroplast-encoded proteins that accumulate reduced radiolabel in the *in vivo* experiment, a 5-min *in organello* pulse-labeling experiment should also be performed.

## Growth of Plants for Pulse-Labeling Experiments

Etioplasts exposed to light for short periods are more translationally active than chloroplasts from plants grown under day–night cycles. It is desirable, therefore, to grow plants to be used in pulse-labeling experiments according to the following regime. Illuminate plants during germination and transfer them to complete darkness after the coleoptiles emerge but before the first leaf has expanded. The short exposure to light inhibits hypocotyl elongation, which simplifies the handling of plants during the labeling experiments. Grow plants in the absence of light for approximately 5 more days until the third leaves have just started to emerge. Transfer plants to an illuminated growth chamber for 24–48 hours. They will accumulate chlorophyll during this period. Seedlings should be no older than 12 days at the time of the labeling experiment. For most mutations, mutant plants are noticeably paler than their wild-type siblings when grown under this regime, and can easily be identified.

## *In vivo* Labeling of Leaf Proteins

Plants should be healthy but not freshly watered to maximize label uptake. Several mutant and several wild-type siblings should be radiolabeled for each experiment.

1. Use a 21-gauge syringe needle to poke leaf approximately 50 times in a 0.5-cm band across a middle segment of a healthy second leaf. Use a backing of soft plastic (e.g., Tupperware) so that the needle does not fully penetrate leaf. For plants with less hardy leaves (e.g., tobacco or *Arabidopsis*), this may destroy the integrity of the leaf. Instead, the leaf surface can be abraded gently with an emery board.

2. Apply 10 $\mu$l of labeling mix to the perforations. The liquid will stay in place due to the waxy cuticle in the surrounding leaf tissue. Labeling

mix consists of 2 $\mu$l 10% bromphenol blue (to visualize the label moving through plant), 50 $\mu$l (at least 500 $\mu$Ci) $^{35}$S-methionine (*trans*-Label [NEN, Wilmington, DE] works well and is inexpensive), and 50 $\mu$l 10 m$M$ sodium phosphate, pH 7. As the liquid is taken up by the plant, apply either more buffer or, for maximum sensitivity, more labeling mix.

3. After desired labeling period (typically between 20 min and 2 hr), excise 1 cm of tissue surrounding the perforations. Immediately extract the proteins according to the standard protein extraction method described earlier, using 300 $\mu$l of homogenization buffer.

4. Assay the incorporation of radiolabel by spotting 5 $\mu$l of each sample on DE81 filter disks. Wash the filters in 0.5 $M$ sodium phosphate dibasic (three times for 3 min), water (two times for 3 min) and ethanol (one time for 30 sec). Add to a scintillation cocktail that is suitable for aqueous samples and measure bound radiolabel in a scintillation counter. Bulk proteins will remain bound to the filter and free amino acids will be washed off. Filters can be labeled with pencil and washed together in the same container. A good 20-min labeling will yield 10,000 cpm/$\mu$l. A yield of 3500 cpm/$\mu$l is adequate for most purposes.

5. Store protein at $-80°$ in aliquots to avoid repeated freeze–thaw cycles.

*Notes*

The shortest pulse period that gives adequate incorporation into proteins such as cytochrome $f$ and the *petD* gene product is 20 min. Labeling periods of 40 min give much better incorporation and are convenient for trial experiments. Proteins that are radiolabeled to near normal levels in a 40-min labeling need not be examined further for defects in synthesis.

To compare rates of synthesis of individual proteins between wild-type and mutant plants, immunoprecipitation reactions must be adjusted so that the mutant and wild-type samples involve the same total amount of the precipitated protein. Failure to do this will lead to an overestimate of the rate of synthesis of the protein in the mutant, since the mutant sample will have a higher ratio of antibody to target protein. It is useful to harvest a bit of nonradiolabeled leaf from a wild-type sibling to use as a source of "carrier" protein for the radiolabeled mutant protein extracts. Quantitative immunoblots can be used to determine how much nonradioactive carrier protein should be added to mutant samples to equalize the amount of the protein of interest between samples.

*In organello* Pulse-Labeling of Maize Chloroplast Proteins

This protocol does not involve the separation of intact from broken chloroplasts. We have found that results obtained with this method are comparable to those obtained following the isolation of intact chloroplasts. All solutions should be ice cold and chloroplasts should be kept on ice until the labeling reaction is started.

1. Remove healthy leaves from 15 seedlings, avoiding stem tissue. Cut into ~1-cm segments and place in a Waring blender or Virtis homogenizer outfitted with razor blades. Add 150 ml GM. Blend at medium speed in two 3-sec bursts. Filter through one layer of Miracloth. Squeeze gently. Transfer fibers back into blender, add another 150 ml GM and repeat grinding. Filter through fresh Miracloth. Squeeze gently.

2. Distribute the filtrate into six 50-ml round-bottom centrifuge tubes. Centrifuge in a swinging bucket rotor (e.g., Sorvall JS13.1) at 4000g for 1 min. The pellets will be small and the supernatants will be green.

3. Drain pellets. Resuspend gently but competely with a soft paintbrush in a total of 1 ml GM. Transfer to a single microfuge tube. Pellet chloroplasts by microcentrifugation for 5 sec at 12,000g at 4°.

4. Resuspend pellet gently in 95 μl RM. Add the following mixture and mix gently: 15 μl 100 mM ATP, 3 μl 50× amino acid mix lacking methionine, 3 μl 0.5 M DTT, 6 μl (≈60 μCi) $^{35}$S-methionine (e.g., NEN trans-Label), 27 μl 2× RM.

5. Place in 28° water bath and incubate 5 min. To maintain similar conditions inside wild-type and mutant chloroplasts, do not illuminate samples since the mutants will be nonphotosynthetic.

6. To stop the reaction, add 150 μl ice-cold GM and 7 μl 200 mM methionine. Pellet chloroplasts in a microfuge at 4° for 15 sec. Discard the supernatant. Resuspend pellet in 50 μl GM and 2 μl 40 mM PMSF.

7. Solubilize membranes by the addition of 5.5 μl 10% SDS and incubate at 55° for 10 min. Add 550 μl RIPA. Centrifuge at maximum speed in a microfuge for 3 min to pellet insoluble material. The pellet will not contain any chlorophyll if solubilization is complete. Spot 5 μl of the supernatant on DE81 filter disks and assay the incorporation of radiolabel as described earlier in the in vivo labeling protocol. Immediately divide the remainder of the supernatant into four aliquots and freeze at −80°. Between 200,000 and 300,000 cpm should have been incorporated. A typical labeling reaction provides enough material for five or six immunoprecipitation reactions to visualize translation products such as cytochrome f and the petD gene product.

*Buffers*

GM: 330 mM Sorbitol, 50 mM HEPES-KOH, pH 8, 1 mM EDTA. Autoclave and store at 4°. RM: 330 mM Sorbitol, 50 mM HEPES-KOH, pH 8, 10 mM MgCl$_2$. Autoclave and store at 4°. 50× amino acid mix lacking methionine: 2 m$M$ of each amino acid, except methionine. Store at $-80°$. Use within 1 month.

RIPA: 0.15 $M$ NaCl, 1% Triton X-100, 0.5% deoxycholate, 0.1% SDS, 20 m$M$ Tris-HCl, pH 7.5, 5 mg/ml aprotinin.

## Immunoprecipitation Protocol

To identify individual radiolabeled proteins unambiguously, it is necessary to purify them by immunoprecipitation. If membranes were not previously solubilized, add 1/10 volume of 10% SDS and heat to 55° for 5 min. Dilute with nine volumes of RIPA (described earlier). Centrifuge in a microfuge at maximum speed for 5 min and remove supernatant to a new tube (even if the pellet is not visible). If solubilization is complete, no chlorophyll will be visible in the pellet. Store the sample in aliquots at $-80°$.

For accurate comparison of protein synthesis rates between mutant and wild-type plants, it is essential that the immunoprecipitation reactions involve the same total amount of the precipitated protein (as described earlier). Prior to assembly of the immunoprecipitation reactions, quantitative immunoblots should be used to determine the amount of nonradioactive wild-type carrier protein to add to radiolabeled mutant samples.

### Antibody Binding

For material labelled *in vivo*, 200,000 cpm per immunoprecipitation is ideal. For material labeled *in organello*, 50,000 cpm per immunoprecipitation is sufficient. Immediately prior to antibody addition, centrifuge protein extract for 5 min at maximum speed in a microfuge to remove any insoluble material that would otherwise cause background signal. Use the supernatant for immunoprecipitation.

Add antiserum to lysate and mix gently. Incubate on ice for between 4 and 14 hr. The amount of antiserum must be determined empirically, although 15 $\mu$l of a high-titer crude serum is usually sufficient. Affinity purified sera are ideal, if available. While several antisera can be mixed in one immunoprecipitation, avoid using more than 20 $\mu$l of crude serum in any single reaction, since the immunoglobulins will overload the gel.

### Bind IgGs to Staph A Cells

Formalin-fixed Staph A cells such as Immuno-Precipitin (BRL, Gaithersburg, MD) provide an inexpensive and effective reagent for the precipitation of IgGs. Because any unbound protein A in the cell suspension will titrate out the antibody, the cells should be washed shortly before use as follows. Shake the bottle of cells to a homogeneous suspension and remove a volume sufficient for several days' experiments. To facilitate pellet resuspension, divide cells between several microfuge tubes with no more than 0.8 ml per tube. Mark cell volume on the tube. Centrifuge at maximum speed in a microfuge for 2 min. Resuspend to original volume in RIPA buffer. Pellet the cells and resuspend in RIPA two more times. Combine the aliquots and store at 4° for up to 1 week.

Add 250 $\mu$l of washed Staph A cells to each immunoprecipitation reaction. This should be sufficient to bind nearly all of the IgG in 20 $\mu$l of crude serum. If using affinity-purified antibody, less Staph A is needed. Mix gently. Store on ice for 1–2 hr. Pellet cells in a microfuge at maximum speed for 2 min. Discard the supernatant.

Wash the cells once in 0.5 ml of high-salt RIPA (RIPA with 0.4 $M$ NaCl) and twice in standard RIPA, by resuspension and microcentrifugation. The high-salt RIPA wash may be useful for decreasing the background of proteins that are weakly associated with the IgG/Staph A/antigen complex.

*Elute Antigen from Cells and Analyze on Gel*

Resuspend final Staph A pellet in 30 $\mu$l 1.5× SDS–PAGE sample buffer. Heat cells to 70° for 10 min to denature proteins. Pellet cells for 5 min at maximum speed in a microfuge. Load supernatant onto an SDS polyacrylamide gel of appropriate composition for the protein of interest. For the immunoprecipitation of cytochrome $f$, do not include urea in the gel because cytochrome $f$ comigrates with the heavy IgG chain in the presence of urea.

Radioactive proteins are detected by electrophoretic transfer to a nitrocellulose membrane, complete drying of the membrane, and direct exposure to X-ray film. Expect exposure times of between 3 and 7 days.

Analysis of Association of Chloroplast mRNAs with Polysomes

If antibodies are not available for immunoprecipitation, defects in translation initiation can nonetheless be detected by a reduced association of an mRNA with ribosomes. For those proteins whose synthesis appears to be reduced in pulse-labeling experiments, association of the corresponding mRNAs with polysomes should be assessed to gain insight into the translational step that is disrupted. Reduced association of an mRNA with ribosomes (1) confirms that the defect is a translational one and (2) indicates a block in initiation or early elongation.

All solutions should be ice cold unless otherwise stated.

1. Grind 0.2 g leaf tissue to a fine powder in liquid nitrogen in a mortar and pestle. Add 1 ml polysome extraction buffer and grind further until thawed. Remove debris by passage through a 3-ml syringe plugged with a

small amount of glass wool. Collect liquid in a microfuge tube on ice. Incubate for 10 min to solubilize membranes.

2. Centrifuge at maximum speed for 5 min in a microfuge at 4°. Remove the supernatant to a new tube. Add 1/20 volume of 10% sodium deoxycholate. Incubate on ice for 5 min to complete solubilization of microsomal membranes. Pellet remaining insoluble material in a microfuge at maximum speed for 15 min at 4°.

3. Layer 0.5 ml of this mixture onto 4.4 ml sucrose gradients (prepared as described below). Balance tubes carefully by the addition of polysome extraction buffer. Centrifuge in an SW50.1 rotor (Beckman, Fullerton CA) at 45,000 rpm for 65 min at 4°.

During this centrifugation, prepare tubes for gradient fractions. Label thirteen 1.5-ml microfuge tubes per gradient, and add to each tube 50 $\mu$l of a solution of 5% SDS and 0.2 $M$ EDTA (pH 8). Collect 0.41- ml fractions by gentle pipeting from the top of the gradient and add to the prepared tubes.

4. To each fraction, add 0.4 ml phenol/chloroform/isoamyl alcohol (25:25:1), vortex thoroughly, and separate phases by microcentrifugation for several minutes. Transfer aqueous phases to new tubes.

5. To each fraction, add 1 ml of *room temperature* 95% ethanol. Completely invert tubes several times to mix thoroughly the sucrose solution with the ethanol. Centrifuge for 15 min in a microfuge at maximum speed at room temperature to pellet the RNA. The warm temperature is essential to prevent the sucrose from pelleting in the lower gradient fractions. Pour off the supernatants. Drain pellets until almost dry. Resuspend each pellet in 30 $\mu$l ice-cold TE (10 m$M$ Tris-HCl, 1 m$M$ EDTA, pH 8). Store samples at −80°.

4 $\mu$l of each fraction is then analyzed by Northern hybridization. The rRNA bands should be easily detected on the filter by methylene blue staining. The distribution of the cytosolic rRNAs in the gradient should be similar between samples. If the distribution of an mRNA is shifted toward the top of the gradient in a mutant sample (relative to a wild-type sample), this provides evidence for a defect in the translation of that mRNA.

*Puromycin Control*

Rapid sedimentation of an mRNA in a sucrose gradient could be due either to polysome association or to association with some other large ribonucleoprotein. Only if the rapid sedimentation is due to polysome association will the sedimentation rate be reduced by puromycin treatment in high salt, conditions under which puromycin releases ribosomes from the mRNA.

After pelleting nuclei, but before adding the sodium deoxycholate, bring the KCl concentration to 0.5 $M$. Add puromycin to 500 $\mu$g/ml. (Puromycin stock is 3 mg/ml in 1.5 $M$ KCl.) Incubate at 37° for 10 min. Add sodium deoxycholate to 0.5%, centrifuge for 15 min to pellet insoluble material, and layer on sucrose gradients as described earlier.

*Buffers*

Polysome extraction buffer: 0.2 $M$ Tris-HCl, pH 9, 0.2 $M$ KCl, 35 m$M$ MgCl$_2$, 25 m$M$ EGTA, 0.2 $M$ sucrose, 1% Triton X-100, 2% polyoxy-ethylene-10-tridecyl ether. Filter through 0.2-$\mu$m pore sterile nitro-cellulose filtration unit and store in aliquots in sterile, disposable plastic tubes at −20°. Immediately prior to use add heparin to 0.5 mg/ml (from a 100 mg/ml stock), 2-mercaptoethanol to 100 m$M$, chloramphenicol to 100 $\mu$g/ml, and cycloheximide to 25 $\mu$g/ml.

10X polysome gradient salts: 0.4 $M$ Tris-HCl, pH 8.0, 0.2 $M$ KCl, 0.1 $M$ MgCl$_2$. Make solution "RNase-free" by treatment with diethyl pyrocarbonate followed by autoclaving, or by filtration through a 0.2-$\mu$m pore nitrocellulose sterile filtration unit into an RNase-free container. Store at −20°.

Gradient preparation: Step gradients can be prepared in bulk by using the following method and stored for long periods at −80°. To smooth out the steps, gradients must be thawed overnight at 4° prior to use. Prepare stock solutions containing 15, 30, 40, and 55% sucrose (see following table). Pipette 1.1 ml of the 55% solution into the bottom of 5-ml ultracentri-fuge tubes and freeze at −80°. Pipette 1.1 ml of the 30% solution on top of the 55% layer and freeze at −80°. Repeat with the 30 and 15% steps.

Recipe to prepare 25 4.4-ml gradients:

|  | 15% | 30% | 40% | 55% |
| --- | --- | --- | --- | --- |
| 70% sucrose (ml) | 6.4 | 13 | 17 | 24 |
| 10× Polysome gradient salts (ml) | 3 | 3 | 3 | 3 |
| DEPC-treated water (ml) | 20.5 | 14 | 10 | 3.3 |
| 100 mg/ml heparin ($\mu$l) | 150 | 150 | 150 | 150 |
| chloramphenicol (50 mg/ml in ethanol) ($\mu$l) | 60 | 60 | 60 | 60 |
| cycloheximide (10 mg/ml in water) | 75 | 75 | 75 | 75 |

Other Methods for Probing Mutant Phenotypes

Other standard methods for analyzing chloroplast gene expression and protein targeting can be successfully scaled down for use with limited mutant tissue. For example, to assess transciption rates in mutants that fail to accumulate specific mRNAs, we have successfully used published procedures for chloroplast transcription run-on assays[26] with as few as fifteen mutant seedlings.[27] Similar numbers of mutant seedlings are sufficient for chloroplast fractionation experiments to localize mistargeted protein precursors in protein targeting mutants.[17]

## Molecular Cloning of Genes Identified by Mutations and Targeted Nuclear Gene Knockouts

Although mutant phenotypes can provide important insights into gene function, detailed understanding of the role of a gene requires that it be cloned. Cloning of *Mu*-tagged alleles can provide a challenge because most *Mu*-active maize plants contain numerous *Mu* elements. To identify the single insertion that is the cause of the mutant phenotype, a simple strategy is to identify *Mu* insertions that are common to all mutant plants arising in many different mutant lineages. It is possible to generate sufficiently diverse lineages in just two generations following the initial mutant identification, if heterozygous plants from the original mutant isolate are used to pollinate numerous (e.g., 50–100) normal inbreds, each of which is then self-pollinated to regenerate ears segregating homozygous mutants. *Mu* insertions that are not tightly linked to the mutation will segregate in these populations. In most cases, the *Mu* elements in the targeted gene will be found in all mutant seedlings and will be absent from closely related homozygous wild-type plants. Occasionally, this relationship does not hold true because of the appearance of a derivative allele during mutant propagation or because of phenotypic suppression.[28]

Southern blots bearing DNA from mutant seedlings arising from diverse branches of the pedigree are probed with each member of the *Mu* transposon family.[15] A *Mu*-containing restriction fragment is sought that is in common to all mutants and absent in closely related homozygous wild-type plants. It is ideal if this analysis is restricted to seedlings exhibiting revertant sectors to ensure that the DNA samples contain a transposon at the locus of interest and are not from stable deletion derivatives. Restriction fragments of mutant DNA that comigrate with the linked *Mu* insertion are purified from agarose gels and a size-selected library is prepared from this DNA by cloning into a plasmid vector. Standard electroporation protocols can provide the high transformation efficiencies ($10^9$ colonies/$\mu$g) that are critical for the success of this method. Colony lifts are screened by hybridization with a probe for the *Mu* family member identified from the Southern

---

[26] R. Klein and J. Mullet, *J. Biol. Chem.* **265,** 1895 (1990).
[27] B. Till and A. Barkan, unpublished data (1996).
[28] R. Martienssen, A. Barkan, W. Taylor, and M. Freeling, *Genes Dev.* **4,** 331 (1990).

analysis. The DNA sequences flanking the cloned *Mu* element are used to confirm that the desired gene has, in fact, been cloned. If multiple transposon-induced alleles are available, each should exhibit a transposon insertion in the same transcription unit. Alternatively, the observation that the cloned *Mu* element has excised from somatic revertant sectors provides strong evidence that the desired gene has been cloned.[18,29]

Many proteins of potential importance in chloroplast gene expression and protein targeting are being identified in biochemical assays or by virtue of homology to proteins of known function. Knowledge of the loss-of-function phenotypes will be critical for establishing the *in vivo* roles of such proteins. A method has been developed in maize for the identification of individuals with a *Mu* transposon in any gene of known sequence.[30,31] This method relies on the polymerase chain reaction to identify, from among a large population of transposon-carrying individuals, those harboring a transposon within the gene of interest. At the present time, Pioneer Hybrid Inc, developers of this technology, will screen their "library" of *Mu*-containing genomes as a service to the research community. This "reverse genetics" technology provides a powerful and critical complement to the more traditional genetic approach described earlier.

### Acknowledgments

I wish to thank all members of my laboratory, and especially Rodger Voelker and Macie Walker, for help with developing and optimizing the protocols described here. I would also like to acknowledge the work of Don Miles, who laid the foundation for a fruitful genetic analysis of chloroplast biogenesis in a higher plant.

[29] R. Martienssen, A. Barkan, M. Freeling, and W. Taylor, *EMBO J.* **8,** 1633 (1989).
[30] R. Meeley and S. Briggs, *Maize Gen. Cooperation Newsletter* **69,** 67 and 82 (1995).
[31] L. Das and R. Martienssen, *Plant Cell* **7,** 287 (1995).

## [5] Obtaining Crystal Structures from Bacterial Photosynthetic Reaction Centers

*By* Günter Fritzsch

### Introduction

Two decades ago it was argued that integral membrane proteins cannot be crystallized. Contrary to their soluble counterparts, membrane proteins possess large hydrophobic surface domains that interact with the alkane

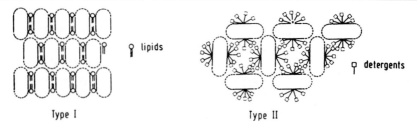

Type I        Type II

FIG. 1. The two basic types of membrane protein crystals. Type I: Stacks of membranes each containing highly ordered "two-dimensionally crystallized" proteins. Type II: Solubilized membrane proteins crystallized within the detergent micelles. The polar surface parts of the proteins are indicated by broken lines. [Taken from H. Michel, *Trends Biochem.* **8**, 56 (1983).]

chains of the lipids. Addition of detergents is required to solubilize membrane proteins in aqueous solutions. The detergent molecules form micelles around the hydrophobic proteins yielding bulky protein–detergent complexes. These complexes represent nonidentical particles not even qualified to serve as building blocks for a well-ordered three-dimensional crystal lattice. Furthermore, the size of these complexes is increased compared to the detergent-free protein. As a consequence, the interactions between individual protein molecules are less likely so that crystal formation seems to be even more unfavorable for those proteins.

Nevertheless, sophisticated techniques have been developed to control the size and shape of the detergent micelles in such a way that membrane proteins can now be prepared as monodisperse solutions and then crystallized despite all counterstatements. A first success was the crystallization of bacteriorhodopsin from *Halobacterium halobium*[1,2] in 1980; the quality of the crystals (diffraction to 4.2-Å resolution), however, was unsatisfactory. The ice was finally broken in 1982 when the photosynthetic reaction center (RC) from the non-sulfur purple bacterium *Rhodopseudomonas viridis* (*Rps. viridis*) was crystallized by Michel[3] who had proposed the existence of two principal types of crystals that can be obtained from membrane proteins[4,5]:

> *Type I* (Fig. 1, left): Proteins in membranes are surrounded by lipids. At high concentrations they can form two-dimensional patterns with hydrophobic interactions in the plane of the membrane. Ordered

[1] H. Michel and D. Oesterhelt, *Proc. Natl. Acad. Sci. U.S.A.* **77**, 1283 (1980).
[2] H. Michel, *EMBO J.* **1**, 1267 (1982).
[3] H. Michel, *J. Mol. Biol.* **158**, 567 (1982).
[4] H. Michel, *TIBS* **8**, 56 (1983).
[5] J. Deisenhofer and H. Michel, *EMBO J.* **8**, 2149 (1989).

stacks of these membrane protein layers may form a three-dimensional crystal-like lattice. Polar interactions dominate the binding in the direction perpendicular to the membrane. Consequently, two contrasting interactions must be increased simultaneously during the growth of those crystals, a requirement that is difficult to fulfill.

*Type II* (Fig. 1, right): The protein is solubilized and expected to crystallize within its detergent micelle. The advantage of this type of crystal is that the intermolecular forces are of polar character in all directions. The main problem of this procedure, however, is the large size of the detergent micelle that must be reduced far enough to facilitate the nucleation process. For this reason, membrane protein with large and stable polar domains that tower out of the membrane are more suitable for this crystallization method than proteins without extramembrane domains.

An essential innovation in preparing type II crystals from membrane-bound proteins such as bacteriorhodopsin and photosynthetic RCs was introduced when small amphiphilic molecules were added to the protein-detergent complex.[2,3] These amphiphiles are too small to form micelles themselves, inserted into existing detergent micelles, however, they facilitate the nucleation process by several effects: (1) They create mixed detergent–amphiphile micelles with a reduced size due to a changed critical micellar concentration (MCC). This effect has been shown by small-angle neutron scattering[6] and by determining the detergent concentration at which resolubilization occurs as a function of the RC concentration.[7] (2) The amphiphiles allow a rearrangement of the detergent molecules in such a way that the detergent–amphiphile complex accommodates unevenesses of the protein surface better than the detergent alone. (3) The polar headgroups of the amphiphilic molecules are relatively small and allow more of the protein's polar groups to interact with neighboring protein molecules. Besides these effects, the suitable application of amphiphiles can prevent phase separation.

As shown by neutron diffraction with $H_2O/D_2O$ contrast variation, the detergent–amphiphile phase in type II crystals is arranged as rings filling the space close to the hydrophobic membrane-spanning helices of the RC. All of these rings are interconnected by cylindrical detergent bridges, forming a detergent network throughout the crystal.[8] The multilateral interac-

[6] P. A. Timmins, J. Hauk, T. Wacker, and W. Welte, *FEBS Lett.* **280,** 115 (1991).

[7] P. Gast, P. Hemelrijk, and A. J. Hoff, *FEBS Lett.* **337,** 39 (1994).

[8] M. Roth, A. Lewit-Bentley, H. Michel, J. Deisenhofer, R. Huber, and D. Oesterhelt, *Nature* **340,** 659 (1989).

tions of protein, precipitant, detergent, and amphilphile have been discussed in detail recently.[9]

In this article, the preparation of two bacterial RCs for crystallization is described. Apart from these RCs several other proteins have been prepared as type II crystals: The *Escherichia coli* outer membrane porins Ompf and PhoE,[10–12] the prostaglandin $H_2$ synthase-1,[13] the light-harvesting antenna complexes (LH2) from *Rps. acidophila*[14] and *Rhodospirillum molischianum*,[15] the beef heart mitochondrial cytochrome $bc_1$ complex,[16] the photosystem I reaction center,[17] another crystal form from bacteriorhodopsin of *Halobacterium halobium*,[18] and, more recently, the *Staphyloccus aureus* α-hemolysin that forms a heptameric transmembrane pore.[19] The membrane protein cytochrome *c* oxidase has been crystallized from two different species in two different ways: the oxidase from bovine heart was treated according to the conventional procedure by a suitable choice of physicochemical crystallization conditions,[20,21] whereas the crystallization of the oxidase from *Paracoccus denitrificans* was mediated by cocrystallization with an antibody $F_v$ fragment.[22,23] The hydrophilic antibody fragment possesses charged residues on its surface that function as sites for electrostatic interactions with neighboring molecules.

The procedures developed for the crystallization of soluble proteins can also be applied to membrane proteins, e.g., vapor diffusion and dialysis

[9] R. M. Garavito, D. Picot, and P. J. Loll, *J. Bioenerg. Biomembr.* **28**, 13 (1996).

[10] R. M. Garavito and J. P. Rosenbusch, *J. Cell Biol.* **86**, 327 (1980).

[11] S. W. Cowan, T. Schirmer, G. Rummel, M. Steiert, R. Ghosh, R. A. Pauptit, J. N. Jansonius, and J. P. Rosenbusch, *Nature* **358**, 727 (1992).

[12] S. W. Cowan, R. M. Garavito, J. N. Jansonius, J. A. Jenkins, R. Karlsson, N. König, E. F. Pai, R. A. Pauptit, P. J. Rizkallah, J. P. Rosenbusch, G. Rummel, and T. Schirmer, *Structure* **3**, 1041 (1995).

[13] D. Picot, P. J. Loll, and M. Garavito, *Nature* **367**, 243 (1994).

[14] G. McDermott, S. M. Prince. A. A. Freer, A. M. Hawthornwaite-Lawless, M. Z. Papiz, R. J. Cogdell, and N. W. Isaacs, *Nature* **374**, 517 (1995).

[15] J. Koepke, X. Hu, C. Muenke, K. Schulten, and H. Michel, *Structure* **4**, 581 (1996).

[16] C. A. Yu, J. Z. Xia, A. M. Kachurin, L. Yu, D. Xia, H. Kim, and J. Deisenhofer, *Biochim. Biophys. Acta* **1275**, 47 (1996).

[17] N. Krauß, W.-D. Schubert, O. Klukas, P. Fromme, H. T. Witt, and W. Saenger, *Nature Struct. Biol.* **3**, 965 (1996).

[18] G. F. X. Schertler, H. D. Bartunik, H. Michel, and D. Oesterhelt, *J. Mol. Biol.* **234**, 156 (1993).

[19] L. Song, M. R. Hobaugh, Ch. Shustak, S. Cheley, H. Bayley, and J. E. Gouaux, *Science* **274**, 1859 (1996).

[20] T. Tsukihara, H. Aoyama, E. Yamashita, T. Tomizaki, H. Yamaguchi, K. Shinzawa-Itoh, R. Nakashima, R. Yaono, and S. Yoshikawa, *Science* **269**, 1069 (1995).

[21] T. Tsukihara, H. Aoyama, E. Yamashita, T. Tomizaki, H. Yamaguchi, K. Shinzawa-Itoh, R. Nakashima, R. Yaono, and S. Yoshikawa, *Science* **272**, 1136 (1996).

[22] C. Ostermeier, S. Iwata, B. Ludwig, and H. Michel, *Nature Struct. Biol.* **2**, 842 (1995).

[23] S. Iwata, C. Ostermeier, B. Ludwig, and H. Michel, *Nature* **376**, 660 (1995).

with salts or polymers as precipitating agents. The hanging drop method, however, is less favorable for drops of greater than 10 $\mu$l because the presence of detergent in the crystallization buffer decreases the surface tension of the drop considerably, thereby inhibiting the formation of a well-shaped hanging drop.

Two-dimensional crystals from membrane proteins have also been grown. The structure of these proteins can be determined by applying electron crystallography. Examples are the bacteriorhodopsin from *Halobacterium halobium*,[24] the plant light-harvesting chlorophyll *a/b* complex,[25] and the microsomal glutathion transferase.[26]

## The Crystallization of Bacterial Reaction Centers

So far, crystals from two purple bacterial RCs diffracting to a resolution of better than 3 Å have been grown:

1. The RC of *Rps. viridis*.[3-5,27,28] Only tetragonal crystals from this RC have been obtained. They diffract to high resolution (<3 Å). The crystallization and structural analysis of this RC have expanded the understanding of photosynthesis in particular and of the vectorial processes in biomembranes generally.
2. The RC from *Rhodobacter sphaeroides* (*Rb. sphaeroides*). Three kinds of well-diffracting crystals can be obtained: orthorhombic,[29-35] trigonal,[36,37] and tetragonal ones.[38] A brief overview of the three

[24] R. Henderson, J. M. Baldwin, T. A. Ceska, F. Zemlin, E. Beckmann, and K. H. Downing, *J. Mol. Biol.* **213,** 899 (1990).
[25] W. Kühlbrandt, D. N. Wang, and Y. Fujiyoshi, *Nature* **367,** 614 (1994).
[26] H. Hebert, I. Schmidt-Krey, and R. Morgenstern, *EMBO J.* **14,** 3864 (1995).
[27] J. Deisenhofer, O. Epp, K. Miki, R. Huber, and H. Michel, *J. Mol. Biol.* **180,** 385 (1984).
[28] J. Deisenhofer, O. Epp, K. Miki, R. Huber, and H. Michel, *Nature* **318,** 618 (1985).
[29] J. P. Allen and G. Feher, *Proc. Natl. Acad. Sci. U.S.A.* **81,** 4795 (1984).
[30] J. P. Allen, G. Feher, T. O. Yeates, D. C. Rees, J. Deisenhofer, H. Michel, and R. Huber, *Proc. Natl. Acad. Sci. U.S.A.* **83,** 8589 (1986).
[31] J. P. Allen, G. Feher, T. O. Yeates, H. Komiya, and D. C. Reeds, *Proc. Natl. Acad. Sci. U.S.A.* **84,** 6162 (1987).
[32] T. O. Yeates, H. Komiya, A. Chirino, D. C. Rees, J. P. Allen, and G. Feher, *Proc. Natl. Acad. Sci. U.S.A.* **85,** 7993 (1988).
[33] C.-H. Chang, M. Schiffer, D. Tiede, U. Smith, and J. Norris, *J. Mol. Biol.* **186,** 201 (1985).
[34] A. Ducruix and F. Reiss-Husson, *J. Mol. Biol.* **193,** 419 (1987).
[35] H. A. Frank, S. S. Taremi and J. R. Knox, *J. Mol. Biol.* **198,** 139 (1987).
[36] S. K. Buchanan, G. Fritzsch, U. Ermler, and H. Michel, *J. Mol. Biol.* **230,** 1311 (1993).
[37] U. Ermler, G. Fritzsch, S. K. Buchanan, and H. Michel, *Structure* **2,** 925 (1994).
[38] J. P. Allen, *Proteins* **20,** 283 (1994).

crystallizations is given in Fritzsch *et al.*[39] Although this RC is of an extreme hydrophobic character, it can be prepared to crystallize according to type II.

The RCs from *Rps. viridis* and from *Rb. sphaeroides* have molecular weights of 145,000 and 102,000, respectively. Both consist of the three similar subunits L, M, and H; the RC from *Rps. viridis* has an additional 40-kDa cytochrome *c* subunit that protrudes into the periplasm. This subunit is stably folded and sufficiently hydrophilic to play a crucial role in the nucleation process of this RC. Since the RC from *Rb. sphaeroides* is missing this subunit, it is much more difficult to crystallize and so the initial conditions for crystallization are different for both RCs.

To avoid heterogeneity of the RCs due to damage by light, all steps of isolation must be performed under low-light conditions, e.g., in a dim lab without direct illumination of the sample.

Isolation of *Rps. viridis* RC

*Cell Growth and Harvesting*

The procedure usually performed for the isolation of highly purified RCs from *Rps. viridis* has been developed by H. Michel.[3] A detailed protocol is given by Buchanan.[40] Cells of *Rps. viridis* grow photoheterotrophically in sodium succinate medium 27[41] (*Rhodospirillaceae* medium) that has been autoclaved at 120° for 45 min. Transparent 25-liter Nalgene bottles are inoculated with about 2 liters of culture, sealed with silicon stoppers and Parafilm. To consume the remaining oxygen in the bottles, the cultures are allowed to grow for 12–24 hr in the dark. After this, the bottles are illuminated by light with an intensity of about 300 $W/m^2$, e.g., by three lamps (Osram Concentra 120V PAR 38 Cool). The distance from lamp to bottle should be large enough (>50 cm) to ensure a temperature of ≤27° inside the bottles.

After about 8 days the cells have reached the late log phase and are harvested by centrifugation (15 min at 6000$g$ at 4°) or by using a continuous flow centrifuge like CEPA (Carl Padberg, Lahr, Germany) and washed with TEA8 buffer [10 m$M$ Tris-HCl, pH 8.0, 0.1% ethylenediaminetetraacetic acid (EDTA), 0.03% NaN$_3$]. The yield amounts to 2.5–4.0 g wet weight

[39] G. Fritzsch, U. Ermler, M. Merckel, and H. Michel, *in* "The Reaction Center of Photosynthetic Bacteria" (M. E. Michel-Beyerle, ed.), p. 3. Springer, Berlin, Germany, 1996.
[40] S. K. Buchanan, Ph.D. Thesis, Johann Wolfgang Goethe-Universität, Frankfurt (1992).
[41] "German Collection of Strains 1989," 4th Ed. DSM—Deutsche Sammlung von Mikroorganischen und Zellkulturen GmbH, Braunschweig, p. 280. 1989.

per liter of culture. At this stage the cells can be frozen in liquid nitrogen and stored at $-20°$ for months or even years. They have to be kept in the dark.

## Ultrasonication, Density Gradient, and Solubilization

The bacteria are disrupted by ultrasonication. Thirty-five-gram cells are suspended in 200 ml TEA8 buffer and sonicated in a 400-ml rosette glass cooled with a water–ice bath (Brand sonifier: duty cycle 50%, output 10). The suspension is centrifuged at 24,000$g$ for 20 min. The supernatant is stored while the pellet is resuspended in TEA8 and once more treated by ultrasonication. Both supernatants are collected and photosynthetic membranes are isolated by differential centrifugation at 40,000$g$ and 4° for 3 hr. The pellet contains the crude membranes. Debris is removed by centrifugation (22,000$g$, 20 min). The crude membranes can be frozen in liquid nitrogen and stored at $-20°$. Alternatively to ultrasonication a French pressure cell (2.9 kPa) can be used for cell disruption.

The crude membranes are homogenized by resuspending in TEA8 (1 g in 2 ml), stirred for 15 min in an ice-cold water bath, and ultrasonicated for 5 min (all procedures in the dark). The suspension is gently layered onto 30–60% (w/w) sucrose gradients, about 3 g (6 ml) membranes per 250-ml gradient. Isopycnic centrifugation is performed at 100,000$g$ for at least 12 hr. Slow acceleration and deceleration should be used. The RC-containing fraction forms dark green-brown bands in the middle of the centrifugation tubes. These bands are collected and washed free from sucrose. The tubes containing the membrane bands are filled up with TEA8 buffer, mixed, and centrifuged at 10,000$g$ for 3 hr. The pellet is then washed three times with TEA8 buffer (40 min at 150,000$g$). The pellet from the last washing step should be as dry as possible. It is weighed and can be frozen in liquid nitrogen and stored at $-20°$.

The purified chromatophores including the RC are isolated from the membranes by solubilization with 6% of the nonionic detergent $N,N$-dimethyldodecylamine-$N$-oxide (LDAO). According to Weyer[42] and Buchanan[40] the following procedure can be applied: 5 g membranes are suspended in 21.6 ml distilled water, 6.7 ml of a 30% v/v LDAO solution (Fluka, Neu-Ulm, Germany) is added slowly. To solubilize the RC completely, stir at room temperature for 20 min. All steps must be performed in the dark. The solution is centrifuged at 8000$g$ and 4° for 20 min. The supernatant contains the solubilized RC.

---

[42] K. A. Weyer, "Isolierung und Sequenzierung der Proteinuntereinheiten des photosynthetischen Reaktionszentrums von *Rhodopseudomonas viridis.*" Wissenschaftliche Forschungsbeiträge Biologie, Biochemie, Chemie, München (1988).

*Molecular Sieve Chromatography*

Pure RCs are obtained by gel filtration. After solubilization the RCs are transferred to a pair of Fractogel columns [TSK HW-55 (S), Merck, Darmstadt] at 4° without any delay. To protect the RCs against light, the columns must be kept in the dark. They are preequilibrated with LDAO buffer (20 mM $NaH_2PO_4$, pH 6.0, 0.01% LDAO, 0.02% $NaN_3$). Because large quantities of this buffer are required in the following, the preparation of 1 liter buffer stock solution (10-fold concentrated) is recommended. A short column (4 cm diameter, 30 cm long) is followed by a long column (100 cm) with the same diameter. Both columns are connected by a thin tube (diameter: 1–2 mm). The flow rate is 100 ml/hr. The eluates are collected in test tubes, 5 min per tube. The filtration process lasts 15–20 hr, the first fractions are colorless, followed by yellow ones. Dark brownish-green bands appear after about 10 hr. They contain the RCs, which are examined spectroscopically between 250 and 900 nm. Fractions with an absorption ratio of $A_{280}/A_{830} \leq 3.0$ are collected and concentrated to 5–10 ml in an Amicon cell (Millipore Corp., Bedford, MA) using a PM 10 membrane.

The concentrated RCs are loaded onto another column (2.5 cm in diameter and 90 cm long) containing the same Fractogel. This column is preequilibrated with LDAO buffer, pH 6.0, and runs again in the dark, but at room temperature. The flow rate is 35–40 ml/hr. After 8–10 hr the RC bands appear. Fractions with an absorption ratio of $A_{280}/A_{830} \leq 2.5$ are collected and again concentrated to 5–10 ml. In the meantime, the column is washed with 10 ml of 20 mM $NaH_2PO_4$, pH 6.5, and 10% LDAO and again equilibrated ($\geq 5$ bed volumes) with the LDAO buffer. The concentrated RCs are applied to the room temperature column for a second time. Fractions with an absorption ratio of $A_{280}/A_{830} \leq 2.1$ are pooled together. This is the last step of RC purification, therefore the spectroscopic determination of purity must be performed carefully. The yield amounts to about 8 mg RC per gram of purified membranes. The correlation between mass and absorbance is given by the extinction coefficient of *Rps. viridis* RC at 830 nm of 280,000 $mol^{-1}cm^{-1}$.

Crystallization of *Rps. viridis* RC

The RC solution is concentrated in an Amicon cell (PM10 membrane) and subsequently in Centricon 30 concentration tubes (Millipore Corp., Bedford, MA) at 6000g to a final $OD_{830} = 40$, i.e., to a concentration of 18 mg/ml. The volume has to be measured. Usually, a 4 M $(NH_4)_2SO_4$ stock solution (without LDAO) is used to prepare the *viridis* mother liquor for crystallization, which is composed of 20 mM $NaH_2PO_4$, 0.063% LDAO,

1.5 $M$ (NH$_4$)$_2$SO$_4$, 3% 1,2,3-heptanetriol, and RC with an OD$_{830}$ = 25. The
amount of required 1,2,3-heptanetriol is calculated as shown later. It is
added as the first component to the concentrated RC solution at OD$_{830}$ =
40 in solid form. It must be dissolved completely, e.g., by vortexing. From
the 4 $M$ (NH$_4$)$_2$SO$_4$ stock solution as much as 60% of the obtained RC
solution volume is taken and mixed with the RC solution so that the final
RC concentration has an OD$_{830}$ = 25. To prevent denaturation of the RC
by the highly concentrated (NH$_4$)$_2$SO$_4$ the mixing must be performed slowly
(1–2 min) to avoid denaturation of the protein. From the volumes mixed
together the final volume of the *viridis* mother liquor can be calculated,
which is necessary to determine the amount of the required 1,2,3-heptane-
triol. The final absorbance of the RC, OD$_{830}$ = 25, corresponds to a protein
concentration of about 11 mg/ml. If necessary, the pH can be adjusted to
other values than 6.0. The dilution effect can be disregarded. After this,
the RC solution is centrifuged with a table centrifuge for 2 min (e.g.,
Eppendorf centrifuge 12,000 rpm). Denatured protein is pelleted.

The RCs are successfully crystallized by the vapor diffusion method
with sitting drops. Up to six drops of about 50 $\mu$l protein solutions are
equilibrated against 8 ml of the *viridis* reservoir solution of 1.8–2.1 $M$
(NH$_4$)$_2$SO$_4$ at 18°. Crystals grow within a few days or a few weeks depending
mainly on the concentration of (NH$_4$)$_2$SO$_4$ in the reservoir. Because the
size of the obtained crystals depends basically on the number of competing
crystal nuclei in the crystallization well, the concentration of (NH$_4$)$_2$SO$_4$
controls not only the number but also the size of the crystals. Many small
crystals appear usually at 2.1 $M$ (NH$_4$)$_2$SO$_4$ whereas few and big ones (or
none at all) grow at 1.8 $M$ (NH$_4$)$_2$SO$_4$. To obtain large crystals, reservoir
concentrations should be applied in 0.05 $M$ steps. The crystallization temperature may range from 12 to 20° without affecting the quality of the
obtained crystals.

The crystals are rod shaped and of a tetragonal space group. They can
become 2 mm long and about 0.7 mm in diameter (Fig. 2). If no crystals
appear after 3 weeks, the nucleation process can be provoked by increasing
the concentration of (NH$_4$)$_2$SO$_4$ in the reservoir.

### Isolation of *Rb. sphaeroides* RCs

*Cell Growth and Harvesting*

Cells of wild-type strain ATCC 17023 are grown photoheterotrophically
in the same medium as *Rps. viridis* (see earlier discussion). They tolerate
temperatures up to 35° and grow faster than cells of *Rps. viridis*. After 7
days the cells have reached the late log phase and are then harvested

FIG. 2. Tetragonal crystals of the *Rps. viridis* RC grown from ammonium sulfate in a sitting drop. The large crystal on the left side has grown parallel to the "surface" of the drop and is almost 1 mm long. The smaller crystal on the right side has grown vertically from below and shows the quadratic cross section that is typical for tetragonal crystals.

by centrifugation as described previously for *Rps. viridis* cells. Cells are concentrated to an $OD_{850} = 50$. They can be frozen in liquid nitrogen and stored at $-20°$.

*Ultrasonication, Density Gradient, and Solubilization*

Disrupture of cells and sucrose gradient are performed by the same procedures given for *Rps. viridis*. The purified chromatophores are concentrated to an $OD_{800} = 50$. At this stage they can be frozen in liquid nitrogen and stored at $-20°$.

After thawing the frozen samples of the RC should be purified and crystallized in a few (max. 5) days without delay. During this time most of the procedures may be carried out at room temperature. Any prolonged interruption of the process or even another freezing of the protein endangers the formation of good crystals.

The *Rb. sphaeroides* RCs are solubilized with 0.5% LDAO in Tris buffer, pH 8.0, 100 m$M$ NaCl, 1 m$M$ sodium ascorbate, 830 $\mu M$ benzamidine, and 0.5 m$M$ phenylmethylsulfonylfluoride (PMSF). The suspension is stirred for 60 min at $4°$ in the dark and then centrifuged at 30,000$g$ and $4°$ for 20 min. The supernatant is centrifuged at 150,000$g$ and $4°$ for 1 hr. The RCs are in the supernatant that is transferred to the anion exchange column (see below).

*Alternative procedure.* The harvested RCs from *Rb. sphaeroides* can also be purified without sucrose gradient. This procedure is based on a method by Feher and Okamura[43] modified by Meyer.[44] The cells are suspended (1 ml/g) in Tris buffer (20 m$M$, pH 8.0) and sonicated for 15 min in an ice-water bath (Brand sonifier: duty cycle 50%, output 7). Undisrupted cells are pelleted by centrifugation (30 min at 16,500$g$ at 4°). The chromatophores in the supernatant are pelleted at 200,000$g$ for 1 hr and resuspended in Tris buffer. The RC is concentrated to an OD$_{860}$ = 50. LDAO and NaCl are added to final concentrations of 0.3% and 100 m$M$, respectively. The suspension is stirred in the dark at room temperature for 30 min and centrifuged at 200,000$g$ and 4° for 1 hr. The supernatant is checked for RCs (band at 860 nm). If no RCs are solubilized, the pellet is resuspended as described earlier and mixed with 0.3% LDAO and 100m$M$ NaCl. Usually, this step must be repeated a third time to solubilize the RCs completely. The RC-containing supernatants are transferred to a light protected DEAE-cellulose column (Whatman DE 52) of 25 cm in length and 2.5 cm in diameter (room temperature). The protein is washed with four concentrations of NaCl (60, 80, 90, and 100 m$M$) to remove protein-free pigments. The elution is performed with 500 m$M$ NaCl in TEAL8 buffer (10 m$M$ Tris-HCl buffer, pH 8.0, 0.1 m$M$ EDTA, 0.02% NaN$_3$, 0.1% LDAO).

*Anion Exchange Chromatography*

Purified chromatophores from 35-g cells of an NaCl concentration of less than 60 m$M$ are transferred to a light-protected 2.5-cm × 12-cm column of DEAE-Sepharose (fast flow) at room temperature and washed with TEAL8 buffer at increasing concentrations of NaCl (80, 100, and 120 m$M$). Omission of NaN$_3$ has no consequences provided purification and crystallization are performed without delay. The concentration of LDAO may vary from 0.06 to 0.18% without detectable influence on the results. Elution is performed with 250 m$M$ NaCl and the brownish fractions are pooled. The spectra of the collected fractions should show the features specific for *Rb. sphaeroides* RC: maxima at 850 and 800 nm, a maximum or at least a shoulder at 760 nm, the 600-nm band, and the carotenoid bands at 455, 480, and 510 nm. The usually used absorbance ratio $A_{280}/A_{800}$ for purity can be misleading at this step of purification because the absorbance of the RC at 800 nm is most likely superimposed by the absorbance of light-harvesting complexes that may still be present in the sample. The eluate

---

[43] G. Feher and M. Y. Okamura, "Chemical composition and properties of reaction centers," *in* "The Photosynthetic Bacteria" (R. K. Clayton and W. R. Sistrom, eds.), pp. 349. Plenum Press, New York, 1978.

[44] M. Meyer, diploma thesis, Ludwig-Maximilians-Universität, München (1992).

is concentrated to $\leq 5$ ml. It is diluted with Na-free TEAL8 buffer $1:4$ and again transferred to a light-protected 2.5-cm $\times$ 12-cm column of DEAE-Sepharose (fast flow) at room temperature. The procedure described previously is repeated once with the difference that the RCs are now eluted with 180 m$M$ NaCl. This elution takes several hours because the separation of the RC–light-harvesting complex proceeds slowly. Fractions with an absorbance ratio $A_{280}/A_{800} \leq 1.3$ are collected and concentrated to an optical density $OD_{800} = 60$. If the majority of the RC containing fractions has an absorbance ratio of $A_{280}/A_{800} \leq 1.25$, no further purification is required. For the growth of trigonal crystals, an even lower purity can be tolerated (see below).

*Molecular Sieve Chromatography*

If necessary, a molecular sieve chromatography step can be applied in addition to both anion-exchange chromatographies or instead of the second one. The concentrated RCs (about 5 ml with an $OD_{800}$ of about 20 to 30) are transferred to a light-protected 2.5-cm $\times$ 70-cm column filled with Fractogel TSK HW-55 (S). Before filtration the columns must be thoroughly equilibrated ($\geq 5$ bed volumes) with TEAL8. The flow rate should be adjusted to about 0.5 ml/min. The filtration works optimally at room temperature. Fractions with an absorbance ratio of $A_{280}/A_{800} < 1.25$ are pooled.

The total mass of *Rb. sphaeroides* RC can be calculated by using the extinction coefficient of *Rb. sphaeroides* RC at 800 nm that amounts to 288,000 mol$^{-1}$ cm$^{-1}$.

Crystallization of *Rb. sphaeroides* RC

The RC solution is concentrated in an Amicon cell (PM10 membrane) and subsequently in Centricon 30 concentration tubes at 6000$g$ to a final $OD_{800} = 55 - 65$, i.e., to a concentration of about 200 $\mu M$. The volume is to be measured. The RCs from *Rb. sphaeroides* are crystallized by the vapor diffusion method with sitting drops. Up to six drops of about 50 $\mu$l mother solution are equilibrated against 8 ml of the reservoir solution.

*Orthorhombic Crystals*

A description for crystallization of *Rb. sphaeroides* RCs in orthorhombic form is given by Allen and Feher.[45] The precipitant to obtain orthorhombic crystals is polyethylene glycol 4000 (PEG$^{4000}$). The RCs must be of high

[45] J. P. Allen and G. Feher, *in* "Crystallization of Membrane Proteins" (H. Michel, ed.), p. 137. CRC Press, Boca Raton, FL, 1991.

purity and crystallized immediately, otherwise they will be denatured. The *sphaeroides* 1 mother liquor for orthorhombic crystals is prepared as follows: 7–7.8% (w/v) 1,2,3-heptanetriol is added to the RC solution in solid form and completely dissolved by vortexing. The suspension is mixed 1 : 1 (v/v) with a solution of 10–15 m$M$ Tris, pH 8.0, 0–0.1% LDAO, 20–24% PEG$^{4000}$, 500–540 m$M$ NaCl. The mixing must be done slowly (1–2 min) under steady stirring to prevent denaturation of the RCs. After mixing, the *sphaeroides* 1 mother liquor consists of ~100 $\mu M$ RC, 3.5–3.9% 1,2,3-heptanetriol, 10–12% PEG$^{4000}$, 0.05–0.1% LDAO, and 340–360 m$M$ NaCl. This is equilibrated against the *sphaeroides* 1 reservoir solution consisting of 10–15 m$M$ Tris-HCl, 500–600 m$M$ NaCl, 18–24% PEG$^{4000}$, 1 m$M$ EDTA, and 0.1% sodium azide. The suitable range of temperature for crystallization is 13–22°. Examples of successful experiments (big orthorhombic crystals) are as follows:

1. 15 m$M$ Tris-HCl, pH 8.0, 360 m$M$ NaCl, 3.9% 1,2,3-heptanetriol, 12% (w/v) PEG$^{4000}$, 0.06% LDAO, 1 m$M$ EDTA, 0.1% sodium azide against 15 m$M$ Tris-HCl, 600 m$M$ NaCl, 22% PEG$^{4000}$, 1 m$M$ EDTA, 0.1% sodium azide[30]
2. 10 m$M$ Tris-HCl, pH 8.0, 300 mM NaCl, 10% PEG$^{4000}$, 0.8% *n*-octyl-ß-D-glucopyranosid, 1 m$M$ EDTA against 10 m$M$ Tris-HCl, 300 m$M$ NaCl, 25% PEG$^{4000}$, 1 m$M$ EDTA[33]
3. 10 m$M$ Tris-HCl, pH 8.0, 340 m$M$ NaCl, 3.5% 1,2,3-heptanetriol, 10% PEG$^{4000}$, 0.06% LDAO, 1 m$M$ EDTA, 0.1% sodium azide against 15 m$M$ Tris-HCl, 550 m$M$ NaCl, 20% PEG$^{4000}$, 1 m$M$ EDTA, 0.1% sodium azide[46]

Conditions for a suitable soak buffer that keeps orthorhombic crystals stable for a few days are 10–15 m$M$ Tris, pH 8.0, 0.06% LDAO, 20% PEG$^{0004}$, 450 m$M$ NaCl, 3.5% 1,2,3-heptanetriol. LDAO may be replaced by 0.8% octyl-$\beta$-D-glucopyranoside or nonyl-$\beta$-D-glucopyranoside.

The procedure to get orthorhombic crystals from *Rb. sphaeroides* is relatively reproducible. These needles appear, together with more or less amorphous precipitate, in almost each crystallization experiment. However, the crystals are frequently thin and poorly ordered. To get large crystals suitable for X-ray crystallography (>2 mm long, >0.5 mm in diameter, and ≤3 Å resolution) is a more difficult task and several attempts are required. Screening within the concentration limits of *sphaeroides* 1 mother liquor as given earlier is recommended.

Under optimal conditions the orthorhombic crystals can become up to 5 mm in length and up to 1 mm in the other directions. In most cases

---

[46] G. Fritzsch, unpublished.

they are partially hollow, i.e., they contain solvent inside the crystal. The mechanical stability is low, and they tend to break easily during manipulations. Temperatures above 23° are harmful. The X-ray stability of the large crystals is sufficiently high to collect one data set with one large crystal (<2 mm) that is shifted along its long axis up to 10 times during exposure.

The space group of orthorhombic crystals is $P2_12_12_1$. The best resolution is 2.8 Å in the direction of the long crystal axis, but worse in the other directions. The number of observed unique reflections ($n_{obs}$) is less than the number of parameters necessary to define the model ($n_{par}$), the ratio $n_{obs}/n_{par}$ is at best 0.81 for orthorhombic crystals.[47] Polarized absorption spectroscopy with these crystals shows a strong dichroism mainly in the region of the special pair absorption around 860 nm.

*Trigonal Crystals*

With potassium phosphate as precipitant, the RC of *Rb. sphaeroides* can form trigonal crystals. They usually coexist with orthorhombic ones in the same crystallization well. As discussed later, the purity of the protein needs not to be extremely high. The standard *sphaeroides* 2 mother liquor for trigonal crystals is prepared as follows: 3.6–6% (w/v) 1,2,3-heptanetriol is added to RC solution in solid form and vortexed until dissolved completely. The suspension is mixed 1:1 (v/v) with a soluton of 1.0–2.0 $M$ $K_2H/KH_2$ $PO_4$, pH 6.5–7.5, 0.02–0.2% LDAO, 0–220 m$M$ NaCl, and 4–10% dioxan. The mixing must be done slowly under steady stirring in order to prevent denaturation of the RC by the high concentration of $K_2H/KH_2PO_4$. The final *sphaeroides* 2 mother liquor consists of ~100 $\mu M$ RC, 0.5–1.0 $M$ $K_2H/KH_2PO_4$, pH 6.5–7.5, 1.8–3.0% 1,2,3-heptanetriol, 0.06–0.15% LDAO, 90–200 m$M$ NaCl, and 2–5% dioxane. The replacement of 1.8–3.0% 1,2,3-heptanetriol by a mixture of 2% 1,2,3-heptanetriol and 1% 1,2,3-hexanetriol reduces the amount of denatured precipitant and yields RC crystals of unchanged quality (see below). This mother solution is equilibrated against the *sphaeroides* 2 reservoir solution consisting of 1.4–1.7 $M$ $K_2H/KH_2PO_4$. The suitable range of temperature for crystallization is 13–24°.

Conditions for a suitable soak buffer to store the crystals stably for a few days are 2.2 $M$ $K_2H/KH_2$ $PO_4$, pH 7.0, 2.0% 1,2,3-heptanetriol, 0.01% LDAO, and 3% dioxane.

The procedure for getting trigonal crystals is not very reproducible. For a well-diffracting crystal usually many experiments are required. In about 50% of the attempts, the RCs aggregate as amorphous precipitate. Screening

[47] C. R. D. Lancaster, U. Ermler, and H. Michel, *in* "Anoxygenic Photosynthesic Bacteria" (R. E. Blankenship, M. T. Madigan, and C. P. Bauer, eds.), p. 503. Kluwer Academic Publishers, Dordrecht, 1995.

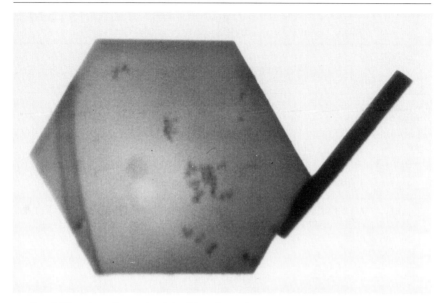

FIG. 3. Flat trigonal crystals of the *Rb. sphaeroides* RC grown from potassium phosphate on a glass slide. The large crystal has grown from below and shows the hexagonal cross section. Its edge-to-edge distance is 2.5 mm, the "long" axis of this crystal, i.e., its thickness, is about 50 μm.

tests within the concentration limits of *sphaeroides* 2 mother liquor as given earlier are recommended, but commonly a crystallization experiment gives good crystals in almost all of the screening trials or nothing at all. Most likely, the main reason for success or failure has to do with the preparation of the protein.

The trigonal crystals can be 4 mm long and up to 1.5 mm in diameter. They form well-shaped hexagonal rods (Fig. 3). At variance with orthorhombic crystals they do not contain empty domains. The mechanical stability is higher than for the orthorhombic form, and as such they are much easier to manipulate. Temperatures of up to 25° are tolerated. The X-ray stability is sufficiently high to collect one data set with one big crystal by shifting it several times along its long axis.

The space group of the trigonal crystals is $P3_121$. The best crystals diffract to beyond 2.4 Å.[48] Contrary to the orthorhombic needles, the trigonal crystals diffract in all directions with the same resolution. The number of independent reflection used to determine the structure at 2.65-Å resolution is 56,141 compared with 21,992 for orthorhombic crystals, and

---

[48] G. Fritzsch, unpublished.

TABLE I
REFINED STRUCTURES FROM BACTERIAL REACTION CENTERS[a,b]

| Resolution | No. of unique reflections | Completeness (%) | $n_{obs}/n_{par}$ | R factor (%) | Mean coordinate error |
|---|---|---|---|---|---|
| *Rps. viridis* | | | | | |
| 1PRC[5] 2.30 | 95,762 | 75.4 | 2.38 | 19.3 | 0.26 |
| *Rb. sphaeroides* | | | | | |
| 2PCR 3.1[62] | 13,493 | 50.8 | 0.63 | 22.0 | 0.5 |
| 4RCR 2.8[32] | 21,992 | 60.0 | 0.81 | 22.7 | 0.4 |
| 1PSS 3.0[63] | 21,518 | 68.9 | 0.79 | 22.3 | 0.5 |
| 1PCR 2.65 (2.4)[37,(64)] | 56,141 (87,000) | 90.4 | 1.91 (2.96) | 18.6 | 0.3 |

[a] After C. R. D. Lancaster *et al.*, *in* "Anoxygenic Photosynthetic Bacteria" (R. E. Blankenship, M. T. Madigan, and C. Bauer, eds.), p. 503. Kluwer Academic Publishers, Dordrecht, 1995.

[b] The structures are named according to their entry code in the Brookhaven data bank (PDB). The completeness was calculated for the number of reflections actually used in the refinement and may differ from those given in the references. The mean coordinate errors are estimates of the upper limit of the mean coordinate error from a Luzzati plot.

the ratio $n_{obs}/n_{par}$ = 1.91 (see Table I). Polarized absorption spectroscopy shows a strong dichroism mainly in the region of the special pair absorption around 850 nm. This is the same behavior seen with orthorhombic crystals, although the space groups and symmetry relationships for both types of crystals are different.

*Tetragonal Crystals*

Another type of well-diffracting crystal from *Rb. sphaeroides* RCs is of tetragonal structure.[38] In this case the mother liquor consists of 10 mg/ml RC, 10 m*M* Tris buffer, pH. 8.0, 6.0% PEG[4000] as precipitant, 0.85% β-octylglucoside as detergent, as well as 0.4% benzamidine hydrochloride and 2.5% 1,2,3-heptanetriol as amphiphiles. Sitting protein drops (50 μl) are equilibrated against a reservoir solution containing 32% PEG[4000]. Crystals grow after approximately 2 weeks with a maximal size of 1.5 × 1.5 × 0.3 mm³. They belong to tetragonal space group P4₃2₁2 with two RCs per assymetric unit.

*Optimal Composition of Amphiphiles for Trigonal Crystals*

The effect of amphiphiles depends on the length of their carbon chain. As a rule, to achieve the same effect by adding amphiphiles with one carbon less in the carbon chain, the three-fold concentration of the additive is needed. The replacement of the additive 1,2,3-heptanetriol by 0.4–1% 1,2,3-octanetriol (which corresponds to 1.2–3% heptanetriol) in the *sphaeroides* 2 mother liquor yields amorphous precipitate and no crystals, not even orthorhombic ones. Application of 1,2,3-hexanetriol in concentrations of 3.6–6% instead of 1,2,3-heptanetriol yields a rather clear crystallization solution over weeks or even months. But this amphiphile inhibits nucleation and crystals appear in less than 25% of experiments. If crystals grow, there are only a few in one crystallization well, therefore they can become really large. Some good results have been obtained by mixing heptanetriol and hexanetriol with final concentrations of 1.5%/1%, and in the range of 1%/2% to 2%/10%, respectively.

The advantages of the heptanetriol–hexanetriol mixture are as follows: (1) The equilibrium between trigonal and orthorhombic crystals is shifted toward trigonal ones. (2) The crystals are usually large, which means suitable for the acquisition of room temperature data sets. (3) Very little amorphous precipitate appears in the crystallization wells.

Since the nucleation rate is decreased with mixed micelles consisting of LDAO, heptanetriol, and hexanetriol, the effect of hexanetriol is most obviously not a further spatial reduction of the detergent–protein complex; rather it is a better fit of the more flexible detergent micelle with the hydrophobic surface domains of the *Rb. sphaeroides* RC.

A few trials with 1 and 2% methyl-2,4-pentanediol (MPD) instead of hexanetriol yielded only in one case small crystals and only on one glass slide. With 2% MPD an increased amorphous precipitation is observed.

### Role of 1,4-Dioxane

The presence of organic solvents such as 1,4-dioxane reduces the dielectric constant in the mother liquor. Therefore electrostatic interactions between RCs are enhanced. The probability for twinned crystals is decreased at low dielectric constant. Omission of 1,4-dioxane, however, not only inhibits the growing of single trigonal crystals of *Rb. sphaeroides* RC, it also leads more frequently to amorphous aggregates. Addition of 2–5% is optimal, although even 10% 1,4-dioxane (with corresponding dilution of the other components in the *sphaeroides* 2 mother liquor) has yielded good results in one experiment.[49]

### Temperature and pH

The optimal temperature range for the growth of trigonal *Rb. sphaeroides* RC crystals is 13–24°. At lower temperatures the crystals grow rather

---

[49] T. Bundschuh, diploma thesis, Johann Wolfgang Goethe-Universität, Frankfurt (1995).

slowly, however, with no detectable loss in quality. At higher temperatures the crystals diffract to a lower resolution, most likely due to a less well ordered crystal lattice. The pH may vary between 6.0 and 7.5. Low pH (6.5) suppresses the appearance of orthorhombic needles and favours trigonal ones. However, this fact should not be utilized to grow trigonal rather than orthorhombic crystals, since experience has shown that trigonal crystals grown at pH 6.5 diffract X-rays to less than 3.0 Å.

*Reproducibility*

The procedure for growing trigonal crystals is less reproducible than that for orthorhombic ones. Only each second trial is successful. The X-ray quality is good in most cases, i.e., the resolution is $\leq 3$ Å. The yield of large crystals can be controlled by the crystallization conditions. Concentrations in the reservoir of $\leq 1.6$ $M$ KHP yield preferentially many small crystals with maximal lengths of axes of about 0.5 mm. They appear after a few days. On the other hand, few large crystals or no crystals at all grow at reservoir concentrations of $\leq 1.5$ $M$ KHP. This growth frequently lasts many weeks or even months. At a temperature of 21° the crystals are usually larger than at 18°.

*Impurity as a Controlling Parameter*

Although in the early days of protein crystallization the crystals grew out of relatively impure preparations, there is a dogma that the highest purity possible for the protein is imperative for getting good crystals.[50] However, the trigonal crystals from *Rb. sphaeroides* RC also grow out of less pure mother liquors. RCs with purity ratios of $A_{280}/A_{800} > 1.6$ can yield big trigonal crystals that diffract well. Even the presence of incompletely separated light-harvesting B800-850 (LH2) complexes does not prevent the growth of trigonal crystals. Because orthorhombic crystals are less tolerant of impurities, the equilibrium between both types of crystal can be "controlled by impurity." Purity ratios $A_{280}/A_{800}$ between 1.6 and 2.0 (no light-harvesting complexes present) yielded preferentially trigonal crystals with unrestricted X-ray qualities. The physicochemical effect of impurities is not understood so far.

*Seeding*

The application of small RC crystals and different minerals does not improve the result of crystallization, either for orthorhombic or for trigonal

---

[50] A. McPherson, *in* "Crystallization of Membrane Proteins" (H. Michel, ed.), p. 1. CRC Press, Boca Raton, FL, 1991.

crystals. Although the RCs accept the seeded material sometimes, the quality of the obtained crystals is not improved compared with the conventionally grown crystals.

### RC from Strain R26, Mutants, and Chemically Modified RCs

RCs from the carotenoid-free *Rb. sphaeroides* strain R26 crystallize as well as the native RC and show the same resolution in the X-ray beam (2.6 Å).[51] Also crystals of different mutants of *Rb. sphaeroides* RCs have been grown. Three mutations in the binding pocket of the primary quinone do not affect the overall structure of the protein and yield crystals that diffract to 2.9 Å.[52]

### Preparation of Thin Crystals for Spectroscopy

For spectroscopic measurements (e.g., polarized absorption spectroscopy) sufficiently thin crystals are needed. The absorption should not exceed 2.0. For the formation of those crystals 2- to 5-$\mu$l droplets of mother solution are placed on glass microscopic slides covers (24 mm × 24 mm × 1 mm) and equilibrated with the respective reservoir solution.[53] Crystals with a thickness of a few microns grow after one or a few days. Trigonal crystals from *Rb. sphaeroides* RCs and tetragonal crystals from *Rps. viridis* RC stick on the glass surface and can easily be treated with other buffer solutions. Orthorhombic crystals from *Rb. sphaeroides*, however, do not adhere to the glass and their use is therefore not recommended for spectroscopic investigations.

### Structure Determination and Refinement

For both soluble and membrane proteins, data collection and crystallographic procedures follow the same principles. If data collection is performed at room temperature, the crystals are mounted in glass capillaries and exposed to the X-ray beam. In most cases, synchrotron radiation yields significantly better resolution than conventional X-ray sources.

The structure of the RC from *Rps. viridis* was determined by applying the method of multiple isomorphous replacement with heavy atom compounds.[27] The phase problem has been solved using five different heavy atom derivatives and further improvement of phases was achieved with

[51] G. Fritzsch, M. Meyer, U. Ermler, and H. Michel, unpublished.
[52] G. Fritzsch *et al.*, in preparation.
[53] G. Fritzsch, S. Buchanan, and H. Michel, *Biochim. Biophys. Acta* **977,** 157 (1989).

solvent flattening.[54] Model building was performed with the FRODO package[55] where use was made of the real-space-refinement facility to place long stretches of helical structures correctly into the electron density. The program packages PROTEIN, EREF,[56,57] TNT,[58] and FRODO were used for refinement.

The structure of the *Rb. sphaeroides* RC from orthorhombic crystals was determined by molecular replacement using the coordinates from *Rps. viridis* RC.[30,59] For the determination of the structure using the trigonal *Rb. sphaeroides* crystals the coordinates of an R-26 model[60] were used.[37] The refinement was performed using a native data set to a resolution of 2.65 Å, and the overall completeness was 90.4%. For positional refinement the X-PLOR program[61] was employed and the final crystallographic R factor was 18.6%. With the trigonal crystals up to 87,000 X-ray reflections[61] have been detected so far. They are the first crystals from this RC that yielded more reflections than parameters ($x,y,z$ coordinates and B factors) that are required for structure determination. Crystallographic data for both RCs are listed in Table I.

Further Developments

The crystallization of membrane-bound proteins is still a challenging task. A strategy for effective initial conditions has not yet been developed,[9] not even for the rather small group of photosynthetic RCs from nonsulfur purple bacteria. Further developments should not only be based on conventional mixtures of buffers, precipitants, detergents, small amphiphiles, and other additives, it should explore new concepts, e.g., cocrystallization with

[54] B. C. Wang, *Methods Enzymol.* **115,** 90 (1985).
[55] T. A. Jones, *J. Appl. Crystallogr.* **11,** 268 (1978).
[56] A. Jack and M. Levitt, *Acta Crystallogr. A* **34,** 931 (1978).
[57] J. Deisenhofer, S. J. Remington, and W. Steigemann, *Methods Enzymol.* **115,** 303 (1985).
[58] D. E. Tronrud, L. F. Ten Eyck, and B. W. Matthews, *Acta Crystallogr. A* **43,** 489 (1987).
[59] C.-H. Chang, D. Tiede, J. Tang, U. Smith, J. Norris, and M. Schiffer, *FEBS Lett.* **205,** 82 (1986).
[60] J. P. Allen, G. Feher, T. O. Yeates, H. Komiya and D. C. Rees, *Proc. Natl. Acad. Sci. U.S.A.* **48,** 5730 (1987).
[61] A. T. Brünger, J. Kuriyan, and A. M. Karplus, *Science* **235,** 458 (1987).
[62] C.-H. Chang, O. El-Kabbani, D. Tiede, J. Norris, and M. Schiffer, *Biochemistry* **30,** 5352 (1991).
[63] A. J. Chirino, E. J. Lous, M. Huber, J. P. Allen, C. C. Schenck, M. L. Paddock, G. Feher, and D. C. Rees, *Biochemistry* **33,** 4584 (1994).
[64] L. Kampmann, G. Fritzsch, U. Ermler, and H. Michel, unpublished.
[65] H. Michel, *Trends Biochem.* **8,** 56 (1993).

antibodies as have been successfully applied for the crystallization of the cytochrome *c* oxidase from *Paracoccus denitrificans.*[22]

## Acknowledgment

I thank Dr. H. Michel for numerous discussions, Mrs. Z. Bojadzijev for technical assistance, and Dr. A. Gardiner for reading the manuscript.

# Section II

# Photosynthetic Complexes: Function/Structure

# [6] Using Genetics to Explore Cytochrome Function and Structure in *Rhodobacter*

*By* HANS-GEORG KOCH, HANNU MYLLYKALLIO, and FEVZI DALDAL

## Introduction

Facultative phototrophic bacteria in general, and *Rhodobacter* (*Rb.*) species in particular, provide excellent experimental models for studying the structure and function of electron transfer components. These organisms have highly branched electron transfer pathways that allow them to grow under a variety of conditions.[1] Under phototrophic growth conditions one or two different *c*-type cytochromes (cyt), the soluble cyt $c_2$ and the membrane-bound cyt $c_y$, facilitate light-driven cyclic electron transfer between the photochemical reaction center and the cyt $bc_1$ complex in these species.[2–4] Under respiratory growth conditions one of the pathways used is similar to the mitochondrial electron transfer chain in that it involves a cyt $bc_1$ complex, a cyt *c* and a cyt *c* oxidase.[5,6] In *Rb. sphaeroides* when cells are grown at high oxygen tensions the predominant cyt *c* oxidase is of $aa_3$ type,[7] whereas under microaerophilic conditions a $cbb_3$-type cyt *c* oxidase,[8] with a presumably higher oxygen affinity, is synthesized. On the other hand, *Rb. capsulatus* is devoid of *a*-type cytochromes,[9] thus only a $cbb_3$-type cyt *c* oxidase is present under both high and low oxygen concentrations.[10] An additional aerobic energy transduction mode, called the *alternate respiratory pathway,* also exists in both organisms. It is independent of the cyt $bc_1$, cyt *c*, and cyt *c* oxidase and uses a quinol oxidase.[6] Furthermore, a variety of compounds, such as dimethyl sulfoxide, trimethylamine oxide, or nitrous

---

[1] D. Zannoni, *in* "Anoxygenic Photosynthetic Bacteria" (R. E. Blankenship, M. T. Madigan, and C. E. Bauer, eds.), p. 949. Kluwer Academic Publishers, Dordrecht, 1995.

[2] F. Daldal, S. Cheng, J. Applebaum, E. Davidson, and R. C. Prince, *Proc. Natl. Acad. Sci. U.S.A.* **83,** 2012 (1986).

[3] T. J. Donohue, A. G. McEwan, S. van Doren, A. R. Crofts, and S. Kaplan, *Biochemistry* **27,** 1918 (1988).

[4] F. E. Jenney and F. Daldal, *EMBO J.* **12,** 1283 (1993).

[5] R. B. Gennis, R. P. Casay, A. Azzi, and B. Ludwig, *Eur. J. Biochem.* **125,** 189 (1982).

[6] B. Marrs and H. Gest. *J. Bacteriol.* **114,** 1045 (1973).

[7] J. P. Shapleigh, J. J. Hill, J. O. Alben, and R. B. Gennis, *J. Bacteriol.* **174,** 2338 (1992).

[8] J. A. Garcia-Horsman, E. Berry, J. P. Shapleigh, J. O. Alben, and R. B. Gennis, *Biochem.* **33,** 3113 (1994).

[9] J.-H. Klemme and H. G. Schlegel, *Arch. Microbiol.* **68,** 326 (1969).

[10] K. A. Gray, M. Grooms, H. Myllykallio, C. Moomaw, C. Slaughter, and F. Daldal, *Biochemistry* **33,** 3120 (1994).

oxide, can also serve as terminal electron acceptors under dark, anaerobic growth conditions.[11] This metabolic versatility and the availability of various molecular and genetic tools obviously make *Rhodobacter* species an excellent model system for performing multidisciplinary approaches to study the structure, function and biogenesis of various electron transfer components.

Cytochromes with various heme prosthetic groups are major electron transfer proteins (Table I).[12] Heme A, heme D, and the more recently discovered heme O are chemical variants of protoporphyrin IX, which forms heme B.[13,14] Among these proteins, *c*-type cytochromes are unique in that the heme moiety is covalently attached to the apoprotein via two thioether linkages between the cysteine residues of a highly conserved heme attachment site (**CxxCH**) and the two vinyl groups of protoporphyrin IX.[13] It now appears that this motif is not exclusive to *c*-type cytochromes, because bacterial genomics has recently uncovered several cytoplasmic proteins with an identical motif, but they are unlikely to contain an attached heme group (e.g., DNA polymerase). Furthermore, some eukaryotic microbes such as *Euglena* and *Critidia* have a modified heme attachment site (***AAQCH***).[15,16] Finally, cytochromes with a noncovalently attached heme group (such as cyt *a, b, d,* and *o*), have no conserved attachment sites, although multiple sequence alignments often reveal conserved amino acids such as histidine as heme ligands.

## Molecular Recognition of Cytochromes

Cytochromes have traditionally been recognized and classified by using optical spectroscopy because they show characteristic absorption spectra that change reversibly with the redox state of their heme Fe atom (Table I). While this technique is a powerful tool to detect the presence of cytochromes in complex biological materials, usually various members of a given type of cyt have closely related absorption properties, thus making their molecular identification impossible. For example, many known *c*-type cytochromes absorb around 550–552 nm (Table I). Furthermore, heme O, in which the vinyl group of the pyrrole ring A of heme B is substituted by a hydroxyethylfarnesyl group and has a methyl group at pyrrole ring 3,[17]

[11] A. G. McEwan, *Ant. van Leeuwenhoek* **66,** 151 (1996).
[12] G. Palmer and J. Reedijk, *J. Biol. Chem.* **267,** 665 (1992).
[13] G. W. Pettigrew and G. R. Moore, "Cytochromes c." Springer-Verlag, Berlin, 1987.
[14] A. Puustinen and M. Wikström, *Proc. Natl. Acad. Sci. U.S.A.* **88,** 6122 (1991).
[15] G. W. Pettigrew, *Nature* **241,** 531 (1973).
[16] G. W. Pettigrew, *FEBS Lett.* **22,** 64 (1972).
[17] W. Wu, C. K. Chang, C. Varotsis, G. T. Babcock, A. Puustinen, and M. Wikström, *J. Am. Chem. Soc.* **114,** 1182 (1992).

TABLE I
GENERAL PROPERTIES OF DIFFERENT CYTOCHROMES[a]

| Cytochrome | Characteristic features | Typical example |
|---|---|---|
| a | *Absorption maximum:* 605 nm | cyt $aa_3$ |
|  | *Prosthetic group:* Protoheme IX derivative, vinyl group at pyrrole ring A is substituted by a 17 carbon hydroxyethylfarnesyl group (heme A, iron chelate of protoporphyrin derivative) | oxidase |
|  | *α-Band of pyridine ferrohemochrome:* 580–590 nm |  |
| o | *Absorption maximum:* 557 nm | cyt *bo* |
|  | *Prosthetic group:* Protoheme IX derivative, vinyl group at pyrrole ring A is substituted by a 17-carbon hydroxyethylfarnesyl group. In comparison to heme A formyl group at pyrrole ring 3 is replaced by methyl group (heme O, iron chelate of protoporphyrin IX derivative | quinol oxidase |
|  | *α-Band of pyridine ferrohemochrome:* 552 nm |  |
| b | *Absorption maximum:* 555–565 nm | cyt $b_{562}$ |
|  | *Prosthetic group:* Protoheme IX (heme B, iron chelate of protoporphyrin IX) |  |
|  | *α-Band of pyridine ferrohemochrome:* Around 555–558 nm |  |
| c | *Absorption maximum:* 550–555 nm | cyt $c_1$ of the |
|  | *Heme group* (substituted protoheme IX) covalently attached via thioether linkages between either or both of the vinyl side chains of protoheme and the protein. Conseved heme-attachment site (**CxxCH**). Penta- (high-spin) or hexa- (low spin) coordinated | $bc_1$ complex |
|  | *α-Band of pyridine ferrohemochrome:* 549–551 nm (two thioether links); 553 nm (single thioether link) |  |
| d | *Absorption maximum:* 625–660 nm | cyt $cd_1$-nitrite |
|  | Degree of conjugation of double bonds is less than in porphyrins, e.g., dihydroporphyrin (heme D), tetrahydroporphyrin (heme $D_1$) | reductase |
|  | *α-Band of pyridine ferrohemochrome:* 600–620 nm |  |

[a] Note that novel prenylated hemes have been identified as cofactors of cytochrome oxidases in archaea. Data gathered from G. W. Pettigrew and G. R. Moore, "Cytochromes c." Springer Verlag, Berlin, 1987; G. Palmer and J. Reedijk, *J. Biol. Chem.* **267**, 665 (1992); A. Puustinen and M. Wikström, *Proc. Natl. Acad. Sci. U.S.A.* **88**, 6122 (1991).

is not readily distinguished from heme B by optical spectroscopy. Hence, additional analytical techniques, combining separation of proteins by standard electrophoretic methods with detection of the heme groups via their intrinsic peroxidase activity, are commonly used. Sodium dodecyl sulfate–polyacrylamide gel electrophoresis (SDS–PAGE) and lithium dodecyl sulfate–PAGE followed by peroxidase staining are especially well suited for the detection of cytochromes with covalently and noncovalently attached

A

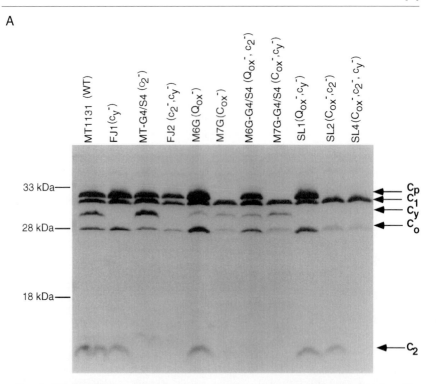

FIG. 1. (A) Cytochrome $c$ (cyt $c$) profile of chromatophore membranes of various Rb. capsulatus strains grown under respiratory conditions on enriched MPYE medium. Two hundred fifty micrograms of total protein was loaded into each lane on an SDS–16.5% polyacrylamide gel like that described by Schägger and von Jagow [Anal. Biochem. **166,** 368 (1987)]. After electrophoresis c-type cytochromes were revealed by TMBZ staining.[18] Molecular weight markers are shown on the left, and the bands corresponding to the cytochromes $c_p$, $c_1$, $c_y$, $c_o$, and $c_2$ are indicated on the right. (B) Optical difference spectra of chromatophore membranes and soluble fractions of various Rb. capsulatus strains grown under respiratory conditions. Ascorbate-reduced *minus* ferricyanide-oxidized difference spectra of soluble fractions (upper panel), ascorbate-reduced *minus* ferricyanide-oxidized difference spectra of chromatophore membranes (middle panel) and dithionite-reduced *minus* ferricyanide-oxidized difference spectra of chromatophore membranes (lower panel). [H. Schägger and G. von Jagow, Anal Biochem. **166,** 368 (1987)].

heme groups, respectively. Peroxidase activity associated with the heme group can be reversibly detected using either 3,3′,5,5′-tetramethylbenzidine (TMBZ),[18] diaminobenzidine,[19] or luminol (5-amino-2,3-dihydro-1,4-phthalazinedione)[20,21] as a substrate. Alternatively, a hemoprotein can also

[18] P. E. Thomas, D. Ryan, and W. Levin, Anal. Biochem. **75,** 168 (1976).
[19] A. McDonnel and L. A. Staehlin, Anal. Biochem. **117,** 40 (1981).
[20] C. Bonfils, S. Charasse, J. Bonfils, and C. Larroque, Anal. Biochem. **226,** 302 (1995).
[21] C. Vargas, A. G. McEwan, and J. A. Downie, Anal. Biochem. **209,** 323 (1993).

B

R. capsulatus

MT1131    FJ1    MT-G4/S4    FJ2    pFJ63/FJ2

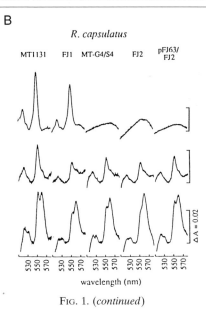

wavelength (nm)

FIG. 1. (*continued*)

be detected after elimination of the heme Fe atom by its intrinsic fluorescence emission.[22] Note that stains relying on peroxidase activity are extremely sensitive to sulfhydryl reducing reagents like 2-mercaptoethanol, and hence special care is required to remove them after gel electrophoresis.

In some cases, cytochromes that are part of enzyme complexes can be recognized directly using specific stains. For example, the presence of an active cyt *c* oxidase can be revealed *in vivo* using the *NADI reaction*.[23] This stain is simply performed by overlaying colonies with a 1:1 (v/v) mixture of 35 m$M$ $\alpha$-naphthol and 30 m$M$ $N,N$-dimethyl-*p*-phenylenediamine dissolved in ethanol and water, respectively. Colonies that contain an active cyt *c* oxidase turn blue within 30 sec on exposure to these chemicals due to the production of indophenol blue. This stain has been extremely useful for exploring the structure, function, and biogenesis of cyt complexes in various species, including *Rb. capsulatus*.[24] Using the NADI reaction, a number of *Rb. capsulatus* mutants lacking specific *c*-type cytochromes have been isolated and characterized (Fig. 1). Note also that the metabolic versatility of these species is clearly reflected in their complex cyt *c* profile (Table II).

[22] M. B. Katan, *Anal. Biochem.* **75,** 132 (1976).
[23] D. Keilin, "The History of Cell Respiration and Cytochrome." Cambridge University Press, Cambridge, Massachusetts, 1966.
[24] H-G. Koch, O. Hwang, M. Grooms, and F. Daldal, *J. Bacteriol.* **180,** 969 (1998).

TABLE II
DIFFERENT C-TYPE CYTOCHROMES IN RHODOBACTER

| Cytochrome | Organisms[a] | MW[b] | $E_{m,7}{}^{c}$ (mV) | Relevant characteristics | Reference[d] |
|---|---|---|---|---|---|
| **A. Soluble** | | | | | |
| Cyt $c_2$ (cycA) | R.c, R.s | 13,000 | +360 | Electron carrier in photosynthesis and respiration | 1,2 |
| Iso-cyt $c_2$ (cycI) | R.s | 15,000 | +290 | Possibly involved in formaldehyde oxidation, sufficient to support Ps growth on overproduction | 3 |
| Cyt c' | R.c, R.s | 14,000 | +30 to +50 | Usually dimeric high-spin cyt c, function unknown | 4 |
| cyt $c_{554}$ (cycF) | R.s | 14,000 | +205 | Induced under aerobic conditions, function unknown | 5 |
| Cyt $c_{551.5}$ | R.s | 16,000 | −254 | Isolated from Ps grown cells | 6 |
| Cyt $c_{552}$ | R.s | 13,500 | +250 | Induced under denitrifying conditions | 7 |
| Cyt $c_{553}$ | R.s. | 25,000 | +120 | Function unknown | 8 |
| napB | R.s | 15,000 | NA[d] | Part of a periplasmic nitrate reductase, predicted to be a diheme cytochrome | 9 |
| **B. Membrane bound** | | | | | |
| Cyt $c_1$ (petC/fbcC) | R.c, R.s | 30,000 | +290 | Subunit of ubihydroquinone : cyt c oxidoreductase, anchored to the membrane by a carboxy-terminal α-helix, interacts with cyt $c_2$ and $c_y$ | 10 |
| Cyt $c_y$ (cycY) | R.c, R.s | 22,000 (R.c) 18,000 (R.s) | +360 | R.c cyt $c_y$ functions as an electron carrier both in photosynthesis and respiration; R.s cyt $c_y$ only in respiration | 11 |
| Cyt $c_p$ (ccoP) | R.c, R.s | 32 | +265 | Predicted to be a diheme cyt c subunit of cyt $cbb_3$ oxidase | 12, 13 |
| Cyt $c_o$ (ccoP) | R.c, R.s | 27 | +320 | Subunit of cyt $cbb_3$ oxidase | 12, 13 |
| dmsC | R.s | 44 | NA | Induced by anaerobic respiration, predicted to be a pentaheme cyt c | 14 |
| napC | R.s | 25 | NA | Subunit of periplasmic nitrate reductase, predicted to be a tetraheme cyt c | 9 |
| cycG | R.s | 32 | NA | Predicted to be a diheme cyt c, forms an operon with cycF, function unknown | 5 |

TABLE II (*continued*)

[a] *R.c,R. capsulatus; R.s,R. sphaeroides.*
[b] For soluble proteins, estimated from SDS–PAGE; for membrane-bound cytochromes calculated from primary sequence when available.
[c] Approximate $E_{m,7}$ values.
[d] NA, information not available.
[e] References:
1. F. Daldal, S. Cheng, J. Applebaum, E. Davidson, and R. C. Prince, *Proc. Natl. Acad. Sci. U.S.A.* **83**, 2012 (1986).
2. T. J. Donohue, A. G. McEwan, S. van Doren, A. R. Crofts, and S. Kaplan, *Biochemistry* **27**, 1918 (1988).
3. M. A. Rott, V. C. Witthuhn, B. A. Schilke, M. Soranno, A. Abdulfatah, and T. J. Donohue, *J. Bacteriol.* **175**, 358 (1993).
4. R. G. Bartsch, *in* "The Photosynthetic Bacteria" (R. K. Clayton and W. R. Sistrom, eds.), p. 249. Plenum Press, New York, 1978.
5. J. E. Flory and T. J. Donohue, *J. Bacteriol.* **177**, 4311 (1995).
6. T. E. Meyer and M. A. Cusanovich, *Biochim. Biophys. Acta* **807**, 308 (1984).
7. W. P. Michalski, D. J. Miller, and D. J. D. Nicholas, *Biochim. Biophys. Acta* **849**, 304 (1986).
8. J. Orr, *Biochim. Biophys. Acta* **57**, 373 (1962).
9. F. Reyes, M. D. Roldán, W. Klipp, F. Castillo, and C. Moreno-Vivián, *Mol. Microbiol.* **19**, 1307 (1996).
10. K. A. Gray and F. Daldal, *in* "Anoxygenic Photosynthetic Bacteria" (R. E. Blankenship, M. T. Madigan, and C. E. Bauer, eds.), p. 747. Kluwer Academic Publishers, Dordrecht, 1995.
11. H. Myllykallio, F. E. Jenney, C. Moomaw, C. Slaughter, and F. Daldal, *J. Bacteriol.* in press.
12. K. A. Gray, M. Grooms, H. Myllykallio, C. Moomaw, C. Slaughter, and F. Daldal, *Biochemistry* **33**, 3120 (1994).
13. J. A. Garcia-Horsman, E. Berry, J. P. Shapleigh, J. O. Alben, and R. B. Gennis, *Biochemistry* **33**, 3113 (1994).
14. T. Ujiiye, I. Yamamoto, H. Nakama, A. Okubo, S. Yamazaki, and T. Satoh, *Biochim. Biophys. Acta* **1277**, 1 (1996).

## Genetic Tools to Analyze Cytochrome Function and Structure in *Rhodobacter* Species

### General Genetic Techniques

Nearly all standard tools used in organisms amenable to genetic analyses are also available in *Rhodobacter* species. Many of those used for investigating different biological processes, including the structure and function of cytochromes, have been described previously by Williams and Taguchi,[25] Donohue and Kaplan,[26] and Scolnik and Marrs.[27] Among the DNA transfer

---

[25] J. C. Williams and A. K. W. Taguchi, *in* "Anoxygenic Photosynthetic Bacteria" (R. E. Blankenship, M. T. Madigan, and C. E. Bauer, eds.), p. 1029. Kluwer Academic Publishers, Dordrecht, 1995.
[26] T. J. Donohue and S. Kaplan, *Methods Enzymol.* **204**, 459 (1991).
[27] P. A. Scolnik and B. L. Marrs, *Ann. Rev. Microbiol.* **41**, 703, (1987).

methods into *Rhodobacter* species, *intra-* and *inter*species conjugation is the most powerful. However, note that conjugation frequency is strain dependent, and varies between $10^{-4}$ and $10^{-8}$ for both *Rb. sphaeroides* and *Rb. capsulatus*.[28,29] The relatively small ($\approx$10-kb) derivatives (such as pRK404 and pRK415) of the broad host range plasmid RK2, containing multiple unique restriction sites and the lacZ$\alpha$ region, are available as mobilizable cloning vectors.[30] Counterselection against the *Escherichia coli* donor strains is often achieved by using the prototrophic properties of recipient *Rhodobacter* species by growth on minimal medium.[31] Counterselection on enriched medium is also possible in the presence of high concentrations of tellurite ($K_2TeO_3$, 200 $\mu$g/ml for aerobic and photosynthetic growth conditions).[32] Although tellurite resistance selection has not been used extensively, several *Rhodobacter* species, such as *Rb. capsulatus* strain B10 or *Rb. sphaeroides* strains 2.4.1, WS8, 2.4.7 and RS2, are known to exhibit high-level resistance to this metal oxyanion.[32] Alternatively, an *E. coli*-specific lytic phage (such as phage $T_4D$) can also be used to prevent growth of the donor strain, independently of the nature of the medium.

Chemical transformation, which has the advantage of not requiring counterselection against the donor strain, is of limited use for *Rhodobacter* due to its low efficiency, although electroporation has significantly improved it for various *Rhodobacter* species. A protocol that has been used successfully for *Rb. capsulatus* is described next:

Inoculate 1 liter MPYE (0.3% peptone, 0.3% yeast extract, 1.6 m$M$ $MgCl_2$, 1.0 m$M$ $CaCl_2$, pH 7.0) medium with 10 ml of seed culture, and grow at 35° under semiaerobic conditions until an $OD_{630}$ of 0.4–0.5 is obtained (10–12 hr).

Harvest cells at room temperature (RT) by centrifugation at 10,000 rpm for 10 min, and resuspend pellet in 1 liter doubly distilled $H_2O$ at RT.

Repeat washing step with 500 ml doubly distilled $H_2O$ at RT, and resuspend the pellet in 20 ml of $H_2O$ with 10% glycerol. Repeat centrifugation step and resuspend cells in 1 ml of 10% glycerol in water. Aliquot competent cells (100 $\mu$l each), quick freeze them using ethanol/dry ice bath, and store at $-80°$.

[28] C. Moreno-Vivian, M. D. Roldan, F. Reyes, and F. Castillo, *FEMS Microbiol. Lett.* **115,** 279 (1994).
[29] D. J. Kelly, D. J. Richardson, S. J. Ferguson, and J. B. Jackson, *Arch. Microbiol.* **150,** 138 (1988).
[30] G. Ditta, T. Schmidhauser, E. Yacobson, P. Lu, X.-W. Liang, D. R. Finlay, D. Guiney, and D. R. Helinski, *Plasmid* **13,** 149 (1985).
[31] W. R. Sistrom, *J. Gen. Microbiol.* **22,** 778 (1960).
[32] M. D. Moore and S. Kaplan, *J. Bacteriol.* **174,** 1505 (1992).

Electroporation is performed using 40 $\mu$l competent cells under the following conditions [Bio-Rad (Richmond, CA) Gene Pulser apparatus, 0.2-cm gap-cuvette]: 400 ohms, 25 $\mu$F, 2.5 kV. Under these conditions the time constant should be around 8 ms. The recommended DNA amount is about 100 ng/40 $\mu$l of competent cells. Immediately after pulsing, add 1 ml MPYE + 10 m$M$ MgSO$_4$ + 10 m$M$ MgCl$_2$, and incubate for 4–6 hr at 35°. Harvest transformed cells by centrifugation, resuspend in 100 $\mu$l MPYE, and spread out on appropriate plates. The transformation frequency observed is about 10$^5$ colonies/$\mu$g DNA for a plasmid of about 10-kb size.

Note that the use of both electroporation and chemical transformation is limited to DNA extracted from *Rhodobacter* species, most likely due to the presence of restriction and modification systems. Restriction endonucleases have been characterized so far only from *Rb. sphaeroides* strains[33,34] although restriction-less strains of both *Rhodobacter* species have been described.[35,36]

Isolation of knockout mutants is essential for defining rigorously the biological function of individual cytochromes. Two different tools are available for allelic exchange between a mutant allele of a gene of interest and its wild-type chromosomal copy:

1. The pSUP series of plasmids, which are unable to replicate in *Rhodobacter* species, are frequently used as suicide vectors.[37] Integration of the plasmid into the chromosome is detected by monitoring the pSUP-encoded antibiotic resistance and allelic exchange by selecting for the appropriate selectable marker located in the gene to be knocked out. In addition to these ColE1-based vectors, Penfold and Pemberton[38] constructed a suicide vector, pJP5603, based on the R6K replicon, which replicates only if the R6K *pir* gene is provided *in trans*. This vector contains the multiple cloning site of pUC19 and harbors the *kan* gene of Tn5, which contains unique *Bgl*II and *Nco*I sites useful in replacing it with other selectable markers, as needed. Another system based on the R6K replicon is described by Roberts and co-workers.[39] It consists of two plasmids, the R6K-based helper plasmid pUX-BF13, which depends for its replication on the *pir* product

[33] S. P. Lynn, L. K. Cohen, J. F. Gardner, and S. Kaplan, *J. Bacteriol.* **138**, 505 (1979).
[34] S. P. Lynn, L. K. Cohen, S. Kaplan, and J. F. Gardner, *J. Bacteriol.* **142**, 505 (1980).
[35] D. P. Taylor, S. N. Cohen, W. G. Clark, and B. L. Marrs, *J. Bacteriol.* **154**, 580 (1983).
[36] T. J. Donohue, J. Chory, T. E. Goldsand, S. P. Lynn, and S. Kaplan, *J. Virol.* **55**, 147 (1985).
[37] R. Simon, U. Priefer, and A. Pühler, *Bio/Technology* **1**, 784 (1983).
[38] R. J. Penfold and J. M. Pemberton, *Gene* **118**, 145 (1992).
[39] Y. Bao, D. P. Lies, H. Fu, and G. P. Roberts, *Gene* **109**, 167 (1991).

provided by pUX-BF5, a pUC19 derivative that replicates only in enteric bacteria. The plasmid pUX-BF13 also contains the mini transposon Tn7-Kan into which foreign DNA can be cloned.

2. A unique system for allele replacement in *Rb. capsulatus* uses the *gene transfer agent* (GTA), which appears to be a defective phage specific for this species.[40] This tool has been extremely useful for isolating many knockout mutants (interposon mutagenesis), including those lacking all major *c*-type cytochromes (Fig. 1). Preparation of the GTA and the GTA-mediated gene transfer have been described by Donohue and Kaplan.[26] Note that due to the limited packing capability of the GTA this method is not suitable for transfer of DNA fragments larger than 4.5–5 kb. Further, in order to use this technique the gene of interest containing an integrated selectable marker should be available beforehand. This selectable allele is first transferred by conjugation into a *Rb. capsulatus* GTA overproducer strain such as R121 or Y262. Due to the instability of GTA overproduction often several transconjugants are pooled and grown by photosynthesis in enriched medium to reassure optimal production. The interrupted allele is then incorporated into the chromosome of the recipient strain by homologous recombination via a double crossover using the selectable marker. GTA is also useful for determining the physical distance between a genetic marker such as a restriction site and various spontaneous mutations located in a given gene. Daldal *et al.*[41] have used this technique, called interposon tagging, to map spontaneous myxothiazol-resistant mutants of *Rb. capsulatus* to the structural genes *fbcFBC/petABC* of the cyt *bc*$_1$ complex.

Although a negative screen (for example, the inability to grow under photosynthetic conditions) is often sufficient for isolation of appropriate mutants, obviously a positive selection is more effective. Sabaty and Kaplan[42] developed a positive selection vector, pPS400, using the promotorless *sacB* gene of *Bacillus subtilis* fused to the upstream regulatory region of the *puc* operon coding for the light-harvesting complex B800-850. In the presence of sucrose the *sacB* gene product, levansucrase, hydrolyzes sucrose and forms fructosyl residues, which are lethal to many gram-negative bacteria, including *Rhodobacter* species. Because the *sacB* gene is under the control of the *puc* promotor, only mutants with reduced *puc* operon transcription survive in the presence of 0.05% fructose.

[40] B. Marrs, *Proc. Natl. Acad. Sci. U.S.A.* **71**, 971 (1974).
[41] F. Daldal, M. K. Tokito, E. Davidson, and M. Faham, *EMBO J.* **8**, 3951 (1989).
[42] M. Sabaty and S. Kaplan, *J. Bacteriol.* **178**, 35 (1996).

## Genetic Modifications of Cytochromes

### Random Mutagenesis

*Chemical Mutagenesis.* Among the chemical mutagens, nitrosoguanidine (NTG) and the alkylating agent ethyl methane sulfonate (EMS) are commonly used for photosynthetic bacteria, including *Rhodobacter* species. In combination with powerful selection and screening methods, like the above-described NADI reaction or the inability to grow in the presence of sucrose, chemical mutagenesis is capable of generating a large number of mutants. A typical protocol for EMS mutagenesis is as follows: A 10-ml aliquot of an overnight culture of the appropriate strain (grown in MPYE to an $OD_{630} \approx 1.0$) is centrifuged, and the cell pellet resuspended in 1 ml of 0.1 $M$ sodium phosphate buffer (pH 7.4), containing 12 mg of EMS. The suspension is incubated at 37° for 30 min, washed three times with phosphate buffer, and finally resuspended in 1 ml buffer. Ten microliters of the mutagenized cells is added to 1 ml of MPYE broth and incubated overnight under appropriate conditions. Overnight cultures are diluted ($10^{-7}$–$10^{-8}$) and plated on MPYE plates. For isolation of mutants unable to grow photosynthetically mutagenized cells are incubated for 2 days under photosynthetic growth conditions, the colonies they form are marked, and the plates are transferred to respiratory growth conditions. Newly arising putative respiratory competent and photosynthetic incompetent cells are purified further and their phenotypes characterized.

*Transposon Mutagenesis.* Tn5 is the most commonly used transposon for transposon mutagenesis in *Rhodobacter* species, with the above-described pSUP plasmids often used as delivery vectors.[37] Among others, Tn5 mutagenesis has been successful in isolating cyt $c_p$ mutants,[24,43] general cytochrome biogenesis mutants,[44] and mutants affecting heme biosynthesis.[45] In comparison to Tn7, which transposes to specific locations in photosynthetic and other bacteria,[46,47] Tn5 apparently has fewer preferences for its insertion sites in *Rhodobacter,* and therefore can be used for "random" mutagenesis. It displays a low probability of genome rearrangements upon

[43] H.-G. Koch, O. Hwang, M. Grooms, and F. Daldal, *in* "Photosynthesis: From Light to Biosphere" (P. Mathis ed.), Vol. II, p. 819. Kluwer Academic Publishers, Dordrecht, 1995.
[44] R. G. Kranz, *J. Bacteriol.* **171,** 456 (1989).
[45] S. W. Biel and A. J. Biel, *J. Bacteriol.* **172,** 1321 (1990).
[46] M. Rogers, N. Ekaterinaki, E. Nimmo, and D. Sherrat, *Mol. Gen. Genet.* **205,** 550 (1986).
[47] D. C. Youvan, J. T. Elder, D. E. Sandlin, R. Zserbo, D. P. Alder, N. J. Panopoulos, B. L. Marrs, and J. E. Hearst, *J. Mol. Biol.* **162,** 17 (1982).

transposition, and is highly stable once integrated into the chromosome.[48] Note that the successful use of Tn5 mutagenesis is strain dependent in both *Rb. capsulatus* and *Rb. spheroides*. Considering a transposition frequency of $10^{-5}$ in *Rhodobacter* species, only strains with a high conjugation frequency for the pSUP plasmids, (such as the *Rb. sphaeroides* strains 2.4.1 and NCIB 8253 and the *Rb. capsulatus* strains B10 and, to a lower frequency, MT1131) yield large pools of "random" Tn5 mutants. Transposition is also effective for localized mutagenesis (i.e., mutagenesis of a cloned gene in *E. coli* as host strain, see below). However, for this purpose pSUP vectors are of limited use as they replicate in *E. coli,* although screening is possible on media containing elevated concentrations of antibiotics (i.e., kanamycin 50–100 $\mu$g/ml) to differentiate between Tn5 insertions in the chromosome and Tn5 insertion in the plasmid. In addition, this approach requires counterselection against the Tn5 delivering vector. On the other hand, *E. coli*-specific phages (such as $\lambda$28 and others) provide better delivery tools, since they are unable to replicate within, or to lysogenize, the host strain. The Tn5 derivative Tn*phoA*, where the insertion element $IS50_L$ is replaced with the *E. coli* alkaline phosphatase gene (*phoA*), lacking the signal peptide, has been used successfully to generate fusion proteins, linking the N terminus of the target gene to *phoA*.[49] Since alkaline phosphatase is only active in the periplasm,[50] this system is especially useful for identifying soluble periplasmic cytochromes and for analyzing the topology of membrane-bound cytochromes, as has been the case for cyt $c_y$[51] and the cytochrome biogenesis protein *CycH*.[52] A similar system for studying secreted proteins was developed by Tadayyon and Broome-Smith[53] by fusion of the *blaM* gene, encoding a $\beta$-lactamase, to Tn5. By using Tn*blaM*, the desired mutants can be isolated by direct selection for ampicillin resistance rather than by phenotypic screening, available when using Tn*phoA*. Additional useful Tn5 derivatives include Tn5-*mob*, which contains the RP4-specific *mob*-site, enabling the target replicon to be efficiently mobilized by the broad host range transfer functions of RP4,[54] and the Tn5-*lacZ* ($\beta$-galactosidase) and Tn5-*luc* (luciferase) that generate gene and operon fusions to monitor gene expression.[55]

[48] W. S. Reznikoff, *Ann. Rev. Microbiol.* **47,** 945 (1993).
[49] C. Manoil and J. Beckwith, *Proc. Natl. Acad. Sci. U.S.A.* **82,** 8129 (1985).
[50] A. I. Derman and J. Beckwith, *J. Bacteriol.* **173,** 7719 (1991).
[51] H. Myllykallio, F. E. Jenney, C. Moomaw, C. Slaughter, and F. Daldal, *J. Bacteriol.* **179,** 2623 (1997).
[52] S. E. Lang, F. E. Jenney, and F. Daldal, *J. Bacteriol.* **178,** 5279 (1996).
[53] M. Tadayyon and J. K. Broome-Smith, *Gene* **111,** 21 (1992).
[54] R. Simon, *Mol. Gen. Genet.* **196,** 413 (1984).
[55] R. Simon, J. Quandt, and W. Klipp, *Gene* **80,** 161 (1989).

*Localized Mutagenesis*

*Use of Mutator Strains.* To study the structure and function of a given cytochrome, production of random mutations within its structural gene is useful. This task has recently been simplified by the development of *E. coli* strains such as XL1-Red (Stratagene, La Jolla, CA) defective in three major DNA repair pathways of *E. coli* (*MutS, MutD*, and *MutT*), which results in 5000-fold higher mutation frequency in comparison to a wild-type strain. Using this system, one base pair change per 2000 nucleotides while propagating a high-copy plasmid in XL1-Red should be expected after overnight growth. This system is easy to use, and has no preference for the kind of mutations (transition, transversion, insertions, or deletions) it induces. It has recently been used to study interactions of membrane proteins VirB9 and VirB10 in *Agrobacterium*,[56] and should be of general use in *Rhodobacter*. Alternatively, the polymerase chain reaction (PCR) may be used for this purpose under nonoptimal amplification conditions with a polymerase defective in proofreading.[57] However, it often requires recloning of the amplified fragment, and to ensure that independent mutations are obtained, separate amplification reactions need to be performed.

*Site-Directed Mutagenesis*

For generating site-specific changes in DNA sequences, oligonucleotide-directed mutagenesis can be utilized. Several methods are commercially available for isolating single-stranded DNA from derivatives of M13 phage or phagemids, which is then used as a template for *in vitro* synthesis primed by a phosphorylated mutagenic primer annealed to this template. Use of uracilated template DNA, routinely isolated from a *dut⁻ ung⁻ E. coli* strain, which incorporates uracil instead of thymine, provides the necessary selection against the wild-type allele after transfering the heteroduplex synthesized *in vitro* into a *dut⁺ ung⁺ E. coli* strain.[58]

Mutations created using site-directed mutagenesis include small and large deletions and insertions, single amino acid replacements, and engineering sites for restriction endonucleases to aid further manipulation of DNA sequences or additions of immunogenic tags to cloned genes. A useful example is the introduction of eight amino acid residues corresponding to the FLAG epitope (Kodak, Eastman Kodak Co., New Haven, CT) into the 3′ end of *cycY* encoding cyt $c_y$ of *Rb. capsulatus*.[50] The addition of this

[56] C. E. Beaupré, J. Bohne, E. M. Dale, and A. N. Binns, *J. Bacteriol.* **179**, 78 (1997).
[57] M. F. Goodman and L. B. Bloom, *in* "Nucleases" (S. M. Linn, R. S. Lloyd, and R. J. Roberts, eds.), 2nd Ed., p. 235. Cold Spring Harbor Laboratory Press, Cold Spring Harbor, New York, 1993.
[58] T. A. Kunkel, *Proc. Natl. Acad. Sci. U.S.A.* **82**, 488 (1985).

epitope was useful to identify the gene product of *cycY* and to purify it using commercially available immunoaffinity columns.

Future Prospectives

Combinatorial mutagenesis described using *Rb. capsulatus* by Goldmann and Youvan[59] has great potential for creating new phenotypes when combined with massive screening of mutants by imaging devices. In the case of photochemical reaction center and light-harvesting I complexes, this approach has already been used to randomize several amino acids simultaneously to probe the correlation between protein sequences and biophysical characteristics of these proteins. Although this approach has not yet been tested with cytochromes, it should be possible to also use it to investigate their structure–function relationship, assuming that screening of mutants is not limiting.

The construction of an ordered cosmid library of *Rb. capsulatus* genomes by Kumar *et al.*,[60] has made it possible to initiate an international *Rb. capsulatus* genome sequencing project. Fonstein *et al.*,[61] are also deleting chromosomal DNA covered by their cosmid collection using GTA, and investigating global changes in the gene expression using this library as a hybridization template.

Due to the advanced genetic tools developed for *Rhodobacter,* these species continue to offer an ideal system for investigating metabolic diversity and defining gene function via disruption of open reading frames identified by direct DNA sequencing. Rapid progress in bacterial genome sequencing projects should make it possible to identify proteomes and hence complete cytochrome complements in several organism. Considering that *Rhodobacter* species are among the most versatile organisms in terms of their growth abilities, it is clear that they continue to be extremely useful for future comparative studies on metabolic diversity and global gene regulation and metabolic engineering.

[59] E. R. Goldman and D. C. Youvan, *in* "Anoxygenic Photosynthetic Bacteria" (R. E. Blankenship, M. T. Madigan, and C. E. Bauer, eds.), p. 1257. Kluwer Academic Publishers, Dordrecht, 1995.

[60] V. Kumar, M. Fonstein, and R. Haselkorn, *Nature* **381,** 653 (1996).

[61] M. Fonstein, E. G. Koshy, T. Nikolskaya, P. Mourachov, and R. Haselkorn, *EMBO J.* **14,** 1827 (1995).

## [7] Comparison of *in Vitro* and *in Vivo* Mutants of PsaC in Photosystem I: Protocols for Mutagenesis and Techniques for Analysis*

*By* JOHN H. GOLBECK

### Background

In this article, an experimental approach that employs the techniques of molecular biology to address biochemical and biophysical problems in Photosystem I (PS I) is outlined. The experimental target is PsaC, a PS I-bound, 8.9-kDa polypeptide that contains two [4Fe–4S] clusters termed $F_A$ and $F_B$.[1] Although the three-dimensional structure of PsaC has not been solved, the main-chain folding pattern is believed to be similar to small bacterial dicluster ferredoxins from *Peptococcus asacharolyticus*[2] (formerly *P. aerogenes*) and *Clostridium acidurici*,[3] which contain two $\alpha$ helices near the iron–sulfur clusters and two regions of two-stranded antiparallel $\beta$ sheet.[4,5] PsaC is likely to possess a pseudo-$C_2$ symmetry axis oriented perpendicular to a vector connecting the two iron–sulfur clusters. The amino acid sequence has two **CxxCxxCxxCP** iron–sulfur binding motifs,[6,7] in which

---

* **Abbreviations:** DCPIP, 1,6-dichlorophenolindophenol; DTT, dithiothreitol; FNR, ferredoxin:NADP⁺oxidoreductase; FWHM, full-width, half-maximum; Gm, Gentamycin; IPTG, isopropylthiogalactose; Km, Kanamycin; LAHG, Light Activated Heterotrophic Growth; PCR, polymerase chain reaction; PMS, phenazine methosulfate; NADP⁺, nicotinamide adenine dinucleotide phosphate; PS I, Photosystem I; SDS, sodium dodecyl sulfate.

 **Definitions:** P700-$F_A$/$F_B$ complex, PS I containing cofactors P700, $A_0$, $A_1$, $F_X$, $F_B$ and $F_A$; P700-$F_X$ core, a minimized PS I where PsaC, PsaD, and PsaE have been removed; P700-$A_1$ core, a further minimized PS I where the $F_X$ cluster has been oxidatively denatured; P700-$A_0$ core, the minimum PS I where the phylloquinone has been extracted; C14D, PsaC with Cys 14 → Asp mutation; C14D-PS I, PS I complex containing C14D; C51D, PsaC with Cys 51 → Asp mutation; C51D-PS I, PS I complex containing C51D; CDK25, a mutant of *Synechocystis* sp. PCC 6803 lacking PsaC, PsaD, and PsaE.

[1] N. Hayashida, T. Matsubayashi, K. Shinozaki, M. Sugiura, K. Inoue, and T. Hiyama, *Curr. Genet.* **12,** 247 (1987).
[2] E. T. Adman, L. C. Sieker, and L. H. Jensen, *J. Biol. Chem.* **251,** 3801 (1976).
[3] E. D. Duee, E. Fanchon, J. Vicat, L. C. Sieker, J. Meyer, and J. M. Moulis, *J. Mol. Biol.* **243,** 683 (1994).
[4] A. Kamlowski, A. van der Est, P. A. Fromme, and D. Stehlik, *Biophys. Acta* **1319,** 185 (1997).
[5] A. Kamlowski, A. van der Est, P. Fromme, N. Krauss, W.-D. Shubert, O. Klukas, and D. Stehlik, *Biochem. Biophys. Acta* **1319,** 199 (1997).
[6] P. P. J. Dunn and J. C. Gray, *Plant Mol. Biol.* **11,** 311 (1988).
[7] H. Oh-oka, Y. Takahashi, K. Kuriyama, K. Saeki, and H. Matsubara, *J. Biochem. (Tokyo)* **103,** 962 (1988).

the first three cysteines in one motif cooperate with the fourth cysteine in the other motif to bind a [4Fe–4S] cluster. Based on analogy with structures of known ferredoxins, the cysteine 14 and cysteine 51 ligands are likely to be surface located, and substitutions in these positions are predicted to introduce minimal changes in the three-dimensional structure of the protein.

*Mutagenesis Strategies*

A two-fold strategy to introduce mutations into PsaC is described: an *in vitro* approach, where recombinant PsaC is rebound onto P700-$F_X$ cores, and an *in vivo* approach where the identical mutations are incorporated into the genome of *Synechocystis* sp. PCC 6803. The *in vitro* approach is founded on the binding of *Esherichia coli*-expressed PsaC, PsaD, and PsaE onto previously prepared P-700-$F_X$ cores to rebuild functional P700-$F_A$/$F_B$ complexes. The *in vivo* approach is founded on the natural competence of the unicellular cyanobacterium *Synechocystis* sp. PCC 6803 to take up DNA and integrate it into its genome by homologous recombination. The idea as originally conceived was to substitute cysteines 14 or 51 with aspartic acid in an attempt to change the spectral, kinetic, and/or thermodynamic properties of $F_B$ or $F_A$. The substitution of oxygen for sulfur ligands would serve two purposes: it would likely preserve the three-dimensional structure of the protein backbone and at the same time provide a ligand to the iron–sulfur cluster. The *in vitro* mutagenesis approach was immediately successful in showing that $F_A$ is ligated by cysteines 48, 51, 54, and 21, and $F_B$ is ligated by cysteines 11, 14, 18, and 58 (Fig. 1).[8] This assignment was later confirmed using the *in vivo* mutagenesis approach, which introduces these same mutations into *Synechocystis* sp. PCC 6803.[9,10] In principle, the *in vitro* and *in vivo* mutants should be complementary, providing the identical PS I phenotype. In practice, the *in vitro* and *in vivo* mutants have shown differences as well as similarities when changes are made to the ligands of the iron–sulfur clusters of PsaC. This richness of information is precisely what makes application of these two independent approaches so useful.

Mutagenesis Methods

In this section the *in vitro* and *in vivo* protocols for introducing cysteine-to-aspartate substitutions into the cysteine 14 and cysteine 51 ligands of PsaC are described. Previous findings on the unbound mutant PsaC protein

---

[8] L. Yu, J. D. Zhao, W. P. Lu, D. A. Bryant, and J. H. Golbeck, *Biochemistry* **32**, 8251 (1993).
[9] J. Yu, Y.-S. Jung, I. Vassiliev, J. H. Golbeck, and L. McIntosh, *J. Biol. Chem.* **272**, 8032 (1997).
[10] Y.-S. Jung, I. Vassiliev, J. Yu, L. A. McIntosh, and J. H. Golbeck, *J. Biol. Chem.* **272**, 8040 (1997).

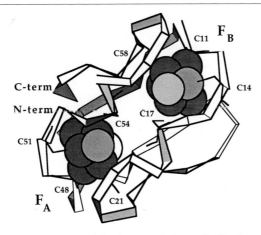

FIG. 1. The location of the $F_A$ and $F_B$ clusters relative to the PsaC protein backbone. $F_B$ is ligated by cysteines 11, 14, 17, and 58 near the N-terminal end of the protein and $F_A$ is shown ligated by cysteines 21, 48, 51, and 54 near the C-terminal end of the protein. The folding of PsaC is based on sequence homology with the known three-dimensional structure of *Peptococcus asacharolyticus*. The difference in mass between the proteins is accommodated by adding 13 amino acids to the C terminus and 8 amino acids between the two iron–sulfur clusters to the *P. asacharolyticus* structure. The center-to-center distance between $F_A$ and $F_B$, and between the two [4Fe–4S] clusters in *P. asacharolyticus*, is ca. 12 Å.

showed that the minimum configuration for folding is two intact iron–sulfur clusters, one of which must be a [4Fe–4S] cluster.[8–11] (Studies on the unbound mutant PsaC proteins are not discussed in detail here.) The minimum structural requirements for PsaC to bind to PS I are determined by reconstituting P700-$F_X$ cores with PsaC, PsaD, and PsaE *in vitro,* and by introducing these same mutations into *Synechocystis* sp. PCC 6803 *in vivo.*

## *In Vitro Mutants of PsaC*

The methodology for constructing *in vitro* mutants of PsaC is outlined. The author's experience is that the biological source of the component proteins is largely immaterial: hybrid P700-$F_A$/$F_B$ complexes have been rebuilt using P700-$F_X$ cores from *Synechococcus* sp. PCC 6301, PsaC from *Synechocystis* sp. 6803, PsaD from *Nostoc* sp. PCC 8009, and PsaE from *Synechococcus* sp. PCC 7002.[12] The only differences are slight alterations in the *g*-values of $F_A$ and $F_B$ when higher plant rather than cyanobacterial

[11] T. Mehari, F. Y. Qiao, M. P. Scott, D. F. Nellis, J. D. Zhao, D. A. Bryant, and J. H. Golbeck, *J. Biol. Chem.* **270,** 28108 (1995).
[12] J. Zhao, W. B. Snyder, U. Mühlenhoff, E. Rhiel, P. V. Warren, J. H. Golbeck, and D. A. Bryant, *Mol. Microbiol.* **9,** 183 (1993).

PsaC is used.[13] The *in vitro* mutagenesis protocol is performed in three discrete stages (Fig. 2). The initial stage includes the preparation of P700-$F_X$ cores by stripping PsaC, PsaD, and PsaE from P700-$F_A$/$F_B$ complexes with the use of chaotropic agents. Alternately, P700-$F_X$ cores can be isolated by growing a *psaC*-minus deletion mutant of *Synechocystis* sp. PCC 6803. (PsaD and PsaE are also missing in the PsaC-minus mutant.) The second stage includes ligating the *psaC, psaD,* and *psaE* genes into appropriate vectors, performing site-directed mutagenesis on *psaC,* moving the altered genes into expression vectors, and overproducing PsaC, PsaD, and PsaE in *E. coli.* The final stage includes rebuilding the P700-$F_A$/$F_B$ complexes by reinserting the iron–sulfur clusters into purified apoPsaC and rebinding holoPsaC, PsaD, and PsaE onto the previously prepared P700-$F_X$ cores. The protocols for the individual steps are outlined below; details are provided in Zhao *et al.*[14]

*Preparation of P700-$F_X$ Cores.* P700-$F_A$/$F_B$ complexes are isolated as PS I trimers from cyanobacterial membranes by treatment with 30 m$M$ *n*-dodecyl-$\beta$-D-maltiside followed by ultracentrifugation in a sucrose density gradient[15] (size exclusion, ion-exchange or affinity chromatography may also be used to purify PS I complexes). Triton X-100 at concentrations $\geq$1% is to be avoided, since it can strip peripheral polypeptides such as PsaF, and to a lesser degree, PsaE and PsaJ, from PS I complexes. P700-$F_X$ cores are isolated by treating P700-$F_A$/$F_B$ complexes at ca. 250 $\mu$g/ml Chl with chaotropic agents such as 6.8 $M$ urea[16] (2 $M$ NaI, 2 $M$ NaSCN, or 2 $M$ NaClO$_4$ are also effective). The mechanism of action is the oxidation of the cubane sulfides, which results in the loss of iron from the clusters; this, in turn, leads to the loss of the three-dimensional structure of PsaC and its ability (as well as that of PsaD and PsaE) to bind to PS I. The removal of PsaC, PsaD, and PsaE is monitored using optical kinetic spectroscopy at 830 nm: when the ca. 30-ms backreaction between P700$^+$ and [$F_A$/$F_B$]$^-$ is replaced by the ca. 1-ms backreaction between P700$^+$ and $F_X^-$, the process is complete [the loss of PsaC, PsaD, and PsaE can be verified by sodium dodecyl sulfate–polyacrylamide gel electrophoresis (SDS–PAGE)]. The reaction is terminated by 10-fold dilution with buffer, the solution is dialyzed against 50 m$M$ Tris buffer, pH 8.3, to remove the residual chaotrope,

[13] T. Mehari, K. G. Parrett, P. V. Warren, and J. H. Golbeck, *Biochim. Biophys. Acta* **1056,** 139 (1991).
[14] J. D. Zhao, N. Li, P. V. Warren, J. H. Golbeck, and D. A. Bryant, *Biochemistry* **31,** 5093 (1992).
[15] J. H. Golbeck, *in* "CRC Handbook of Organic Photochemistry and Photobiology" (W. M. Horspool and P. S. Song, eds.), p. 1407. CRC Press, Boca Raton, Florida, 1995.
[16] K. G. Parrett, T. Mehari, P. G. Warren, and J. H. Golbeck, *Biochim. Biophys. Acta* **973,** 324 (1989).

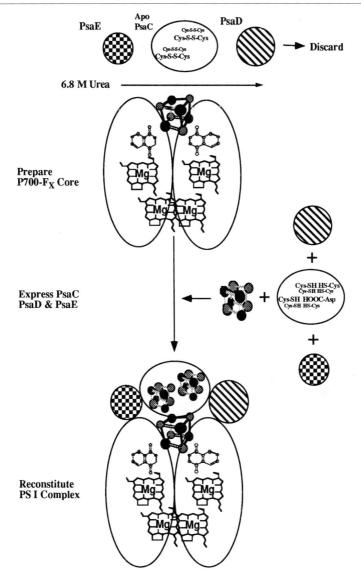

FIG. 2. Strategy for resolving and reconstituting pS I *in vitro* using *Synechococcus* sp. PCC 6301. This schematic diagram depicts the three steps for introducing mutations into PsaC. Step 1 involves treating a P700-$F_A$/$F_B$ complex with 6.8 *M* urea and removing PsaC, PsaD, and PsaE, which are subsequently discarded. The resulting P700-$F_X$ core is purified and reconstituted with *E. coli*-expressed PsaD, mutant apoPsaC, and PsaE concomitant with reinsertion of the iron–sulfur clusters into PsaC. The mutagenesis of PsaC is performed in *E coli* and is described separately in the text. The resulting mutant PS I complex is purified of inorganic reagents and unbound proteins and stored at −80°.

and the P700-$F_X$ cores are purified of unbound PsaC, PsaD, and PsaE by ultrafiltration over a YM-100 membrane (Amicon, Beverly, MA). Alternately, P700-$F_X$ cores can be isolated directly from the *psaC*-minus mutant of *Synechocystis* sp. PCC 6803[17] by treating the cyanobacterial membranes with 30 m$M$ *n*-dodecyl-$\beta$-D-maltoside followed by ultracentrifugation in a sucrose density gradient. The mutant strain still has the *psaD* and *psaE* genes but PsaD and PsaE do not accumulate on the PS I reaction centers in the absence of PsaC.[17] The P700-$F_X$ cores can be stored in 20% glycerol at $-80°$ for an indefinite period without loss of activity.

*Mutagenesis and Expression of PsaC.* Mutagenesis of the *psaC* gene of *Synechococcus* sp. PCC 7002 is performed by the method of Kunkel *et al.*[18] using appropriate synthetic oligonucleotides (24 mers) to prime second-strand synthesis. The mutant *psaC* genes are recloned into plasmid pUC19 as *Xba*I-*Sst*I fragments and subsequently recloned into the *Xba*I and *Nde*I sites of plasmid pET-3a.[14] Overexpression of the mutant *psaC* genes is accomplished in *E. coli* strain BL 21 (DE3) grown in medium supplemented with 50 m$M$ ferric ammonium citrate.[19,20] Expression is initiated by the addition of 0.5 m$M$ IPTG to the growth medium; after 1 hr 20 $\mu$g/ml rifamycin is added and expression is continued for an additional 5–7 hr. The cells are disrupted by passage through a French pressure cell and the inclusion bodies are isolated by centrifugation and resuspended in buffer containing 2 m$M$ DTT. The PsaC inclusion bodies are solubilized in buffer containing 6.5 $M$ urea and 2 m$M$ DTT, and the PsaC apoprotein is purified by gel filtration chromatography on Sephadex G-75 using 50 m$M$ Tris buffer, pH 8.3, as eluant. (A fraction of the protein in the inclusion body contains iron and/or iron–sulfur clusters, but these are destroyed during the suspension in urea.) Typically, 100 mg of protein as inclusion bodies are isolated from a 1-liter culture of *E. coli.*[14]

*Reconstitution of P700-$F_A$/$F_B$ Complexes.* Wild-type or recombinant PsaC, PsaD, and PsaE spontaneously rebind to P700-$F_X$ cores on reincorporating the iron–sulfur clusters into PsaC. The source of protein can be either wild-type PsaC purified from higher plant or cyanobacterial thylakoids,[21] cloned PsaC that is expressed in *E. coli,*[20] or site-modified PsaC that is expressed in *E. coli.*[22] PsaD, but not PsaE, is required for tight binding of PsaC onto P700-$F_X$ cores. PsaE cannot substitute for PsaD, and it

[17] J. P. Yu, L. B. Smart, Y. S. Jung, J. H. Golbeck, and L. McIntosh, *Plant Mol. Biol.* **29**, 331 (1995).

[18] T. A. Kunkel, J. D. A. Roberts, and R. A. Zabour, *Methods Enzymol.* **154**, 367 (1987).

[19] J. Zhao, P. V. Warren, N. Li, D. A. Bryant, and J. H. Golbeck, *FEBS Lett.* **276**, 175 (1990).

[20] N. Li, J. Zhao, P. Warren, J. Warden, D. Bryant, and J. Golbeck, *Biochemistry* **30**, 7863 (1991).

[21] J. H. Golbeck, T. Mehari, K. Parrett, and I. Ikegami, *FEBS Lett.* **240**, 9 (1988).

[22] Y. S. Jung, I. R. Vassiliev, F. Y. Qiao, F. Yang, D. A. Bryant, and J. H. Golbeck, *J. Biol. Chem.* **271**, 31135 (1996).

has no added effect on the stability of the PsaD-reconstituted PS I complex, but it is usually included in the protocol. The reconstitution and rebinding are performed simultaneously by incubating apoPsaC, PsaD, and PsaE at 0.5 $\mu$g/ml with P700-$F_X$ cores at 50 $\mu$g/ml Chl and 1% 2-mercaptoethanol for 10 min under aerobic conditions. An aliquot of $FeCl_3$ is added dropwise to the stirred solution to a concentration of 200 $\mu M$, followed 10 min later by a similar addition of $Na_2S$ to a concentration of 200 $\mu M$. After incubation for 12 hr at room temperature, the solution is dialyzed against 50 m$M$ Tris buffer, pH 8.4, to remove the inorganic cofactors, ultrafiltered over a YM-100 membrane to remove the unbound low molecular mass proteins. Anaerobic conditions are not necessary if the insertion of the iron–sulfur clusters is concomitant with rebinding of PsaC onto P700-$F_X$ cores. Phosphate buffer is to be avoided, because it interacts with the arginines on the $F_X$-loop of PsaA/PsaB and interferes with the binding of PsaC.[23] The reconstitution of the P700-$F_A$/$F_B$ complex is nearly quantitative when wild-type PsaC and PsaD are used.[9]

### *In Vivo Mutants of PsaC*

The methodology for constructing *in vivo* mutants of PsaC is now outlined. *Synechocystis* sp. PCC 6803 cannot normally grow in the total absence of light, even in the presence of a reduced carbon source. Although the mechanism remains obscure, it was discovered that a single 5-min pulse of light (specifically blue-light, 450 nm) will allow *Synechocystis* sp. PCC 6803 to grow heterotrophically in the presence of glucose.[24] This process is termed light-activated heterotrophic growth (LAHG). The *in vivo* mutagenesis protocol is a three-step process (Fig. 3). The initial stage involves construction of a *psaC*-minus deletion strain of *Synechocystis* sp. PCC 6803 and genetic verification. The second stage involves *psaC* mutagenesis and genetic characterization of the mutants. The final stage involves transformation of the recipient strain and confirmation of complete segregation. The complete segregation (no wild-type gene copies present) of mutations in genes encoding PS I proteins requires the selection of strains capable of LAHG. The protocols for the individual steps are outlined later; details are provided in Yu *et al.*[9]

*Construction of a psaC Deletion Strain.* The host strain of *Synechocystis* sp. PCC 6803 is constructed by deleting *psaC* and replacing it in the genome with a bacterial gene conferring kanamycin resistance ($Km^r$).[10] To make this (and later) construction(s) without affecting *psaC* transcription, it is necessary to add synthetic DNA sequences to the native *Synechocystis* sp. PCC 6803 genome with appropriate DNA restriction endonuclease sites.

[23] S. M. Rodday, S. S. Jun, and J. Biggins, *Photosynth. Res.* **36**, 1 (1993).
[24] S. L. Anderson and L. McIntosh, *J. Bacteriol.* **173**, 2761 (1991).

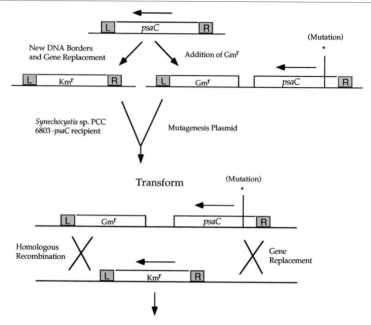

Fig. 3. Strategy for engineering *in vivo* mutations in PsaC using *Synechocystis* sp. PCC 6803 genome. The cloned genes are assembled and/or modified on an *E. coli* bacterial plasmid of choice. The two example genes in this figure are a bacterial drug resistance marker, gentamicin resistance, functional in *Synechocystis* sp. PCC 6803, and the gene encoding the apoprotein PsaC. These genes are, in turn, flanked by *Synechocystis* sp. PCC 6803 DNA. The circular plasmid DNA is added to *Synechocystis* sp. PCC 6803 cells grown under LAHG conditions. The DNA and cells are mixed for approximately 1 hr and then plated on BG-11 growth plates containing gentamicin and glucose. These plates are grown at 34° under LAHG conditions. Homologous recombination is shown by the X's indicating the probable sites of recombination. Transformants are selected for by growth. Figure courtesy of Lee McIntosh, 1995.

This is required due to the relatively small size of the *psaC* gene, which provides few native DNA restriction sites for manipulation. PCR-amplified *psaC* upstream (454-bp) and downstream (208-bp) flanking regions, with linker sequences for *Bam*HI and *Eco*RI at either end, are cloned into pUC118. A Km$^R$ cassette excised from pUC4K using *Eco*RI, is inserted into the *Eco*RI site that separates the upstream and downstream flanking regions to form the plasmid pUC118-ΔC. Km$^r$ colonies are selected under LAHG conditions and verified by Southern hybridization using a *psaC* probe. Once complete segregation of the deletion mutation is confirmed, the deletion strain is grown under LAHG conditions using glucose as carbon source in the presence of Km.

*psaC Mutagenesis.* Plasmids for site-directed mutagenesis are constructed *in vitro* and manipulated in *E. coli.* To perform the selection of transformed cells and accomplish the necessary genetic segregation, a bacterial gene encoding a drug resistance marker is inserted next to the target DNA. In this instance, gentamycin resistance (Gm$^r$) is inserted next to *psaC* as the marker for transformation. A 1.5-kb *Eco*RI–*Nco*I fragment containing *psaC* is excised from the plasmid p6.1S3.5I and cloned into pUC119 at *Sma*I site to create plasmid pC. A 2.0-kb Gm$^R$ cassette, cut with *Bam*HI from pUC119, is inserted at *Bbs*I site downstream of the *psaC* gene on pC to create plasmid pCG. Single-stranded DNA is used as template for site-directed mutagenesis. Oligonucleotides are designed to effect the desired changes in the coding sequence while at the same time destroying a restriction site; this allows verification of the mutations by observing the change in the digestion pattern. A 953-bp *Bgl*II–*Pst*I fragment containing the *psaC* mutation is cut out of pC and ligated with a 5.7-kb partial digestion product from pCG devoid of the corresponding fragment to form pCG with the mutated *psaC* gene.

*Transformation and Genetic Segregation.* A restriction fragment that contains the *psaC* mutation is recloned into an engineered plasmid and used to transform the CDK25 recipient strain of *Synechocystis* sp. PCC 6803.[10] To integrate specific mutations, especially those which may affect growth, it is necessary to segregate genetically the alleles of that gene in the absence of any selection favoring wild-type genomes. If a mutation to a specific reaction important for growth is the goal, then the mutations must be segregated in the absence of that activity. In this example, the activity is photoautotrophic growth using PS I (and PS II). The process of selection in *Synechocystis* sp. PCC 6803 is a case of genome sorting. As in all cyanobacteria, *Synechocystis* sp. PCC 6803 has a large chromosome of which there exist multiple copies per cell. During a transformation/gene replacement event, double homologous recombination leads to the integration of the foreign DNA and/or mutated native gene. Segregation then follows whereby the mutant genome is replicated and positively selected for by the growth conditions conferred by a specific drug resistance. Segregation can take 4–10 weeks under normal LAHG conditions. This genome sorting is followed by Southern blotting/mapping of the DNA. It is possible to use a specific probe, one that differentiates between wild-type and altered DNA, to follow genetic segregation. Such sorting usually takes about five serial platings of individually selected transformants.[25] In this instance, Gm$^r$

---

[25] L. Smart, S. Anderson, and L. McIntosh, *EMBO J.* **10,** 3289 (1991).

colonies are selected and genetically verified using Southern hybridization and growth in the presence of Km. DNA blotting provides a necessary test to determine that no appreciable wild-type gene copies exist in the selected cells line. An even more sensitive test, polymerase chain reaction (PCR), is employed to complete the genetic analysis of transformed cell lines. In these analyses, specific oligonucleotide primers are designed for each mutation and employed in PCR reactions with DNA from isolated transformants.

Spectroscopic Techniques

In this section spectroscopic techniques that are employed to determine of the structural and functional consequences of the cysteine-to-aspartate mutations in PsaC are described. Each method has its inherent strengths and weaknesses, and the full picture of the structural and functional consequences of the mutagenesis emerges only when a combination of techniques is employed.

*Iron–Sulfur Cluster Structure*

EPR spectroscopy is the technique of choice to probe the structure of iron–sulfur clusters, i.e., the type ([2Fe–2S], [3Fe–4S], or [4Fe–4S]) and the spin state ($S = 1/2$, 3/2 or higher). (The less commonplace, but more diagnostic Mössbauer and EXAFS spectroscopies can also be used, but chlorophyll absorption and fluorescence in PS I limit the usefulness of MCD and resonance Raman spectroscopies in this application.) The strength of EPR is that it can distinguish between wild-type $F_A$ and $F_B$ on the basis of the separable principal values of the g-tensors ($g = 2.05$, 1.94, and 1.86 for $F_A$ and $G = 2.07$, 1.92, and 1.88 for $F_B$). The weakness of EPR is that the lack of resonances around $g = 2$ does not necessarily indicate that the iron–sulfur cluster is missing. Iron–sulfur clusters can exist in a variety of spin states due to ferromagnetic and antiferromagnetic couplings between the four iron atoms in the cubane cluster.[26] The $S = 1/2$ state, with resonances centered around $g = 2$, is the most common in biological clusters. This is fortunate: since the $F_A$ and $F_B$ resonances are distributed over less than 60 mT, the linewidths are relatively narrow, and the amplitudes are high [amplitude is proportional to $1/(\text{linewidth})^2$]. The $S = 3/2$ state, which has been observed in cysteine-to-serine mutants of *E. coli* fumarate reductase[27] and proposed in cysteine-to-serine mutants to PsaC,[9] can show reso-

[26] L. Noodleman, D. A. Case, J. M. Mouesca, and B. Lamotte, *J. Biol. Inorg. Chem.* **1**, 177 (1996).

[27] A. T. Kowal, M. T. Werth, A. Manodori, G. Cecchini, I. Schroder, R. P. Gunsalus, and M. K. Johnson, *Biochemistry* **34**, 12284 (1995).

nances which are distributed over as much as 400 mT, with correspondingly low signal amplitudes. An added difficulty is that for highly rhombic systems (i.e., when all of the $g$ factors differ: $g_x \neq g_y \neq g_z$), the low-field resonance can occur near the $g = 4.3$ signal from octahedrally coordinated iron, the midfield resonance can occur around $g = 2$ and may be buried under other $S = 1/2$ resonances, and the high-field resonance can be so broad that it cannot be observed. Nevertheless, the occurrence of high spin states must be considered when any changes are made in the ligands to the iron–sulfur clusters of ferredoxins, including PsaC.

A further complication in EPR is that the rapid spin relaxation properties of paramagnetic $[4Fe–4S]^{1+}$ clusters requires that the measurement be made at cryogenic temperatures, making kinetic studies difficult. In practice, the signal amplitude is optimized by measurement over a range of temperatures and microwave powers. This combination must be determined experimentally for every PS I mutant. The details of the illumination protocol must also be planned with care, because at cryogenic temperatures, charge separation between $P700^+$ and $[F_A/F_B]^-$ occurs irreversibly. Illumination of a dark-frozen, wild-type PS I complex at 15 K results in the transfer of only one electron from P700 and in the partial reduction of $F_A$ and $F_B$ in an ca. 3:1 ratio. Since *either* $F_A$ *or* $F_B$ is reduced in any given reaction center, there is no magnetic interaction between iron–sulfur clusters, and the $g$-values of the two acceptors can be determined by visual inspection or by numerical simulation of the EPR spectra. Freezing a wild-type PS I complex during illumination results in the transfer of two (or more) electrons, leading to the complete reduction of both $F_A$ and $F_B$. Chemical reduction with sodium dithionite at pH 10.5 ($E^0 = -626$ mV) also results in the complete reduction of both $F_A$ and $F_B$ in wild-type PS I complexes. Chemical reduction of the PS I complexes is carried out by adjusting the pH to 10.5 with 1 $M$ glycine–NaOH and adding 1 mg of sodium dithionite per 300 $\mu$l of sample under anaerobic conditions. Freezing in the light and chemical reduction result in the simultaneous reduction of both $F_A$ and $F_B$ in any given PsaC protein, resulting in magnetic interaction between the two reduced iron–sulfur clusters.

## Efficiency of Electron Transfer

Time-resolved optical absorption spectroscopy is the technique of choice to probe efficiency of electron transfer; i.e., the ratio of forward electron transfer from P700 through $A_0$, $A_1$, $F_X$ to $F_B$ and $F_A$ to the recombination with $P700^+$. The premise is that any inefficiency in forward electron transfer will be reflected in an enhanced charge recombination with $P700^+$. The strength of this method is that the backreactions of $A_0^-$, $A_1^-$, $F_X^-$, $F_B^-/F_A^-$

with $P700^+$ occur in well-defined time ranges, and the source of electron transfer inefficiency can usually be located by performing relatively straightforward time-resolved optical measurements in the near infrared, where the cation of P700 is detected as an absorbance increase. The weakness of this method is that it is not possible to distinguish between the backreactions of $F_A^-$ and $F_B^-$ with $P700^+$. A more serious complication is that it is not possible to discriminate between inefficient forward electron transfer and a heterogeneous population of reaction centers. (Quantitative immunoblotting or ELISA assays are the methods of choice to determine sample heterogeneity.) The instrumentation is usually laboratory built and consists of a measuring beam at 830 nm provided by a semiconductor diode laser, and an excitation beam provided by a frequency-doubled, Q-switched Nd-YAG laser operating at 532 nm at a FWHM of 10 ns and a flash energy up to 135 mJ. The signals are usually averaged, and the flashes are spaced in time so that the ground state of P700 is fully recovered prior to the next flash. The measurement in the near infrared allows a strong, nonactinic measuring beam to increase the $S/N$ in spite of the weaker differential extinction coefficient of $P700/P700^+$ at 830 nm ($\varepsilon = 6000\ M^{-1}\ cm^{-1}$) compared to 698 nm ($\varepsilon = 64,000\ M^{-1}\ cm^{-1}$).[28] The short duration of the flash and 10-ns instrument response time allows backreactions from acceptors from $A_0$ through $F_A/F_B$ to be measured. (A xenon flashlamp with a FWMH of 10 $\mu$s can be used if only backreactions from $F_X$, $F_B$, and $F_A$ are of interest.) Samples are suspended under anaerobic conditions in 25 m$M$ Tris-HCl, pH 8.3, 4 m$M$ DCPIP, and 10 m$M$ sodium ascorbate to 50 $\mu$g ml$^{-1}$ Chl and measured in a 10-mm $\times$ 5-mm quartz cuvette.

The identification of the electron acceptor that backreacts with $P700^+$ is based on the kinetics of the absorbance change at 830 nm. The interpretation of the data must take into account multiple kinetic components attributed to a particular acceptor,[29] the problem of species- and preparation-dependent equilibrium between $F_X$ and $A_1$[30] and the broad timescale of backreactions that occur with $P700^+$.[31] The best way to overcome these problems is to monitor $P700^+$ kinetics over many decades of time rather than making separate measurements in relatively narrow time windows. The resulting $\Delta A_{830}$ decay curves are fit to the "sum of several exponentials with baseline" using the Marquardt algorithm. The fit usually includes a baseline offset accounting for long-term phases and/or possible drift of

[28] P. Mathis and P. Sétif, *Israel J. Chem.* **21**, 316 (1981).
[29] K. Brettel and J. H. Golbeck, *Photosynth. Res.* **45**, 183 (1995).
[30] P. Sétif and K. Brettel, *Biochemistry* **32**, 7846 (1993).
[31] I. R. Vassiliev, Y. S. Jung, M. D. Mamedov, A. Y. Semenov, and J. H. Golbeck, *Biophys. J.* **72**, 301 (1997).

signal zero during long timescale acquisition. The benefit of this approach is that since placing the kinetics on the logarithmic timescale does not require using different scales to display traces which have their decay components in different time domains, the results can be evaluated visually even before the multiexponential fit is performed. Analysis of wild-type P700-$F_A$/$F_B$ complexes from *Synechocystis* sp. PCC 6301 shows that the backreaction is biphasic with $t_{1/2}$ of 10 and 80 ms in a ca. 1 : 4 ratio; in P700-$F_X$ cores, the backreaction is biphasic with $t_{1/2}$ of 450 $\mu$s and 1.5 ms in a ca. 5 : 1 ratio; in P700-$A_1$ cores, the backreaction is biphasic with $t_{1/2}$ of 10 and 130 $\mu$s in a ca. 3 : 1 ratio; and in P700-$A_0$ cores, the backreaction is monophasic with $t_{1/2}$ of ca. 30 ns. Because triplet excited states of light-harvesting chlorophyll molecules contribute to the absorption at 830 nm and decay in the microsecond time domain, optimization of the flash energy is essential: the intensity must be high enough to approach saturation of P700$^+$ but not so high as to excite chlorophyll triplets. Because triplets are more difficult to saturate than P700$^+$, the two processes are distinguished by measuring $\Delta$830 nm as a function of flash energy.

### Number of Functional Electron Acceptors

Sequentially timed xenon or laser flashes are employed to probe stoichiometry; i.e., the number of electron acceptors that function after the $F_X$ cluster in PS I. This method is a variant of the technique of time-resolved optical absorption spectroscopy: the same instrumentation is used but a rapid donor to P700$^+$ is employed and the spacing of the flashes is shortened. The trick is to choose a concentration of PMS which donates to P700$^+$ faster than the $F_A^-$ and $F_B^-$ backreactions, but slower than the $F_X^-$ backreaction. The idea is that following the first two flashes, forward electron transfer from PMS to P700$^+$ will outcompete the backreactions, thereby trapping reduced $F_A^-$ and $F_B^-$. The strength of this method is that the number of electron acceptors after $F_X$ can be determined at room temperature.[32] The weakness of this method is that the backreaction kinetics of $F_B^-$ and $F_A^-$ may be altered in the PS I mutants, and this may compromise the distinction between the kinetics on flashes one, two, and three. The PMS concentration must hence be adjusted to take into account the altered backreaction kinetics. (These kinetics are determined using well-spaced single flashes and a slower electron donor to P700$^+$; see previous section on Efficiency of Electron Transfer.) The experiment is conceptually simple. In wild-type PS I complexes, the first flash leads to $F_X[F_A/F_B]^-$ since P700$^+$ is rapidly reduced by PMS. The second flash leads to $F_X$ $F_A^-$/$F_B^-$, since P700$^+$ will again be

---

[32] H. Bottin, P. Sétif, and P. Mathis, *Biochim. Biophys. Acta* **894,** 39 (1987).

reduced by PMS. The third (and subsequent) flash(es) lead to $F_X^- F_A^- F_B^-$, but now the $F_X^-$ P700$^+$ backreaction occurs because it outcompetes forward donation by PMS. In practice, four flashes are usually administered; the third and fourth flash should show similar kinetics if $F_A$ and $F_B$ are indeed reduced. If either $F_A$ or $F_B$ is missing or kinetically incompetent, then the second and all subsequent flashes rather than the third and all subsequent flashes will reflect the kinetics of the P700$^+$ $F_X^-$ backreaction. In practice, one counts the number of flashes, $n$, required to obtain the P700$^+$ $F_X^-$ backreaction. The number of acceptors functioning after $F_X$ is $n - 1$. The method can be verified using wild-type PS I, where $n = 3$ when $F_A$ and $F_B$ are functional, and using Hg-treated PS I, where $n = 2$ when only $F_A$ is present.[33]

*Electron Transfer Throughput*

Steady-state rates of NADP$^+$ reduction are used to probe throughput; i.e., the electron transfer rate through the PS I complexes from cytochrome $c_6$, to ferredoxin or flavodoxin. The strength of this method is that a real-time, room temperature electron transfer rate from a physiologically relevant donor to a physiologically relevant acceptor is measured under continuous illumination, rather than an electron transfer event on a single turnover flash. The weakness of this method is that any change in the electron donor or acceptor binding sites will mimic a bottleneck in electron transfer among the bound acceptors. In my experience, artificial electron donors such as DCPIP or PMS are competent electron donors for higher plant PS I complexes, but they are poor electron donors to cyanobacterial PS I complexes. The rates that can be sustained are typically 40–80 $\mu$mol/mg Chl/hr, which is far below the rates of 800 $\mu$mol/mg Chl/r that can be supported using cytochrome $c_6$.[34] Overexpression of cytochrome $c_6$ in *E. coli* has not yet been accomplished, so the protein must be isolated from a batch culture of cyanobacteria. *Spirulina pacifica* (probably a mixture of *S. maxima* and *S. platensis*), which can be obtained from Cyanotech Corp., a commercial supplier in Hawaii, provides an excellent source of highly active cytochrome $c_6$. Purification is carried out in a manner similar to that described by Lojero and Krogmann.[35] The flavodoxin and ferredoxin are recombinant proteins; the genes are derived from *Synechocystis* sp. PCC 7002, and the proteins are overproduced in *E. coli*.[36] Purification is carried out on the clarified cell lysate by ion-exchange chromatography on Mono Q. Until it

[33] H. Sakurai, K. Inoue, T. Fujii, and P. Mathis, *Photosynth. Res.* **27**, 65 (1991).
[34] Y. S Jung, L. Yu, and J. H. Golbeck, *Photosynth. Res.* **46**, 249 (1995).
[35] C. G. Lojero and D. W. Krogmann, *Photosynth. Res.* **47**, 293 (1996)
[36] U. Mühlenhoff, J. D. Zhao, and D. A. Bryant, *Eur. J. Biochem.* **235**, 324 (1996).

is certain that flavodoxin and ferredoxin interact at the same binding sites on PS I, we usually repeat $NADP^+$ reduction experiments using each electron acceptor. Flavodoxin-mediated $NADP^+$ reduction occurs between the singly and doubly reduced flavin, and ferredoxin-mediated $NADP^+$ reduction measures the single-electron electron transfer to the physiologically relevant acceptor. We also measure flavodoxin reduction without coupling it to $NADP^+$ where electron transfer occurs between the oxidized and singly reduced flavin. Having systematically performed the three assays on all samples, significant differences in rates sustained by these two acceptors in any mutant PS I complexes are yet to be found.

Flavodoxin reduction is measured by monitoring the rate of change in the absorption at 467 nm. The reaction mixture includes 15 $\mu M$ flavodoxin and PS I complexes at 5 $\mu$g ml$^{-1}$ of Chl in 50 m$M$ Tricine, pH 8.0, 50 m$M$ MgCl$_2$, 15 $\mu M$ cytochrome $c_6$ from *S. maxima*, 6 m$M$ sodium ascorbate, and 0.05% $n$-dodecyl-$\beta$-D-maltoside. $NADP^+$ reduction is measured by monitoring the rate of change in the absorption of NADPH at 340 nm. The reaction mixture includes 15 $\mu M$ flavodoxin and 0.8 $\mu M$ spinach ferredoxin : $NADP^+$ oxidoreductase (FNR), and PS I complexes at 5 $\mu$g ml$^{-1}$ Chl in 50 m$M$ Tricine, pH 8.0, 10 m$M$ MgCl$_2$, 15 $\mu M$ cytochrome $c_6$, 6 m$M$ sodium ascorbate, 0.05% $n$-dodecyl-$\beta$-D-maltoside, 0.5 m$M$ $NADP^+$, and 0.1% 2-mercaptoethanol. When ferredoxin-mediated $NADP^+$ photoreduction is measured, the reaction mixture includes 5 $\mu M$ spinach ferredoxin and 0.8 $\mu M$ spinach FNR, and PS I complexes at 5 $\mu$g ml$^{-1}$ Chl in 50 m$M$ Tricine, pH 8.0, 10 m$M$ MgCl$_2$, 15 $\mu M$ cytochrome $c_6$, 6 m$M$ sodium ascorbate, 0.05% $n$-dodecyl-$\beta$-D-maltoside, 0.5 m$M$ $NADP^+$, and 0.1% 2-mercaptoethanol. The instrumentation consists of a dual-beam spectrometer with a sample chamber engineered to hold a four-sided (clear) cuvette that can be side illuminated with two high-intensity, light-emitting diode banks at ca. 660 nm (LS1, Hansatech Ltd., Norfolk, UK). The light intensity must be saturating at the chlorophyll concentration used. If there is any doubt (for example, suspicion that antenna chlorophylls are disrupted in energy transfer), a double reciprocal plot of intensity versus rate is plotted and extrapolated to infinite light intensity. The photomultiplier is protected with appropriate glass filters, which block the scattered actinic light, but pass the 467- or 340-nm measuring beam.

## Case Studies: C14D and C51D in PsaC

These principles are applied by describing two case studies: cysteine 14-to-aspartate and cysteine 51-to-aspartate mutations in PsaC. The results of the *in vitro* and *in vivo* methods of mutagenesis are compared and contrasted, and the strengths and weaknesses of the spectroscopic methods

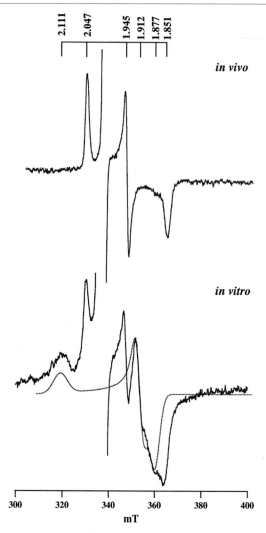

FIG. 4. EPR spectrum of *in vivo* and *in vitro* C14D-PS I complexes. The samples were frozen in darkness and illuminated at 15 K prior to and during measurement. The large resonance near 338 mT is the photochemically generated P700$^+$ radical at $g = 2.002$. The 250-$\mu$l samples contained 1 mg ml$^{-1}$ Chl, 300 $\mu$M DCPIP, and 1 m$M$ sodium ascorbate in 50 m$M$ Tris buffer, pH 8.3, containing 0.04% Triton X-100. Spectrometer conditions microwave power, 20 mW; modulation amplitude; 10 G at 100 kHz. A dual-mode resonator (ER/4116 DM) was used in perpendicular modes at a frequency of 9.6471 GHz for the *in vivo* sample and at a frequency of 9.6459 GHz for the *in vitro* sample. The field legend refers only to the *in vitro* sample; the field legend for the *in vivo* sample is not depicted. The *g*-value legend refers to both samples. The numerical simulation of F$'_B$ (dotted line) was carried out using *g* values of 2.111, 1.912, and 1.877 and linewidths of 100, 50, and 60 MHz. [Figure (top) reprinted

employed for their analysis are described. The properties of PS I complexes from the *in vivo* and *in vitro* C14D-PS I and C51D-PS I mutants are described in detail in Yu *et al.*,[9] Jung *et al.*,[10] and Yu *et al.*[37,38]

## *Cysteine 14 → Aspartate in $F_B$: Disagreement between in Vivo and in Vitro Mutants*

*Structure.* The differences between the *in vivo* and *in vitro* C14D-PS I mutants are best seen by EPR when the samples are frozen to 15 K in darkness and illuminated to promote only one electron to the acceptor side. The *in vivo* C14D-PS I mutant shows a strong set of resonances at $g = 2.047, 1.945$, and $1.851$ characteristic of $F_A$ (Fig. 4, top), and there is no evidence for a minority reduction of $F_B$ as is found in wild-type PS I complexes. The *in vitro* C14D-PS I mutant also shows a strong set of resonances at $g = 2.047, 1.945$, and $1.851$ characteristic of $F_A$ (Fig. 4, bottom, solid line) but here a second species, denoted $F'_B$, is present at $g = 2.111$, $1.912$, and $1.877$ (Fig. 4, bottom, dashed line). Differences are also found when the samples are thawed and frozen under constant illumination to promote more than one electron to the acceptor side. The *in vivo* C14D-PS I mutant shows the same set of resonances depicted in Fig. 4 (top), but they are ca. 20% more intense. In contrast, the *in vitro* C14D-PS I mutant shows a new set of resonances at $g = 2.047, 1.937, 1.919$, and $1.884$ that resembles an "interaction spectrum" from the simultaneous reduction of $F_A$ and $F_B$ in wild-type PS I complexes (not shown). On the basis of the EPR spectra in the $g = 2$ region, it therefore *appears* that $F_B$ is missing in the *in vivo* C14D-PS I mutant, but that $F_B$ is present, albeit in an altered state, in the *in vitro* C14D-PS I mutant.

*Efficiency.* The backreaction kinetics in PS I complexes from the *in vivo* and *in vitro* C14D-PS I mutants are depicted in Fig. 5. The kinetics of both C14D-PS I mutants resemble a mixed "core-" and "complex-type," with backreactions occurring from every electron acceptor after $A_0$. As shown in Fig. 5 (top), a large fraction of the kinetics in the *in vivo* C14D-PS I mutants is derived from faster than tens of microsecond to millisecond decay phases attributed to the $P700^+ F_X^-$ and $P700^+ A_1^-$ backreactions.

[37] L. Yu, D. A. Bryant, and J. H. Golbeck, *Biochemistry* **34,** 7861 (1995).
[38] L. A. Yu, I. R. Vassiliev, Y. S. Jung, D. A. Bryant, and J. H. Golbeck, *J. Biol. Chem.* **270,** 28118 (1995).

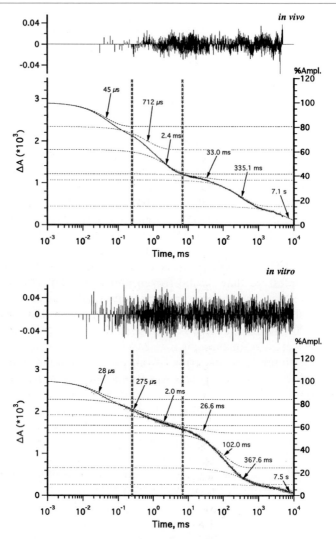

FIG. 5. Kinetics of $\Delta A_{830}$ absorbance changes with *in vivo* and *in vitro* C14D-PS I. The mutant PS I complexes were suspended in 25 m$M$ Tris buffer, pH 8.3, with 0.03% *n*-dodecyl-$\beta$-D-maltoside, 10 m$M$ sodium ascorbate, and 4 $\mu M$ DCPIP at 50 $\mu$g mml$^{-1}$ Chl. Each trace (dots) is an average of 16 measurements taken at 50-sec intervals. The multiexponential fit is overlaid as a solid line; residuals of the fit are shown at the top of the graph. The major individual exponential components are shown as dashed lines, with the percent on the right ordinate, and with offsets equivalent to the relative contribution of the component. The vertical dotted bars separate the time domains in which most of the backreactions of $A_1$ (left), $F_X$ (middle), and $F_A/F_B$ (right) correspondingly occur. The data indicate that while only 42% of the electrons are transferred to $F_A/F_B$ in the *in vivo* C14D-PS I complex, a total of 61%

Analysis of the millisecond and slower decay phases, which are attributed to the $P700^+$ $[F_A/F_B]^-$ backreaction and the slower donation to $P700^+$ by external donors, shows that the quantum efficiency of electron donation to $F_A/F_B$ is 41% on a single turnover flash. Figure 5 (bottom) shows a somewhat smaller fraction of the kinetics in the *in vitro* C14D-PS I complex derived from faster than tens of microsecond to millisecond decay phases attributed to the $P700^+$ $F_X^-$ and $P700^+$ $A_1^-$ backreactions. Analysis of the millisecond and slower decay phases shows that the quantum efficiency of electron donation to $F_A/F_B$ is 61% on a single turnover flash. However, in both the *in vivo* and *in vitro* C14D-PS I mutants, the kinetics cannot distinguish between efficiency of electron transfer or sample heterogeneity (i.e., whether the samples consist of physical mixtures of intact PS I complexes and PS I cores). This result indicates that, if $F_B$ is indeed missing as is implied by the EPR data, then its absence has a relatively minor effect on the efficiency of electron transfer to $F_A$.

*Number of Acceptors.* The hypothesis that $F_B$ is absent in the *in vivo* C14D-PS I mutant is shown to be suspect by performing multiple flash studies which show the number of electron acceptors present at room temperature. In the *in vivo* C14D-PS I mutant, the reduction of $P700^+$ on the first and second flashes occurs with multiple kinetic phases; only the slower phases are relevant in this analysis. As shown in Fig. 6 (top), these slower kinetic phases occur with $t_{1/2}$ of 3.47 and 0.47 ms, which represents PMS forward donation and the $F_X^-$ backreaction, respectively. On the third and fourth flashes, the phase due to PMS forward donation is missing, and only the $P700^+$ $F_X^-$ backreaction is present. The minimum $n$ value of 3 indicates the presence of two ($n$ minus 1) electron acceptors function after $F_X$ in the *in vitro* C14D-PS I mutant. In the *in vitro* C14D-PS I mutant (Fig. 6, bottom), the pattern of reduction of $P700^+$ on the first and second flashes is similar to that of the *in vivo* C14D-PS I mutant. (The lower proportion of faster kinetic phases may be related, in part, to the higher quantum efficiency of electron transfer to $F_A/F_B$ in the *in vitro* C14D-PS I mutant and, in part, to the different flash intervals and concentrations of PMS used in the two experiments.) On the third and fourth flashes, the kinetic phase due to PMS forward donation is missing, and only the $P700^+$ $F_X^-$ backreaction is present. Thus, two acceptors are present in *both* the *in vivo* and *in vitro* C14D-PS I complexes. These mutants are distinctly different from an $F_B$-less sample after $HgCl_2$ treatment, where the amplitude

---

of the electrons are transferred to $F_A/F_B$ in the *in vitro* C14D-PS I complex. [Figure (top) reprinted with permission from J. Yu, L. Vassiliev, Y.-S. Jung, J. H. Golbeck, and L. McIntosh, *J. Biol. Chem.* **272,** 8032 (1997); (bottom) unpublished data from Ilya Vassiliev, 1996.]

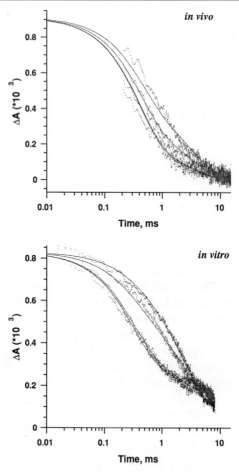

Fig. 6. Multiple flash experiments with *in vivo* and *in vitro* C14D-PS I. The protocol for the *in vivo* C14D-PS I complexes involved excitation of with a train of four consecutive flashes separated by 15 ms and administered in the presence of 10 $\mu M$ PMS and 2 m$M$ sodium dithionite. The protocol for the *in vitro* C14D-PS I complexes involved excitation a train of four consecutive flashes separated by 100 ms and administered in the presence of 100 $\mu M$ PMS and 3 m$M$ sodium dithionite. The experimental data are depicted as the dotted lines; the two-exponential fits of the kinetics are overlaid as the solid lines. If $F_B$ were missing from the mutant C14D-PS I complexes, the third and fourth flashes would be similar to the second lash. Instead, the kinetics of the third and fourth flashes indicate the presence of two acceptors functioning after $F_X$. The *in vitro* measurement was carried out by Dr. Klaus Brettel. [Figure (top) reprinted with permission from Y.-S. Jung, I. Vassiliev, J. Yu, L. McIntosh, and J. H. Golbeck, *J. Biol. Chem.* **272,** 8040 (1997); (bottom) adapted with permission from L. Yu, D. A. Bryant, and J. H. Golbeck, *Biochemistry* **34,** 7861 (1995). Copyright 1995 American Chemical Society.]

TABLE I

IN VIVO CYT C₆: NADP⁺ REDUCTASE ACTIVITIES OF MUTANT
PSAC-PS I COMPLEXES[a]

| Preparation | Flavodoxin reduction[b] | NADP⁺ reduction[b] | |
|---|---|---|---|
| | | Flavodoxin | Ferredoxin |
| PS I | 900 | 930 | 820 |
| PS I Core[c] | 40 | 70 | 40 |
| PsaC-PS I[d] | 460 | 700 | 540 |
| C14D$_{PsaC}$-PSI | 510 | 700 | 480 |
| C51D$_{PsaC}$-PSI | 460 | 700 | 540 |

[a] Adapted from J. Yu, I. Vassiliev, Y.-S. Jung, J. H. Golbeck, and L. McIntosh, *J. Biol. Chem.* **272**, 8032 (1997).
[b] Rates in $\mu$mol (mg Chl)$^{-1}$ hr$^{-1}$.
[c] Isolated from CDK25 mutant of *Synechocystis* sp. PCC 6803.
[d] After rebinding *E. coli*-expressed PsaC, PsaD, and PsaE.

of the slow component differs between the first and the second flashes.[31,32] This analysis reinforces the principle that the absence of an EPR signal in the $g = 2$ region does not necessarily indicate the absence of an iron–sulfur cluster. Drawing from results on the *unbound* C14D PsaC protein,[37] $F'_B$ in the *in vivo* C14D-PS I mutant may a [4Fe–4S] cluster in a high spin state, probably $S = 3/2$, which is difficult to detect by EPR.

*Electron throughput.* Given that the spin states and ligand environments are different in the *in vivo* and *in vitro* C14D-PS I mutants, the question arises whether this leads to differences in rates of electron transfer. As shown in Table I, P700-$F_X$ cores isolated from the CDK25 mutant of *Synechocystis* sp. PCC 6803 support only 8% (flavodoxin) and 5% (ferredoxin) wild-type rates of NADP⁺ reduction. Reconstitution of a P700-$F_A$/$F_B$ complex using *E. coli*-expressed PsaC, PsaD, and PsaE leads to the recovery of 75% (flavodoxin) and 66% (ferredoxin) wild-type rates, thereby showing that $F_X$ by itself is incapable of reducing these soluble electron transfer proteins.[39] The *in vivo* C14D-PS I mutant supports high rates of electron transfer: 75% (flavodoxin) and 59% (ferredoxin) when compared to the wild-type and 100% (flavodoxin) and 100% (ferredoxin) when compared to the reconstituted P700-$F_A$/$F_B$ complex. As shown in Table II, P700-$F_X$ cores isolated from *Synechococcus* sp. PCC 6301 support only 17% (flavodoxin) and 6% (ferredoxin) wild-type rates of NADP⁺ reduction.

[39] J. Hanley, J. Kear, G. Bredenkamp, G. Li, P. Heathcote, and M. Evans, *Biochim. Biophys. Acta* **1099**, 152 (1992).

TABLE II

In Vitro Cyt $C_6$: NADP$^+$ Reductase Activities of Mutant
PsaC-PS I Complexes[a]

| Preparation | Flavodoxin reduction[b] | NADP$^+$ reduction[b] | |
|---|---|---|---|
| | | Flavodoxin | Ferredoxin |
| PS I | 900 | 930 | 820 |
| PS I Core[c] | 60 | 160 | 50 |
| PsaC-PS I[d] | 475 | 580 | 490 |
| C14D$_{PsaC}$-PSI | 300 | 370 | 280 |
| C51D$_{PsaC}$-PSI | 540 | 476 | 416 |

[a] Adapted from Y.-S. Jung, L. Vassiliev, F. Qiao, F. Yang, D. Bryant, and J. H. Golbeck, *J. Biol. Chem.* **271,** 31135 (1996).
[b] Rates in $\mu$mol (mg Chl)$^{-1}$ hr$^{-1}$.
[c] Prepared from *Synechococcus* sp. PCC 6301 P700-$F_A/F_B$ complexes using 6.8 $M$ urea.
[d] After rebinding *E. coli*-expressed PsaC, PsaD, and PsaE.

Reconstitution of a P700-$F_A/F_B$ complex using *E. coli*-expressed PsaC, PsaD, and PsaE leads to the recovery of 62% (flavodoxin) and 60% (ferredoxin) wild-type rates. The *in vitro* C14D-PS I mutant also supports high rates of electron transfer: 40% (flavodoxin) and 34% (ferredoxin) when compared to the wild-type and 64% (flavodoxin) and 57% (ferredoxin) when compared to the reconstituted P700-$F_A/F_B$ complex. [Rates of flavodoxin reduction (uncoupled to NADP$^+$ reduction) in both the *in vivo* and *in vitro* C14D-PS I mutants are similar those of ferredoxin- and flavodoxin-mediated NADP$^+$ reduction.] These studies show that the *in vivo* and *in vitro* C14D-PS I complexes are capable of high-throughput electron transfer to physiologically relevant electron acceptors regardless of the spin state or the ligand environment of the [4Fe–4S] cluster in the $F_B$ site.

*Hypothesis.* The *in vitro* C14D-PS I mutant contains a sulfur ligand provided by the 2-mercaptoethanol. This conclusion takes into account additional *in vitro* mutagenesis experiments which showed that the identical EPR spectrum of $F_B'$ occurs regardless of whether the replacement amino acid is aspartate, serine, alanine, or glycine.[22] The 2-mercaptoethanol is derived from the reconstitution protocol used to insert iron–sulfur clusters into the mutant apoPsaC proteins, and is retained in the mutant site. (The other three 2-mercapthethanol ligands are displaced by the cysteine residues in the ligand exchange reaction which occurs when the [4Fe–4S] clusters are inserts into apoPsaC.) In contrast, the *in vitro* C14 mutant contains an oxygen ligand provided by the replacement amino acid. This conclusion also takes into account additional *in vivo* mutagenesis experiments which showed that a similar EPR spectrum is found in the serine mutant, but no PsaC is accumulated in the alanine mutant.[10] The lack of a suitable ligand in the alanine mutant leads to a protein which either does not fold or contains a [3Fe–4S] cluster. Previous studies with *in vitro* mutants showed that PsaC requires two [4Fe–4S] clusters to rebind to PS I.

*Cysteine 51 → Aspartate in $F_A$: Agreement between in Vivo and in Vitro Mutants*

*Structure.* In contrast, the *in vivo* and *in vitro* C51D-PS I mutants in the $F_A$ site have remarkably similar EPR spectra. When the samples are frozen to 15 K in darkness and illuminated, the *in vivo* C51D-PS I mutant shows a weak set of EPR resonances at $g = 2.067, 1.934$, and $1.882$ characteristic of $F_B$, and there is no evidence for $F_A$ as is found in wild-type PS I complexes (data not shown). The *in vitro* C51D-PS I mutant shows an identical weak set of resonances at $g = 2.167, 1.934$ and $1.882$ characteristic of $F_B$ (data not shown). When the samples are thawed and frozen to 18 K during illumination, the *in vivo* C51D-PS I mutant shows the same set of resonances but they are now ca. four times greater in amplitude (not shown). When the spectrum is measured at 6 K, the $F_B$ resonances become microwave power saturated and a new set of resonances appears at $g = 2.110$, $1.940$, and $1.851$, presumably from a modified $F'_A$ cluster in the mutant site (Fig. 7, top). (The signal at $g = 1.85$ is due to the high-field resonance of a partially reduced $F_X$ cluster.) The *in vitro* C51D-PS I mutant shows the same set of intense $F_B$ resonances when measured at 18 K (data not shown) and the same new set of resonances attributed to $F'_A$ when measured at 6 K (Fig. 7, bottom). These studies show that in contrast to the C14D-PS I mutant, two EPR species, $F_B$ and a modified $F'_A$, are visible around $g = 2$ (but with different relaxation properties) in both the *in vivo* and *in vitro* mutants.

*Efficiency.* A surprisingly large fraction of the backreaction kinetics in the *in vivo* C51D-PS I mutants is derived from faster than tens of microsecond to millisecond decay phases attributed to the $P700^+ F_X^-$ and $P700^+ A_1^-$ backreactions (Fig. 8, top). Analysis of the millisecond and slower decay phases, which are derived from the $P700 + [F_A/F_B]^-$ backreaction and the slower donation to $P700^+$ by external donors, shows that the quantum efficiency of electron donation to $F_A/F_B$ is only 29% on a single turnover flash. Figure 8 (bottom) shows a much larger fraction of the kinetics in the *in vivo* C14D-PS I complex derived from faster than tens of microsecond to millisecond decay phases attributed to the $P700^+ F_X^-$ and $P700^+ A_1^-$

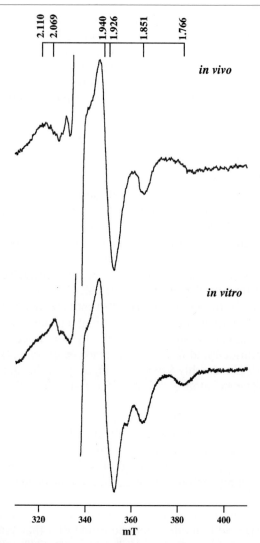

FIG. 7. EPR spectrum of *in vivo* and *in vitro* C51D-PS I complexes. The samples were frozen under continuous illumination and measured at 6 K. The large resonance near 338 mT is the photochemically generated P700$^+$ radical at $g = 2.002$. The 250-$\mu$l samples contained 1 mg ml$^{-1}$ Chl, 300 $\mu$M DCPIP, and 1 m$M$ sodium ascorbate in 50 m$M$ Tris buffer, pH 8.3, containing 0.04% Triton X-100. Spectrometer conditions; microwave power, 20 mW; modulation amplitude; 10 G at 100 kHz. A standard mode resonator (ER/4102 ST) was used at microwave frequency of 9.458 GHz for the *in vivo* sample and 9.448 GHz for the *in vitro* sample. The field legend refers only to the *in vitro* sample; the field legend for the *in vivo* sample is not depicted. The $g$-value legend refers to both spectra. [Figure (top) reprinted with permission from Y.-S. Jung, I. Vassiliev, J. Yu, L. McIntosh, and J. H. Golbeck, *J. Biol.*

backreactions. Analysis of the millisecond and slower decay phases, which are attributed to the P700 + $[F_A/F_B]^-$ backreaction and the slower donation to P700$^+$ by external donors shows that the quantum efficiency of electron donation to $F_A/F_B$ is 60% on a single turnover flash. Kinetics alone cannot distinguish between efficiency of electron transfer and sample heterogeneity. Nevertheless, even though $F_A$ is visible by EPR in both the *in vivo* and *in vitro* C51D-PS I mutants, this study shows that the quantum yield of electron transfer to the terminal electron acceptors is different and considerably less than unity.

*Number of Acceptors.* In the *in vivo* C51D-PS I mutant, the reduction of P700$^+$ on the first and second flashes also occurs with multiple kinetic phases. As shown in Fig. 9 (top), the slower phases which represent PMS forward donation and the $F_X^-$ backreaction contribute a larger proportion of the kinetics than in the C14D-PS I mutants. The kinetics on first and second flashes are typical, containing both kinetic phases, but on the third fourth flash, the kinetic phase due to PMS forward donation is greatly diminished, and on the fourth flash, only the P700$^+$ $F_X^-$ backreaction is present. In the *in vitro* C51D-PS I mutant (Fig. 9, bottom), the reduction of P700$^+$ on the first and second flashes represents largely the PMS forward donation and a small amount of the $F_X^-$ backreaction, but on the third and fourth flashes, the kinetic phase due to PMS forward donation is missing, and only the P700$^+$ $F_X^-$ backreaction is present. (The differences in kinetics of the *in vitro* and *in vivo* mutants are largely due to the different flash intervals and concentrations of PMS used in the two experiments.) The minimum *n* value of 3 indicates the presence of two (*n* minus 1) electron acceptors functioning after $F_X$ in both the *in vivo* and *in vitro* C51D-PS I mutants. Thus, these studies show that $F_A'$ and $F_B$ are functional in the *in vivo* and *in vitro* C51D-PS I complexes.

*Electron Throughput.* The *in vivo* C51D-PS I mutant supports 75% (flavodoxin) and 66% (ferredoxin) of wild-type rates, and the *in vitro* C51D-PS I mutant supports 51% (flavodoxin) and 51% (ferredoxin) wild-type rates (Tables I and II). Moreover, the *in vivo* C51D-PS I mutant supports rates that are identical to the reconstituted P700-$F_A/F_B$ complex using the CDK25 mutant of *Synechocystis* sp. PCC 6803 (Table I). Similarly, the *in vivo* C51D-PS I mutant supports rates that are only slightly lower than a PsaC- (and PsaD-) reconstituted P700-$F_A/F_B$ complex using a urea-stripped PS I core from *Synechococcus* sp. PCC 6301 (Table II). (Rates of flavodoxin

*Chem.* **272**, 8040 (1997); (bottom) reprinted with permission from L. A. Yu, I. R. Vassiliev, Y. S. Jung, D. A. Bryant, and J. H. Golbeck, *J. Biol. Chem.* **270**, 28118 (1995).]

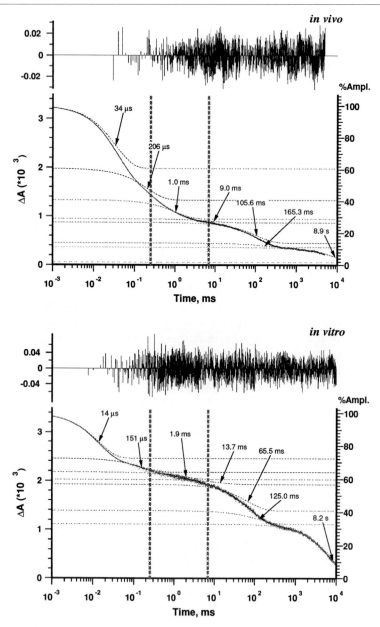

FIG. 8. Kinetics of $\Delta A_{830}$ absorbance changes with *in vivo* and *in vitro* C51D-PS I. The mutant PS I complexes were suspended in 25 m$M$ Tris buffer, pH 8.3, with 0.03% $n$-dodecyl-$\beta$-D-maltoside, 10 m$M$ sodium ascorbate, and 4 $\mu M$ DCPIP at 50 $\mu$g ml$^{-1}$ Chl. Each trace (dots) is an average of 16 measurements taken at 50-sec, intervals. The multiexponential fit

reduction alone generally are similar to those of feredoxin- and flavodoxin-mediated $NADP^+$ reduction.) These results show that the electron transfer throughput is relatively unaffected by the presence of a mixed-ligand $F'_A$ cluster in the *in vivo* or *in vitro* C51D-PS I mutants.

*Hypothesis.* The *in vitro* and *in vivo* C51D-PS I mutants both contain an oxygen ligand provided by the replacement amino acid. This conclusion takes into account the identical EPR spectra in *in vitro* and *in vivo* mutants as well as mutagenesis experiments which show that the *in vitro* C51A-PS I mutant, but not the *in vivo* C51A-PS I mutant, assembles a functional PS I.[10] (It follows that $H_2O$ or $OH^-$ from the solvent is not the adequate "rescue ligand" at the mutant site.) The ligand which occupies the C14 and C51 modified sites in *unbound* PsaC proteins is 2-mercaptoethanol, regardless of whether the mutation is glycine, alanine, serine, or aspartate.[9] The implication is that the environment of the $F_A$ and $F_B$ clusters, which appears similar in the unbound proteins, becomes differentiated when PsaC is bound to the PS I complex. Hence, thiolates provided from cysteine or from 2-mercaptoethanol are the preferred ligands to [4Fe–4S] clusters except when steric or other considerations come into play.

## Assessment of Mutant PsaC Proteins

These case studies show that a combination of mutagenesis methods and spectroscopic techniques is desirable for analysis of PS I mutants. The differences in the EPR spectra of the *in vitro* and *in vivo* C14D-PS I mutants and the similarities in the EPR spectra of the *in vitro* and *in vivo* C51D-PS I mutants provide clues as to the identity of the ligands in the mutant site. Ultimately, the differences can be traced to the use of 2-mercaptoethanol in the iron–sulfur insertion into the *E. coli*-expressed PsaC apoprotein. The external thiolate functions as a "rescue ligand" at the mutated site, compensating for the missing cysteine ligand in PsaC.

---

is overlaid as a solid line; residuals of the fit are shown at the top of the graph. The major individual exponential components are shown as dashed lines, with the percent on the right ordinate, and with offset equivalent to the relative contribution of the component. The vertical dotted bars separate the time domains in which most of the backreactions of $A_1$ (left), $F_X$ (middle), and $F_A/F_B$ (right) correspondingly occur. The data indicate that while only 29% of the electrons are transferred to $F_A/F_B$ in the *in vivo* C51D-PS I complex, a total of 60% of the electrons are transferred to $F_A/F_B$ in the *in vitro* C51D-PS I complex. [Figure (top) reprinted with permission from J. Yu, I. Vassiliev, Y.-S. Jung, J. H. Golbeck, and L. McIntosh, *J. Biol. Chem.* **272,** 8032 (1997); (bottom) unpublished data from Ilya Vassiliev, 1996.]

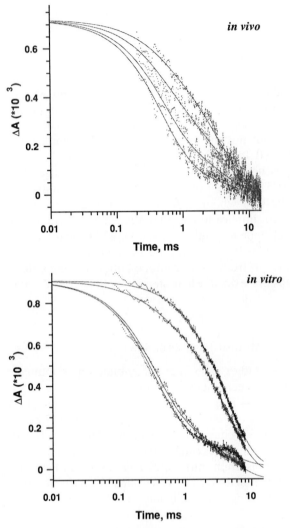

FIG. 9. Multiple flash experiments with *in vivo* and *in vitro* C51D-PS I. The protocol for the *in vivo* C5D-PS I complexes involved excitation of with a train of four consecutive flashes separated by 15 ms and administered in the presence of 10 $\mu M$ PMS and 2 m$M$ sodium dithionite. The protocol for the *in vitro* C51D-PS I complexes involved excitation with a train of four consecutive flashes separated by 100 ms and administered in the presence of 100 $\mu M$ PMS and 3 m$M$ sodium dithionite. The experimental data are depicted as the dotted lines; the two exponential fits of the kinetics are overlaid as the solid lines. If $F_A$ were missing from the mutant C51D-PS I complexes, the third and fourth flashes would be similar to the second flash. Instead, the kinetics of the third and fourth flashes indicate the presence of two acceptors functioning after $F_X$. The *in vitro* measurement was carried out by Dr. Klaus Brettel. [Figure (top) reprinted with permission from Y.-S. Jung, I. Vassiliev, J. Yu, L. McIntosh, and J. H. Golbeck, *J. Biol. Chem.* **272,** 8040 (1997); (bottom) adapted with permission from L. Yu, D. A. Bryant, and J. H. Golbeck, *Biochemistry* **34,** 7861 (1995). Copyright 1995 American Chemical Society.]

The following additional conclusions can be drawn:

1. All of the aspartate mutants contain a functional [4Fe–4S] cluster in the modified C14 and C51 sites. The lack of characteristic EPR signals for [3Fe–4S] clusters, the results of the multiple flash experiment ($n$ = 3), and the equivalent amount of irreversible P700$^+$ produced by low-temperature illumination (not shown; see Jung *et al.*[10]) support this assessment. Oxygen as well as sulfur can apparently support the presence of a [4Fe–4S] cluster in the mutant sites. Note that the identity of the ligand in the mutant site cannot be determined from the EPR spectrum, but is inferred by comparing the *in vivo* with the *in vitro* results.

2. Regardless of the identity of the replacement ligand, the mixed-ligand [4Fe–4S] clusters are capable of electron transfer on a single turnover flash. The quantum yields are less than unity, indicating either an inefficiency in electron transfer or heterogeneity in the population of reaction centers. Quantitative immunoblots and ELISA methods are needed to distinguish between these possibilities. The mutants are also capable of relatively high steady-state rates of electron transfer from cytochrome $c_6$ to NADP$^+$, using ferredoxin or flavodoxin as electron acceptor. The relatively poor correspondence between the quantum yields determined on a single turnover flash and the steady-state rates of NADP$^+$ reduction remains to be explained. It is possible that some of the rate differences observed for the mutant proteins in NADP$^+$ photoreduction results from difference in docking ability of ferredoxin and flavodoxin to the altered sites on the PS I complexes.

## Acknowledgments

Dr. Don Bryant (The Pennsylvania State University) and Dr. Jindong Zhao carried out the *in vitro* mutagenesis of PsaC as well as the expression of PsaC, PsaD, and PsaE in *E. coli*. Dr. Lee McIntosh (Michigan State University) and Jian-Ping Yu carried out the *in vivo* mutagenesis of PsaC in *Synechocystis* sp. PCC 6803. Kevin Parrett and Tetemke Mehari developed the methods for isolating P700-F$_X$ cores and reconstituting P700-F$_A$/F$_B$ complexes; Lian Yu reconstituted and studied the spectroscopic properties of the *in vitro* mutants; and Yean-Sung Jung studies the spectroscopic properties of the *in vivo* mutants. Dr. Ilya Vassiliev developed the optical spectroscopic protocols for analysis of the PS I acceptors. Funding for this work was provided by NSF (DMB-8517391, DMB-8905065, MCB-9205756, MCB-9696179 and MCB-9723661) and USDA (87-CRCR-1-2382 and 96-37306-2632) grants.

# [8] Isolation and Functional Study of Photosystem I Subunits in the Cyanobacterium *Synechocystis* sp. PCC 6803

*By* JUN SUN, AN KE, PING JIN, VAISHALI P. CHITNIS, and PARAG R. CHITNIS

## Introduction

Photosystem I (PS I) is a multiheteromeric pigment–protein complex in thylakoid membranes of cyanobacteria.[1-3] PS I catalyzes the photooxidation of plastocyanin or cytochrome $c_6$ in thylakoid lumen and the photoreduction of ferrodoxin or flavodoxin in cyanobacterial cytoplasm. The PS I complex contains the photosynthetic pigments (chlorophyll *a* and *β* carotene) and five electron transfer centers ($A_0$, $A_1$, $F_X$, $F_A$, and $F_B$) that are bound to the PsaA, PsaB, and PsaC proteins. In addition, PS I complex contains at least eight other polypeptides that are accessory in their functions. Recent X-ray crystallographic analysis of cyanobacterial PS I at 4-Å resolution[4] has revealed its structure and overall organization. Application of molecular genetics has now become indispensable to determine the role of individual amino acids that provide a precise environment for the cofactors to function in efficient transfer of energy and electrons.

In recent years, cyanobacteria have been used increasingly to study structure–function relations in photosynthetic proteins, including PS I. The mesophilic cyanobacterium, *Synechocystis* sp. PCC 6803 is a model system for using molecular genetic approaches to study functions of PS I proteins. Its genome is completely sequenced[5] and the protein components of its photosystem are identified.[1] Targeted mutations in all genes for PS I proteins are available. Here we describe biochemical and molecular genetic methods and resources to study PS I of *Synechocystis* sp. PCC 6803.

[1] P. R. Chitnis, Q. Xu, V. P. Chitnis, and R. Nechushtai, *Photosynth. Res.* **44,** 23 (1995).

[2] P. R. Chitnis, *Plant Physiol.* **111,** 661 (1996).

[3] P. Fromme, *Curr. Opin. Struct. Biol.* **6,** 473 (1996).

[4] N. Krauß, W.-D. Schubert, O. Klukas, P. Fromme, H. T. Witt, and W. Saenger, *Nature Struct. Biol.* **3,** 965 (1996).

[5] T. Kaneko, S. Sato, H. Kotani, A. Tanaka, E. Asamizu, Y. Nakamura, N. Miyajima, M. Hirosawa, M. Sugiura, S. Sasamoto, T. Kimura, T. Hosouchi, A. Matsuno, A. Muraki, N. Nakazaki, K. Naruo, S. Okumura, S. Shimpo, C. Takeuchi, T. Wada, A. Watanabe, M. Yamada, M. Yasuda, and S. Tabata, *DNA Res.* **3,** 109 (1996).

Isolation of Photosystem I Complexes

*Rapid Preparation of Photosystem I Trimers*

Cyanobacteria PS I can exist in trimeric quaternary structure in the membranes.[6] The PS I trimers can be purified rapidly by sucrose gradient ultracentrifugation.[7] The yield of PS I trimers is influenced by several factors: the choice of detergent that is used for solubilization of membranes, the ratio of detergent to membranes, and ionic conditions and pH during solubilization. Among a dozen different detergents tested, dodecyl-$\beta$-D-maltoside has been shown to be the most suitable detergent to dissolve membranes with minimum disturbance to the quaternary structure of PS I. The optimal ratio of chlorophyll to detergent for the photosynthetic membranes of *Synechocystis* sp. PCC 6803 is 1 : 15. The presence of high concentrations of monovalent cations or acidic pH shifts the monomer–trimer balance toward monomers, whereas calcium stimulates trimerization. A typical procedure for isolation of PS I trimers is given next.

Cells are harvested at the late exponential stage of growth, and suspended in SMN solution (0.4 m$M$ sucrose, 10 m$M$ NaCl, 50 m$M$ MOPS, pH 7.0) with the protease inhibitors phenylmethyl sulfonylfluoride (PMSF, 0.2 m$M$) and benzamidine (5 m$M$). The cells are broken in a bead beater and thylakoids are isolated by centrifugation at 50,000$g$ for 60 min. Typically, 3 liters of cells [1.0 optical density (OD) at 730 nm] yield thylakoids containing about 5 mg chlorophyll.

The membranes (1 mg chlorophyll/ml) are incubated for 30 min at room temperature with 1 m$M$ CaCl$_2$ in SMN. They are solubilized by addition of dodecyl $\beta$-D-maltoside to a final concentration of 1.5% and incubation for 15 min on ice.

Insoluble material is pelleted by centrifugation at 20,000$g$ for 15 min and the supernatant is layered on a 10–30% step gradient of sucrose in 10 m$M$ MOPS, pH 7.0, 0.05% dodecyl $\beta$-D-maltoside. Ultracentrifugation at 160,000$g$ for 16 hr resolves the pigmented complexes of the photosynthetic membranes into distinct bands (Fig. 1A). Samples containing 200 $\mu$g chlorophyll can be used in the 13-ml tubes of a SW41 rotor. The heavier green band contains PS I trimers without any significant impurity (Fig. 1B). Typically, 60% chlorophyll in thylakoid can be recovered as PS I trimers by this procedure.

[6] J. Kruip, D. B. Bald, E. Boekema, and M. Rogner, *Photosynth. Res.* **40**, 279 (1994).
[7] V. P. Chitnis and R. P. Chitnis, *FEBS Lett.* **336**, 330 (1993).

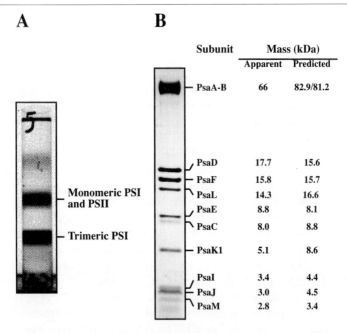

| Subunit | Mass (kDa) | |
|---|---|---|
| | Apparent | Predicted |
| PsaA-B | 66 | 82.9/81.2 |
| PsaD | 17.7 | 15.6 |
| PsaF | 15.8 | 15.7 |
| PsaL | 14.3 | 16.6 |
| PsaE | 8.8 | 8.1 |
| PsaC | 8.0 | 8.8 |
| PsaK1 | 5.1 | 8.6 |
| PsaI | 3.4 | 4.4 |
| PsaJ | 3.0 | 4.5 |
| PsaM | 2.8 | 3.4 |

FIG. 1. Resolution of photosystem I. Monomeric and trimeric forms of PS I are resolved by sucrose gradient ultracentrifugation (A) and the PS I subunits are separated by Tricine-urea sodium dodecyl sulfate–polyacrylamide gel electrophoresis (SDS–PAGE) (B). PsaK2 cannot be detected in the wild-type PS I preparation.

## Purification of Photosystem I by Ion-Exchange Chromatography and Ultracentrifugation

The "Nelson" method for purification of plant PS I[8] can be used for purifying active PS I from *Synechocystis* sp. PCC 6803 membranes. The isolated thylakoid membranes are adjusted to 500 μg chlorophyll/ml with SMN buffer and Triton X-100 is added at a ratio of 1 : 10 (chlorophyll : Triton X-100, w/w). The membranes containing 5–10 mg chlorophyll are solubilized for 15–20 min on ice and centrifuged at 20,000g for 10 min at 4°. After centrifugation, the supernatant is applied to a 2.5 × 10-cm DEAE-cellulose column equilibrated with the MOPS-Triton buffer (10 mM MOPS, pH 7.0, 0.05% Triton X-100). The unbound and loosely bound proteins are removed by washing the column with 50 mM NaCl in MOPS-Triton buffer. The bound proteins are eluted with a 50–200 mM NaCl gradient in MOPS-Triton buffer. The green fraction, containing the PS I complexes, is eluted

---

[8] C. Bengis and N. Nelson, *J. Biol. Chem.* **250**, 2783 (1975).

by ~100 m$M$ NaCl. To concentrate PS I complexes further, this chromatography step can be repeated by diluting the green fraction four times with MOPS-Triton buffer without NaCl and loading this diluted green fraction on a second DEAE-cellulose colume. After washing, PS I complexes can be eluted with 200 m$M$ NaCl in MOPS-Triton buffer and collected as a dark green fraction. Chromatography is performed at 4°.

The dark green fractions are combined for sucrose gradient ultracentrifugation. The fractions are applied to a 5–30% sucrose step gradient in MOPS-Triton buffer and ultracentrifugation at 160,000g for 16–18 hr at 4°. Purified PS I complexes form a distinct green layer in the centrifuge tube and can be collected by puncturing the bottom of the tube or by using a glass Pasteur pipette. If necessary, PS I complexes can be concentrated with a 50K Macrosep centrifugal concentrator (Amicon). Purified PS I is stored at −20° or −80° for further use. Typical yield of PS I using this method is approximately 7 mg chlorophyll in PS I from thylakoids containing 10 mg chlorophyll. The PS I purified by this method is suitable for biochemical and spectroscopic studies. These preparations contain approximately 100 chlorophyll molecules per P700.

*Alternative Methods*

In addition to the methods already described, several other methods have been successfully used to purify PS I complexes from cyanobacterial membranes. Digitonin solubilization, sucrose gradient ultracentrifugation, and preparative PAGE can be used to isolate different types of PS I reaction centers from cyanobacterial thylakoid.[9] Homogeneous PS I complexes can also be purified by preparative HPLC purification: anion exchange chromatography step followed by hydroxyapatite chromatography and gel-filtration chromatography steps.[10]

Purification of Photosystem I Subunits

Purified individual PS I subunits are useful in structural, cross-linking, and topographical studies. They are also used in resolution–reconstitution experiments to understand function of PS I proteins. In addition, they have been used as antigens to raise subunit-specific antibodies. For purpose of raising subunit-specific antibodies, individual PS I subunits can be separated by Tricine-urea SDS–PAGE[11] and stained with Coomassie Blue. The gel

[9] S. Katoh, *Methods Enzymol.* **167,** 263 (1988).
[10] M. Rogner, P. J. Nixon, and B. A. Diner, *J. Biol. Chem.* **265,** 6189 (1990).
[11] Q. Xu, L. Yu, V. P. Chitnis, and P. R. Chitnis, *J. Biol. Chem.* **269,** 3205 (1994).

slices can be excised and used directly as antigens for subunit-specific antibodies.

### From Isolated Complexes

PsaC, PsaD, and PsaE are peripheral subunits on the reducing side of
PS I complex. They can be separated and purified from the PS I complexes.
To extract these subunits from the PS I core, solid ultrapure urea is added
to a final concentration of 7.5 $M$. The final PS I concentration is adjusted
to 200 $\mu$g chlorophyll/ml and 1 $M$ Tris-HCl buffer (pH 7.5) is added to
make a final concentration of 20 m$M$. Urea is dissolved by slow shaking
at room temperature for about 30 min. Most of the PsaC, PsaD, and PsaE
subunits should dissociate from the PS I core by urea treatment. Extracted
proteins are separated by ultrafiltration with a 100K Microsep centrifugal
concentrator at 3000g for 2 hr at 10–15°. The eluted fraction contains
PsaC, PsaD, and PsaE and can be concentrated by ultrafiltration with a
3K Microsep centrifugal concentrator.

The individual subunits from the mixture of PsaC, PsaD, and PsaE
can be separated by ion-exchange chromatography. An Econo-Pac high S
cartridge (Bio-Rad Laboratories, Hercules, CA) is equilibrated with 20
m$M$ Tris-HCl (pH 7.5) buffer, the urea extracted peripheral proteins are
loaded, and the column is washed with 2 ml of 20 m$M$ Tris-HCl (pH 7.5)
buffer. The unbound fraction contains the PsaC and PsaE subunits, whereas
the PsaD subunit binds to the column and can be eluted with a 0–500 m$M$
NaCl gradient. To separate PsaC and PsaE subunits, the concentrated
unbound fraction from the Econo-Pac high S cartridge is applied on an
Econo-Pac high Q cartridge (Bio-Rad Laboratories) equilibrated with 20
m$M$ Tris-HCl (pH 7.5) buffer, and the column is washed with 2 ml of 20
m$M$ Tris-HCl (pH 7.5) buffer. The unbound fraction contains PsaE subunit.
PsaC subunit can be eluted with a 0–500 m$M$ NaCl gradient. The purity
of the separated proteins is examined by Tricine-urea SDS–PAGE and
Western blotting.

### From Overexpressed Proteins

The peripheral PS I subunits (as PsaD and PsaE) can also be purified
after overexpression in *Escherichia coli* cells. The genes (*psaD* and *psaE*)
are cloned in pET-21a and the transformed *E. coli* cells are used for overexpression. Typically, a fermentor container with 10 liters LB is prewarmed
to 37° and inoculated with 400 ml overnight culture. Cells are aerobically
grown at 37° for 2–3 hr until the $OD_{600}$ of the culture reaches 0.6–1.0. Then
1 m$M$ of isopropylthiol-$\beta$-D-galactoside (IPTG) is used to induce the
protein overexpression. Cells are harvested at 3 hr after induction and

suspended in minimal TS buffer (20 m$M$ Tris-HCl, pH 7.0) containing 5 m$M$ PMSF and 5 m$M$ benzamidine

To extract proteins, the cells are broken by sonication and centrifuged at 20,000$g$ for 30 min to pellet inclusion bodies and membranes. The pellets are resuspended in 5 volumes of wash buffer [100 m$M$ Tris-HCl (pH 7.0) 5 m$M$ ethylenediaminetetraacetic acid (EDTA), 5 m$M$ dithiotureitol (DTT), 2 M urea, and 2% Triton X-100] and centrifuged. This step is repeated until the supernatant is clear. The pellets are resuspended in 5 volumes of wash buffer without Triton X-100 and urea and centrifuged again. Finally, the pellets are homogenized in extraction buffer (50 m$M$ Tris-HCl, pH 7.0, 5 m$M$ EDTA, 8 $M$ guanidine-HCl, and 5 m$M$ DTT) at the ratio of 1 ml buffer per gram of original wet cells and centrifuged at 120,000$g$ for 1 hr. The supernatant containing overexpressed protein is filtered through a 0.45-$\mu$m filter. PsaE can be purified by DEAE-cellulose chromatography as unbound fraction while impurities remain bound to DEAE-cellulose with 10 m$M$ Tris-HCl buffer (pH 7.5). Purification of PsaD can be performed on an Econo-Pac high S cartridge as described earlier. Purified proteins can be desalted by dialysis against 10 m$M$ Tris-HCl (pH 8.0) at 4° for 16 hr and concentrated by ultrafiltration with 3K Microsep centrifugal concentrator. Protein concentrations are measured from deduced extinction coefficiency 1 A$_{280}$ = 0.97 mg/ml for PsaD and 1 A$_{280}$ = 1.38 mg/ml for PsaE.

The PsaD and PsaE subunits can be overexpressed as fusion proteins with polyhistidine tag at C terminus and purified by affinity chromatography. The $psaD$ gene is inserted into pET-21b(+) vector between $Xho$I and $Nde$I sites. Plasmid containing $psaD$ gene is introduced into BL21 (DE3) strain of $Escherichia\ coli$. The gene is induced by incubating with 1 m$M$ IPTG at 37° for 3 hr. Induced cells are broken by sonication. Most overexpressed PsaD is in the soluble fraction. The soluble overexpressed tagged PsaD protein is purified by His·Bind Resin and Buffer kit (Novagen, Madison, WI). His·Bind Resin (2.5 ml) is prepared in a small polypropylene column and washed by 3 volumes of sterile deionized water. The affinity material Ni$^{2+}$ is immobilized on resin by 5 volumes of charge buffer (50 m$M$ NiSO$_4$). After the column is equilibrated by 3 volumes of binding buffer (5 m$M$ imidazole, 0.5 $M$ NaCl, 20 m$M$ Tris-HCl, pH 7.9), samples are loaded on column. The column is then washed with 10 volumes of binding buffer and 6 volumes of wash buffer (60 m$M$ imidazole, 0.5 $M$ NaCl, 20 m$M$ Tris-HCl, pH 7.9). Bound proteins are finally eluted in elute buffer (1 $M$ imidazole, 0.5 $M$ NaCl, 20 m$M$ Tris-HCl, pH 7.9) and desalted twice using Econo-Pac 10DG column (Bio-Rad) to remove the imidazole completely. Purified PsaD proteins are stored in −20° for further use.

Topographical Characterization of Photosystem I

During the past few years, major advances in X-ray crystallography[4,12] have provided a framework for understanding the overall architecture of PS I. However, due to the lower resolution of these studies, more detailed information about the structure of PS I is not available. Topographical studies using biochemical techniques reveal exposed domains or exposed amino acid residues, the interactions among PS I subunits or between PS I and its electron donor or acceptor, and the roles of individual residues in the interaction. Topographical studies on complex assemblies like PS I are the most convenient means for correlating defects in function due to mutations with perturbations in structural organization. It is prudent to generate information beyond crystallographic studies for a complete and rapid understanding of structure.

*Limited Proteolysis*

Limited proteolysis coupled with domain-specific antibodies and N-terminal amino acid sequencing provides a powerful tool for understanding details of the organization of PS I.[13–15] To study the accessibility of PS I subunits to proteases, purified PS I complexes, thylakoid membranes, or osmotically shocked cells[16] are incubated with proteases at a final chlorophyll concentration of 200 $\mu$g/ml. The proteolysis reaction conditions for eight proteases are listed in Table I.[15]

The PS I complexes are incubated with different concentrations of proteases, denatured with SDS, and separated by modified Tricine-urea SDS–PAGE. After electrotransfer, polyvinylidene difluoride (PVDF) membranes are stained by Coomassie Blue or immunodetected by three domain-specific antibodies: anti-PsaA2 against the N terminus of PsaA, anti-PsaB450 against an extramembrane loop between transmembrane helices VII and VIII, and anti-PsaB718 against the C terminus of PsaB (Fig. 2). The apparent mass of a protein fragment can be determined from the migration of protein molecular weight standards on electrophoresis. The protein fragments that are visible in Coomassie Blue staining and recognized by immunodetection, can be subjected to N-terminal amino acid sequencing. Exposed proteolysis sites can be located from the N-terminal sequences of proteolysis fragments, their apparent mass and reactivity to antibodies.

[12] N. Krauß, W. Hinrichs, I. Witt, P. Fromme, W. Pritzkow, Z. Dauter, C. Betzel, K. S. Wilson, H. T. Witt, and W. Saenger, *Nature* **361,** 326 (1993).

[13] Q. Xu and P. R. Chitnis, *Plant Physiol.* **108,** 1067 (1995).

[14] Q Xu, J. A. Guikema, and P. R. Chitnis, *Plant Physiol.* **106,** 617 (1994).

[15] J. Sun, Q. Xu, V. P. Chitnis, P. Jin, and P. R. Chitnis, *J. Biol. Chem.* **272,** 21793 (1997).

[16] G. S. Tae and W. A. Cramer, *Biochemistry* **33,** 10060 (1994).

## TABLE I
### REACTION CONDITIONS OF PROTEASES TREATMENT[a]

| Proteases | Thermolysin | Glu-C | Chymotrypsin | Papain | Lys-C | Trypsin | C.ostripain | Pepsin |
|---|---|---|---|---|---|---|---|---|
| Specificity | ^[VILWMF] | [E]^ | [YFW]^ | [RKILG]^ | [K]^ | [KR]^ | [R]^ | [FLYWI]^ |
| Final protease concentration | 250 μg/mg chlorophyll | 250 μg/mg chlorophyll | 250 μg/mg chlorophyll | 10 μg/mg chlorophyll | 250 μg/mg chlorophyll | 250 μg/mg chlorophyll | 250 μg/mg chlorophyll | 250 μg/mg chlorophyll |
| Proteolysis condition[b] | 10 m$M$ MOPS (pH 7.0); 5 m$M$ CaCl$_2$ | 50 m$M$ NH$_4$HCO$_3$ (pH 7.8) | 50m$M$ Tris-HCl (pH 8.0); 10 m$M$ CaCl$_2$ | 10 m$M$ MOPS (pH 7.0); 0.5 m$M$ EDTA; 1.5 m$M$ CySH | 25 m$M$ Tris-HCl (pH 7.5); 1 m$M$ EDTA | 50 m$M$ Tris-HCl (pH 8.0); 1 m$M$ CaCl$_2$ | 20 m$M$ Tris-HCl (pH 7.5); 1 m$M$ CaCl$_2$; 2 m$M$ DTT | 20 m$M$ NaH$_2$PO$_4$ (pH 4.0) |
| Incubation | 37°; 30 min | 15°; 30 min | 25°; 1 hr | 25°; 1 hr | 37°; 1 hr | 37°; 1 hr | 37°; 1 hr | 37°; 1 hr |
| Termination | 20 m$M$ EDTA | 20 m$M$ PMSF | 20 m$M$ PMSF | 20 m$M$ PMSF | 20 m$M$ PMSF | 20 m$M$ PMSF | 20 m$M$ EDTA | 20 m$M$ pepstatinA |

[a] The source of these eight proteases are as follows: sequencing grade endoproteinase Lys-C (from *Pseudomonas aeruginosa*; EC 3.4.21.50), clostripain (endoproteinase Arg-C from *Clostridium histolyticum*; EC 3.4.21.4) are purchased from Promega Biotech, Madison, WI. Papain (from *Carica papaya*; EC 3.4.22.2), pepsin (from porcine stomach; EC 3.4.23.1), sequencing grade endoproteinase Glu-C (protease V8 from *Staphylococcus aureus*; EC 3.4.21.19) and chymotrypsin (from bovine pancreas; EC 3.4.21.1) are obtained from Boehringer Mannheim, Indianapolis, IN. Thermolysin (protease type X from *Bacillus thermoproteolyticus*; EC 3.4.24.4) is purchased from Sigma, St. Louis, MO.

[b] 0.05% Triton X-100 is added for purified PS I complexes.

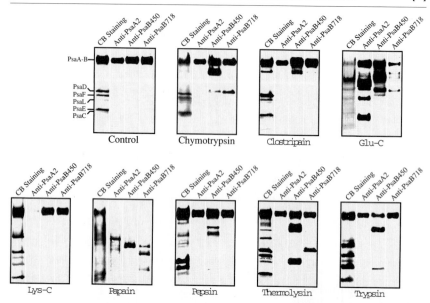

FIG. 2. Limited proteolysis of photosystem I proteins. Purified PS I complexes are partially digested by different proteases and resolved by Tricine-urea SDS–PAGE. After electrotransfer to PVDF membranes, protein fragments are detected by Coomassie Blue (CB) staining or by immunodetection with three domain-specific antibodies against different regions in PsaA or PsaB.

## Modification of Surface-Exposed Residues

To study the topography of PS I, it is important to probe the exposed protein surface by modification of specific residues. Two biotin derivatives can be used to modify surface-exposed residue. N-hydroxysuccinimidobiotin (NHS-biotin) specifically reacts with the N terminus and the ε-amino group of lysyl residues,[17] while biotin-maleimide (M-biotin) modifies the sulfhydryl groups of cysteinyl residues. Both reagents are used to study surface-exposed residues in the PS I complex. Purified PS I complex or PS I subunits are biotinylated at a final protein concentration of 0.5–1.5 mg/ml under the modification conditions listed in Table II.

Biotin-labeled PS I can be further used in protease treatments. The labeled PS I subunits are separated by Tricine-urea SDS–PAGE and electroblotted to Immobilon-P membranes.[18] The blot is probed with peroxi-

[17] E. A. Bayer and M. Wilchek, *Methods Enzymol.* **184,** 138 (1990).
[18] Q. Xu, Y. S. Jung, V. P. Chitnis, J. A. Guikema, J. H. Golbeck, and P. R. Chitnis, *J. Biol. Chem.* **2693,** 21512 (1994).

TABLE II
REACTION CONDITIONS FOR BIOTINYLATION TREATMENT

| Modification reagent | N-hydroxysuccinimidobiotin (NHS-biotin) | Biotin-maleimide (M-biotin) |
|---|---|---|
| Specificity | ε-Amino group of lysine | Sulfhydryl group of cysteine |
| Final concentration | 60 μM | 60 μM |
| Modification condition | 10 mM MOPS (pH 7.0); 0.05% Triton X-100) | 10 mM MOPS (pH 7.0); 0.05% Triton X-100 |
| Incubation | Room temperature 60 min in dark | Room temperature 60 min in dark |
| Termination | 100 mM Glycine | 10 mM DTT |
| Labeled proteins in PS I | PsaAB, PsaD, PsaF, PsaL, PsaE | PsaC |

dase-conjugated Avidin (Cooper Biomedical, Malvern, PA) and then developed with hydrogen peroxide and 4-chloro-naphthol[19] or with enhanced chemiluminescence reagents (Amersham, Arlington Heights, IL).

*Chemical Cross-Linking*

Chemical cross-linking can be used to investigate the interaction of PS I subunits. It can reveal the physical relationship between the PS I subunits. Glutaraldehyde is a bifunctional cross-linking reagent that reacts principally with the amino-group of lysyl residues,[20] and is stored at $-20°$ prior to use. Purified PS I complexes at a concentration of 150 μg chlorophyll/ml are treated with 15 mM glutaraldehyde (Sigma) in the presence of 10 mM MOPS-HCl (pH 7.0) and 0.05% Triton X-100 for 30 min on ice with continuous shaking. The cross-linking reactions are quenched by the addition of glycine to a final concentration of 100 mM. Subsequently, the samples are diluted with an excess of 10 mM MOPS-HCl (pH 7.0), filtered using Centricon-100 (Amicon, Beverly, CA) ultrafiltration, and then applied for analytical gel electrophoresis, Western blotting, and other analysis. Nonspecific cross-linking effects may occur at higher concentrations of glutaraldehyde. The following cross-linked products can be found[21–22]: PsaC–PsaD, PsaC–PsaE, PsaD–PsaL, and PsaE–PsaF.

N-Ethyl-1,3-[3-(diethylamino)propyl]carbodiimide (EDC), a zero-length cross-linker that forms a covalent bond between primary amines

[19] L. K. Frankel and T. M. Bricker, *Biochemistry* **31**, 11059 (1992).
[20] K. Peters and F. M. Richards, *Ann. Rev. Biochem.* **46**, 523 (1977).
[21] T. S. Armbrust, P. R. Chitnis, and J. A. Guikema, *Plant Physiol.* **111**, 1307 (1996).
[22] Q. Xu, V. P. Chitnis, A. Ke, and P. R. Chitnis, *in* "Photosynthesis: From Light to Biosphere" (P. Mathis, ed.), p. 87. Kluwer, Dodrecht, 1995.

and carboxyl,[23] can also be used for studies of the physical relationship between the PS I subunits. Purified PS I at a final concentration of 200 $\mu$g chlorophyll/ml is mixed well with 80 m$M$ fresh-prepared EDC (Sigma) in the presence of 20 m$M$ HEPES, pH 7.0, and 0.05% Triton X-100. After incubation at room temperature for 60 min, the reaction is quenched by the addition of 1 $M$ Tris-HCl, pH 7.0, to a final concentration of 200 m$M$ at room temperature for 15 min. The following cross-linked products have been detected by Western blotting[21]: PsaE–PsaF and PsaC–PsaD. EDC can also be used to cross-link the PsaD of *Synechocystis* sp. PCC 6803 PS I with ferredoxin.[24]

Biochemical Assays for Photosystem I Function

*Oxygen Evolution and Uptake*

Photosynthetic electron transport activities are determined by the rates of oxygen evolution or uptake. Oxygen measurements are made on an oxygen monitoring system (Hansatech, England) at 25° under a light density of 2430 $\mu$mol m$^{-2}$ s$^{-1}$. Total photosynthetic activity is measured as oxygen evolution using whole cells (10 $\mu$g chlorophyll/ml) suspended in 40 m$M$ HEPES, pH 7.0, after adding the final electron acceptor NaHCO$_3$ to the reaction mixture to a final concentration of 10 m$M$. PS I activity can be measured as the rate of oxygen uptake via the Mehler reaction[25,26] using whole cells (in 40 m$M$ HEPES, pH 7.0) or thylakoid membranes (in 10 m$M$ MOPS) or isolated PS I complexes in (10 m$M$ MOPS, pH 7.0, 0.05% Triton X-100) containing 10 $\mu$g chlorophyll in the presence of 50 $\mu$M 3-(3,4-dichlorophenyl)-1,1-dimethylurea (DCMU), 1 m$M$ ascorbic acid, 1 m$M$ 3,6-diaminodurene, and 2 m$M$ methyl viologen.[27] In this reaction, PS II activity is inhibited by DCMU while ascorbate donates electrons to PS I via 3,6-diaminodurene. Methyl viologen accepts electrons from PS I and donates them in turn to oxygen, thereby reducing oxygen concentration in the reaction mixture. After calculating the slopes of lines, the oxygen evolution or uptake rates are calibrated by measuring the oxygen consumption of air-saturated water by sodium dithionite.

[23] S. Bauminger and M. Wilchek, *Methods Enzymol.* **70,** 151 (1980).
[24] C. Lelong, P. Setif, B. Lagoutte, and H. Bottin, *J. Biol. Chem.* **269,** 10034 (1994).
[25] A. H. Mehler, *Arch. Biochem. Biophys.* **34,** 339 (1951).
[26] A. H. Mehler, *Arch. Biochem. Biophys.* **33,** 65 (1951).
[27] V. P. Chitnis, Q. Xu, L. Yu, J. H. Colbeck, H. Nakamoto, D. L. Xie, and P. R. Chitnis, *J. Biol. Chem.* **268,** 11678 (1993).

## $NADP^+$ Photoreduction

The reductase activity of PS I can be estimated in a coupled enzyme assay that monitors flavodoxin-mediated or ferredoxin-mediated $NADP^+$ photoreduction with ferredoxin : $NADP^+$ oxidoreductase. Isolated photosynthetic membranes at a concentration of 5 $\mu$g chlorophyll/ml are used for the measurement. Rates of flavodoxin-mediated $NADP^+$ photoreduction are measured in a 3-ml volume using 15 $\mu M$ flavodoxin in 50 m$M$ Tricine, pH 8.0, 10 m$M$ MgCl$_2$, 15 $\mu M$ cytochrome $c_6$, 6 m$M$ sodium ascorbate, 0.05% $n$-dodecyl-$\beta$-maltoside with 0.5 m$M$ $NADP^+$, and 0.8 $\mu M$ ferredoxin : $NADP^+$ oxidoreductase. Rates of ferredoxin-mediated $NADP^+$ photoreduction are measured under the same conditions except for the substitution of 5 $\mu M$ ferredoxin for 15 $\mu M$ flavodoxin. Both measurements are made by monitoring the rate of change in the absorption of NADPH at 340 nm (the extinction coefficient is 6.22 m$M^{-1}$ cm$^{-1}$). The rates are determined using a Cary 219 or a Shimadzu 160A spectrophotometer fitted with appropriate narrowband and interference filters attached to the surface of the photomultiplier. The samples are illuminated using banks of high-intensity, red light-emitting diodes (LS1, Hansatech Ltd.). The light intensity is saturating at the chlorophyll concentration used. In wild-type membranes, the rates of ferredoxin- and flavodoxin-mediated $NADP^+$ photoreduction are 460 and 480 $\mu$mol/mg chlorophyll per hour, respectively.[18]

## P700 Estimation

The estimation of P700 involves the measurement of absorption changes at 700 or 430 nm induced by chemical oxidation and reduction.[28,29] A Cary 219 or a Shimadzu 160A spectrophotometer is used to measure the oxidized minus reduced spectra. Purified PS I complexes and thylakoid membranes in 10 m$M$ MOPS, pH 7.0, and 0.05% Triton X-100 are adjusted to 25 $\mu$g of chlorophyll/ml (giving an absorbency of about 2 at 680 nm). Two samples in identical cuvettes are used to record the baseline. Then, one sample is oxidized with ferricyanide (1 m$M$) and the other is reduced with ascorbic acid (2 m$M$). After equilibration, a difference spectrum is recorded and the P700 concentration is determined using the extinction coefficients: 64 m$M^{-1}$ cm$^{-1}$ at 700 nm or 44 m$M^{-1}$ cm$^{-1}$ at 430 nm.

In addition to producing an absorbance change at 700 nm, the oxidation of P700 also results in an absorbance increase around 820 nm.[30] This ab-

[28] B. Kok, *Biochim. Biophys. Acta* **48**, 527 (1961).
[29] T. V. Marsho and B. Kok, *Methods Enzymol.* **69**, 280 (1980).
[30] B. Kok, *Nature* **179**, 583 (1957).

TABLE III

SUBUNIT-DEFICIENT PHOTOSYSTEM I MUTANTS AND SUBUNIT-SPECIFIC ANTIBODIES IN *SYNECHOCYSTIS* SP. PCC 6803

| Subunit/gene | Gene features | | | Subunit-specific antibodies | | Subunit-deficient mutants | | | |
|---|---|---|---|---|---|---|---|---|---|
| | Size (bp) | Expression | Location in genome | Rabbit polyclonal antibody against | Ref. | Mutant strain | Genotype | Phenotype | Ref. |
| PsaA/*psaA* | 2256 | With *psaB* | 941686–943941 | Overproduced fusion protein containing *N*-terminal sequence | *a* | PS I-Less | 3' End of *psaA* and 5' end of *psaB* replaced by Cam^r | PS I missing; no photoautotrophic growth | *b* |
| PsaB/*psaB* | 2196 | With *psaA* | 944187–946382 | Overproduced fusion protein containing an extramembrane sequence | *a* | BDK8 | *psaB* gene partially replaced by Kan^r | PS I missing; no photoautotrophic growth | *c* |
| | | | | | | ΔB-RCPT | *psaB* gene partially replaced by Kan^r | PS I missing; no photoautotrophic growth | *d* |
| | | | | | | JUN1 | *psaB* region corresponding to C-terminal domain replaced by Kan^r | PS I missing; no photoautotrophic growth | *e* |
| PsaC/*psaC* | 246 | Monocistronic | 2287579–2287334 | Overproduced protein | *f* | CDK25 | *psaC* interrupted by Kan^r | PsaC, PsaD, PsaE missing; no photoautotrophic growth | *g* |
| PsaD/*psaD* | 426 | Monocistronic | 126639–127064 | Overproduced protein | *f* | ADC4 | *psaD* replaced by Cam^r | No Fd-mediated electron transfer | *h–k* |
| PsaE/*psaE* | 225 | Monocistronic | 1982049–1982273 | Overproduced protein | *f* | AEK2 | *psaE* replaced by Kan^r | Reduced Fd-mediated NADP^+ photoreduction | *k,l* |
| PsaF/*psaF* | 498 | With *psaJ* | 1688053–1687556 | Native protein from polyacrylamide gel | *m* | AFK6 | *psaF* replaced by Kan^r; *psaJ* transcriptionally inactivated | PsaF, PsaJ missing; PsaE decreased; normal rate of P700 reduction by cyt $c_6$ | *f,k,n* |

136

| Gene/protein | No. | Overexpression | GenBank | Protein source | Ref | Mutant | Construct | Phenotype | Ref |
|---|---|---|---|---|---|---|---|---|---|
| PsaI/psaI | 123 | With psaL | 3458023–3458145 | Overproduced fusion protein | o | AIC9 | psaI interrupted by Cam$^r$ | PsaI, PsaL⁻ missing; small decrease in Fd-mediated NADP$^+$ photoreduction | o |
| PsaI/psaI | 123 | With psaF | 1687448–1687326 | Not available | | AJC8 | psaI replaced by Cam$^r$ | PsaF easily lost during detergent treatment | m |
| PsaK1/psaK1 | 261 | ND | 156391–156651 | Native protein from poly-acrylamide gel | p | AK1C | psaK1 interrupted by Cam$^r$ | | p |
| PsaK2/psaK2 | 387 | ND | 3322763–3322377 | | | AK2S | psaK2 replaced by Spec$^r$ | | p |
| PsaL/psaL | 474 | With psaI | 3457459–3457932 | Native protein from poly-acrylamide gel | j | ALC7 | psaL interrupted by Cam$^r$ | No PS I trimers formed | k,q,r |
| PsaM/psaM | 96 | ND | 467201–467296 | Not available | | AMS | psaM interrupted by Strep$^r$ | Less PS I activity, reduced growth at high light intensity | p |

[a] J. Sun, Q. Xu, V. P. Chitnis, P. Jin, and R. P. Chitnis, J. Biol. Chem. 272, 21793 (1997).
[b] S. Boussiba and W. F. J. Vermaas, in "Research in Photosynthesis" (N. Murata, ed.), p. 429. Kluwer, Dordrecht, 1992.
[c] L. B. Smart and L. McIntosh, Plant Mol. Biol. 21, 177 (1993).
[d] L. B. Smart, P. V. Warren, J. H. Golbeck, and L. McIntosh, Proc. Natl. Acad. Sci. U.S.A. 90, 1132 (1993).
[e] J. Sun and P. R. Chitnis, unpublished results.
[f] Q. Xu, L. Yu, V. P. Chitnis, and P. R. Chitnis, J. Biol. Chem. 269, 3205 (1994).
[g] L. Yu, L. B. Smart, Y. S. Jung, J. Golbeck, and L. McIntosh, Plant Mol. Biol. 29, 331 (1995).
[h] P. R. Chitnis, P. A. Reilly, and N. Nelson, J. Biol. Chem. 264, 18381 (1989).
[i] Y. Cohen, V. P. Chitnis, R. Nechushtai, and P. R. Chitnis, Plant Mol. Biol. 23, 895 (1993).
[j] Q. Xu, T. S. Armbrust, J. A. Guikema, and P. R. Chitnis, Plant Physiol. 106, 1057 (1994).
[k] Q. Xu, Y. S. Jung, V. P. Chitnis, J. A. Guikema, J. H. Golbeck, and P. R. Chitnis, J. Biol. Chem. 269, 21512 (1994).
[l] P. R. Chitnis, P. A. Reilly, M. C. Miedel, and N. Nelson, J. Biol. Chem. 264, 18374 (1989).
[m] Q. Xu, W. R. Odom, J. A. Guikema, V. P. Chitnis, and P. R. Chitnis, Plant Mol. Biol. 262, 291 (1994).
[n] P. R. Chitnis, D. Purvis, and N. Nelson, J. Biol. Chem. 266, 20146 (1991).
[o] Q. Xu, D. Hoppe, V. P. Chitnis, W. R. Odom, J. A. Guikema, and P. R. Chitnis, J. Biol. Chem. 270, 16243 (1995).
[p] P. R. Chitnis, unpublished results.
[q] V. P. Chitnis, Q. Xu, L. Yu, J. H. Golbeck, H. Nakamoto, D. L. Kie, and P. R. Chitnis, J. Biol. Chem. 268, 11678 (1993).
[r] V. P. Chitnis and P. R. Chitnis, FEBS Lett. 336, 330 (1993).

sorbance increase around 820 nm can also be produced by the oxidation of P680. The slower P700 reduction and the faster P680 reduction allows discrimination of the absorbance changes by using a system with a time resolution of less than 1 $\mu$s, The Hansatech P700 accessory kit can be used for P700 estimation in isolated membranes or purified complexes by monitoring the absorbance changes at 820 nm by light-induced P700 oxidation.

## Molecular Genetic and Biochemical Resources

Due to its simplicity and similarity to higher plants, *Synechocystis* sp. PCC 6803 has been extensively used as a model system to study photosynthetic proteins. These studies make a rich resource for further studies. Many subunit-deficient PS I mutants have been developed and many subunit-specific, even domain-specific, antibodies have been raised in recent years (Table III). Finally, the completely sequenced genome of *Synechocystis* sp. PCC 6803 makes another great resource for the studies on this cyanobacterium. This achievement will greatly benefit molecular genetics research. The information of the genome can be easily obtained from the web page at http://pka3.kazusa.or.jp/cyano/cyano.html.

### Recipient Strains that Lack Specific Photosystem I Subunits

Subunit-deficient PS I mutants are made by partially or completely replacing the target gene with an antibiotic resistance gene or by inserting an antibiotic resistance gene into the target gene to interrupt the target subunit (Table III). Traditionally, the genes were cloned by construction of the genomic DNA libraries and its screening using heterologous probes or labeled oligonucleotides. Now, with the completely sequenced genome of *Synechocystis* sp. PCC 6803, the required genes can be easily cloned by amplification with polymerase chain reaction (PCR). For gene replacement, a target gene is deleted by digestion and the upstream and downstream sequences are ligated with the antibiotic resistance gene replacing the gene under study. For gene interruption, the antibiotic resistance gene is introduced into a site in the target gene. Constructions are introduced back into the *Synechocystis* sp. PCC 6803 genome through natural transformation and homologous recombination. Recipient strains are tested by PCR, Southern blotting, Northern blotting, and/or Western blotting to ensure that the gene is inactivated and the protein is absent in the mutant strain.

### Antibodies

Subunit-specific rabbit polyclonal antibodies have been raised using native proteins from polyacrylamide gel or overproduced proteins in *Esche-*

*richia coli.* Several domain-specific rabbit polyclonal antibodies in PsaA and PsaB have also been generated using overexpressed fusion proteins as antigens. The specificity of the antibodies is tested by Western blotting using both isolated thylakoid membranes and purified PS I complex resoluted by electrophoresis.

## Acknowledgments

This work is supported part by grants from the National Science Foundation (MCB 9723001), the National Institutes of Health (GM53104), and the U.S. Department of Agriculture–National Research Initiative Competitive Grants Program (USDA-NRICGP 92-37306-7661); Journal Paper J-17318 of the Iowa Agriculture and Home Economics Experiment Station, Ames, Iowa; Projects 3416 and 3431 and supported by Hatch Act and State of Iowa funds.

## [9] Delineation of Critical Regions of the ε Subunit of the Chloroplast ATP Synthase through a Combination of Biochemical and Site-Directed Mutagenesis Approaches

*By* RICHARD E. MCCARTY and JEFFREY A. CRUZ

## Introduction

The chloroplast adenosine triphosphate (ATP) synthase ($CF_1$–$CF_0$) consists of nine different proteins and a total of about 20 polypeptide chains.[1] The catalytic portion of the synthase ($CF_1$) is water soluble once it is removed from the thylakoid membrane and is very amenable to study by biochemical means. One gram or more of $CF_1$ can be prepared from a readily available source—market spinach leaves—in a week's work. Although $CF_1$ in solution cannot catalyze net ATP synthesis, it has properties that mimic $CF_1$ bound to $CF_0$ in thylakoids. Among them is that the rate of ATP hydrolysis by either membrane bound or soluble $CF_1$ is very low.[2]

The rate of ATP synthesis by illuminated thylakoid membranes can be as high as 5 $\mu$mol min$^{-1}$ mg membrane protein$^{-1}$. Because $CF_1$ is about 10% of the thylakoid membrane protein, this rate corresponds to 50 $\mu$mol min$^{-1}$ mg $CF_1^{-1}$. In contrast, the rate of ATP hydrolysis by $CF_1$ in thylakoids in the dark is less than 1% of that of ATP synthesis in the light. Clearly,

---

[1] D. R. Ort and K. Oxborough, *Annu. Rev. Plant Physiol. Plant Mol. Biol.* **43**, 269 (1992).
[2] R. E. McCarty and E. Racker, *J. Biol. Chem.* **243**, 829 (1968).

the ATP synthase is switched on in the light and off in the dark. In this way, wasteful ATP hydrolysis by the synthase in the dark is prevented.[1]

The ATPase activity in $CF_1$ in solution may be activated 50-fold or more by a number of treatments that either promote the dissociation of (alcohols and detergents) or denature (heat) the $\varepsilon$ subunit. Modification of the $\gamma$ subunit also causes ATPase activation. Reduction of the $\gamma$ disulfide decreases $\varepsilon$ binding[3] and trypsin cleavage of $\gamma$ within a regulatory domain abolishes $\varepsilon$ binding.[3,4] Alterations in $\varepsilon-\gamma$ interactions by formation of the electrochemical proton gradient by photoelectron transfer are likely involved in activation in the light.

The $\varepsilon$ subunit of $CF_1$ may be specifically removed from the complex by treating $CF_1$ bound to DEAE-cellulose with a buffered ethanol/glycerol mixture. Both $\varepsilon$ and $CF_1$ deficient in $\varepsilon$ $[CF_1(-\varepsilon)]$ can be obtained in stable, active forms.[5] $CF_1(-\varepsilon)$ has high ATPase activity that may be inhibited to control levels by addition of purified $\varepsilon$. Also, $CF_1(-\varepsilon)$ binds tightly and specifically to $CF_0$, but does not block proton conductance by $CF_0$ unless $\varepsilon$ is added.[5]

The $\varepsilon$ subunit of *Escherichia coli* $F_1$ ($ECF_1$) is also an ATPase inhibitor[6] and has been studied extensively. Truncation and point mutations of $\varepsilon$ were generated and tested for their ability to confer to an $\varepsilon$-deficient strain the ability to grow on a nonfermentable carbon source.[7,8] Although *E. coli* is clearly the organism of choice for genetic studies, the chloroplast system offers the advantages of ease of preparation of $CF_1$ and that $\varepsilon$ is not required for binding to $CF_0$.

The molecular analysis of a protein component of an oligomeric complex can be especially effective when efficient means for depleting the component specifically and reconstituting it have been developed. In this article, we describe methods for the removal of the $\varepsilon$ subunit from $CF_1$ as well as assays of $\varepsilon$ function, ATPase inhibition, and inhibition of proton conductance of $CF_1$ in membranes reconstituted with $CF_1(-\varepsilon)$. Procedures used to subclone and overexpress *atpE,* the gene for $\varepsilon$, in *E. coli* are outlined. A rapid procedure for obtaining active $\varepsilon$ from inclusion bodies is given, as well as the approaches used to obtain several mutant $\varepsilon$. A less effective method for obtaining overexpressed $\varepsilon$ has been reported.[9]

[3] P. Soteropoulous, K-H. Süss, and R. E. McCarty, *J. Biol. Chem.* **267,** 10348 (1992).

[4] K. E. Hightower and R. E. McCarty, *Biochemistry* **35,** 4846 (1996).

[5] M. L. Richter, W. J. Patrie, and R. E. McCarty, *J. Biol. Chem.* **259,** 7371 (1984).

[6] P. C. Sternweis and J. B. Smith, *Biochemistry* **19,** 526 (1980).

[7] M. Jounouchi, M. Takeyama, T. Noumi, Y. Moriyama, M. Maeda, and M. Futai, *Arch. Biochem. Biophys.* **292,** 87 (1992).

[8] D. J. LaRoe and S. B. Vik, *J. Bacteriol.* **174,** 633 (1992).

[9] D. Steinmann, H. Lill, W. Junge, and S. Englebrecht, *Biochim. Biophys. Acta,* **1187,** 354 (1994).

## Methods

### Subcloning of atpE

For both the TA cloning (Invitrogen, Carlsbad, CA) and the subcloning into pET3c (Novagen, Madison, WI) the DH5α strain of *E. coli* is used as the transformation host. Ampicillin at 150 μg/ml is used throughout for selection of plasmid-bearing cells. Plating is done on selective LB agar. Cultures for plasmid purification are grown in selective 2× YT broth.[10]

The *atpE* gene is amplified by polymerase chain reaction (PCR) from pMSD1, a plasmid bearing an intact copy of the gene.[11] The forward primer (5′-TTTCATATGACCTTAAATCTTTGTTACTGACTCCG-3′) includes the start codon as part of an *Nde*I restriction site, whereas the reverse primer (5′-TCGGGATCCTTACGAGGAAATCGTATTGCG-3′) includes a *Bam*HI site immediately following the stop codon. The PCR product is subcloned into the TA cloning vector, pCRII, and the integrity of the *atpE* gene is confirmed by DNA sequencing. The pCRII-*atpE* construct is digested with *Nde*I and *Bam*HI and the fragment containing the *atpE* gene is subcloned in frame into the pET3c expression vector. This plasmid is denoted pETε1.

### Expression of atpE

pETε1 is used to transform the *E. coli* BL21DE3 expression host. Transformed cells are plated on selective LB agar. For all manipulations the selective agent ampicillin (150 μg/ml) is used. A single colony is used to inoculate two 5-ml cultures of selective LB broth. After growth overnight (to stationary phase) at 37°, the two cultures are used to inoculate 1 liter of selective LB broth. The 1-liter culture is incubated at 37° with shaking. After the culture reaches an A at 550 nm of from 0.3 to 0.5, isopropylthiogalactoside is added to a final concentration of 0.3 m$M$. The culture is then incubated at 37° for an additional 2–3 hr. Cells are harvested by centrifugation at 7000$g$ at 4° for 5 min. Pellets are resuspended in 20 ml of 50 m$M$ Tris-HCl, pH 8.0, 100 m$M$ NaCl, 1 m$M$ ethylenediaminetetraacetic acid (EDTA) (TNE). The suspension is stored at −80°.

### Purification of ε Inclusion Bodies

Cell suspensions are thawed in ice-chilled water bath, and the cells are lysed by a single pass through a French press at 20,000 psi and 4°. The

[10] J. Sambrook, E. F. Fritsch, and T. Maniatis, *in* "Molecular Cloning. A Laboratory Manual." Cold Spring Harbor Laboratory Press, New York, 1989.

[11] J. A. Cruz, B. Harfe, C. A. Radkowski, M. S. Dann, and R. E. McCarty, *Plant Physiol.* **109**, 1379 (1995).

lysate is centrifuged for 10 min at 7000g and 4°. Recombinant ε is found in the insoluble fraction, presumably in the form of inclusion bodies. The pellet is resuspended in 20 ml TNE with 0.1% (w/v) sodium deoxycholate. The suspension is centrifuged as described earlier, and the pellet resuspended in 20 ml TNE with 1% (v/v) with IPEGAL (Sigma, St. Louis, MO). The suspension is centrifuged as described earlier and the pellet resuspended in 20 ml TNE. Following a final centrifugation, the inclusion bodies are resuspended in 10 ml of TNE and stored at −20° in 1-ml aliquots. Typical yields of recombinant ε range between 10 and 40 mg per liter of culture.

## Solubilization and Folding of Recombinant ε

Inclusion body suspensions (1 ml) are thawed and centrifuged at 16,000g in a microcentrifuge. The pellet is resuspended in 1 ml of deionized water and the suspension centrifuged as described earlier. The pellet is dissolved in 100–200 μl of 25 mM Tris-HCl, pH 8.0, 8 M urea, 5 mM dithiothreitol (DTT), and a small amount of insoluble material is pelleted by centrifugation, as was done earlier. The urea solution must be freshly prepared. Protein concentration is measured[12] by subtracting the contributions of the urea to the background. This method also underestimates the concentration of ε by 43%. Corrected values are used.

Direct dilution has been used successfully for the refolding of small globular proteins.[13] Rapid dilution of ε in urea solution is more effective in restoring activity to ε than is slow removal of the urea by dialysis. Urea-solubilized ε (0.3–4 mg/ml) is diluted (1:10) rapidly with vortex mixing into an ice-chilled dilution buffer consisting of 5 volumes 50 mM Tris-HCl, pH 8.0, 40% (v/v) ethanol, 3 volumes glycerol, and 1 volume deionized water. For the recombinant wild-type 2-mercaptoethanol is added to a final concentration of 10 mM. Ethanol and glycerol stabilize ε purified from spinach for storage at 4°.[5] Although some work suggests that these components are not absolutely required for folding,[9,14] both appear to enhance folding and stabilize the folded product.[11] Recombinant ε in urea solution stored at room temperature retains its ability to fold upon dilution into ε buffer for at least several hours. After dilution the recombinant ε is stable for at least 24 hr at 4°.

[12] M. M. Bradford, *Anal. Biochem.* **72**, 248 (1976).
[13] J. L. Cleland, S. E. Builder, J. R. Swartz, M. Winkler, J. Y. Chang, and D. I. C. Wang, *Biotechnology* **10**, 1013 (1992).
[14] C. A. Radkowski, J. A. Cruz, and R. E. McCarty, unpublished observations (1997).

*Preparation of $CF_1(-\varepsilon)$ and $\varepsilon$*

(Procedure adapted from Richter *et al.*)[5] Preswollen DEAE-cellulose from Sigma (about 10 g) is defined twice with deionized water and then once with the equilibration buffer (25 m$M$ Tris-HCl, pH 8.0, 0.5 m$M$ ATP, 5 m$M$ DTT). A 1.5- × 2.5-cm column is equilibrated at room temperature with 5 volumes of the equilibration buffer. $CF_1$ is prepared as described[3] with modifications.[11] An ammonium sulfate precipitate of $CF_1$ (20–30 mg) is centrifuged for 10 min at 12,000$g$ and 4°. The pellet is dissolved in 20–30 ml of the equilibration buffer and solid DTT was added to a final concentration of 50 m$M$. After 2 hr at room temperature the solution was diluted to 80–100 ml with the equilibration buffer.

The solution is loaded slowly (approximately 1 drop per second) onto the column. The column is washed with 5 volumes of the equilibration buffer. Then the column is washed with cold $\varepsilon$ extraction buffer [25 m$M$ Tris-HCl, pH 8.0, 20% (v/v) ethanol, 30% (v/v) glycerol, 40 m$M$ NaCl, 0.5 m$M$ ATP, 5 m$M$ DTT]. The first 45 ml of the eluted $\varepsilon$ is collected in 5-ml fractions and stored on ice. After passing approximately 200 ml of the $\varepsilon$ extraction buffer through the column, the column is then washed with 10 volumes of 25 m$M$ Tris-HCl and 0.5 m$M$ ATP at room temperature. $CF_1$ now lacking the $\varepsilon$ subunit [$CF_1(-\varepsilon)$] is eluted with 25 m$M$ Tris-HCl, pH 8.0, 0.4 $M$ NaCl, and 0.5 m$M$ ATP.

*Notes on Storage and Handling*

The purified $\varepsilon$ can be stored at 4° in the buffer used for its extraction without much loss in activity over a period of weeks.[5] The $\varepsilon$ solutions can be concentrated by ultrafiltration, although during storage aggregates may form at concentrations of greater than 0.3–0.5 mg/mL.[3]

$CF_1(-\varepsilon)$ can be stored at 4° as an ammonium sulfate precipitate for many weeks, using conditions identical to that of the storage of $CF_1$. Prior to use in assays or for reconstitution with NaBr treated thylakoids, $CF_1(-\varepsilon)$ was desalted, usually in 5 m$M$ Tris-HCl, pH 8.0, or 5 m$M$ Tricine-NaOH, pH 8.0, using the method of Penefsky.[15] Desalted $CF_1(-\varepsilon)$ is stable for days at room temperature in the presence of 1 m$M$ ATP.

*Preparation of $\varepsilon$-Deficient Thylakoids*

$CF_1(-\varepsilon)$ may be reconstituted specifically with thylakoids depleted of $CF_1$.[5] This ability enables measurement of two of the activities $\varepsilon$ has as a part of the fully assembled ATP synthase: (1) inhibition of the ATPase activity of the ATP synthase and (2) restriction of proton flow through $CF_0$.

[15] H. S. Penefsky, *J. Biol. Chem.* **252**, 2891 (1977).

Treatment of thylakoid membranes with NaBr dissociates $CF_1$ and leaves $CF_0$ and much of the electron transport chain intact. The procedure for the preparation of NaBr-treated thylakoids given next is modified from that reported by Nelson and Eytan.[16] To spinach thylakoid membranes [100 mg of chlorophyll in 60 ml of cold 0.4 $M$ sucrose, 0.02 $M$ Tricine-NaOH (pH 8.0), 0.05 $M$ NaCl, and 0.005 $M$ DTT] add 40 mL of cold 5 $M$ NaBr with gentle stirring. The suspension is incubated on ice and in the dark for 30 min with occasional stirring. An equal volume of cold deionized water is added and the mixture centrifuged for 30 min at 15,000$g$ and 4°. The pellet is completely resuspended in cold 10 m$M$ Tricine-NaOH (pH 8.0) in a glass-Teflon homogenizer. The suspension is filtered through a single layer of Pellon grade 930 interfacing (available at fabric stores) to remove particles resistant to resuspension and then diluted to 1 liter with cold 10 m$M$ Tricine-NaOH (pH 8.0). The suspension is centrifuged for 40 min at 15,000$g$ and 4° and the membrane pellets resuspended using a glass-Teflon homogenizer and no more than 20 ml of cold buffered sucrose-DTT solution. The NaBr-treated thylakoids are stored at $-80°$ in small aliquots at 1–2 mg of chlorophyll/mL in the same solution. The membranes remain stable for months.

Reconstitution of $CF_1(-\varepsilon)$ with NaBr-treated thylakoids as described by Cruz et al.[11] gives less consistent results than the modified procedure reported here. NaBr-treated thylakoids equivalent to 600 $\mu$g of chlorophyll are mixed with 3 mg of $CF_1(-\varepsilon)$ in 5 m$M$ Tricine-NaOH, pH 8.0, and with 150 $\mu$l of 0.1 $M$ ATP, pH 7.0. The volume is adjusted to 1.45 ml with 5 m$M$ Tricine-NaOH, pH 8.0. After 5 min at room temperature, 45 $\mu$l of 0.5 $M$ $MgCl_2$ is added. The mixture is kept at room temperature for 5 min and then an equal volume of cold STN [0.4 $M$ sucrose, 0.02 $M$ Tricine-NaOH (pH 8.0), 0.05 $M$ NaCl] with 2 m$M$ ATP and 7 m$M$ $MgCl_2$ is added. After 5 min on ice, the mixture is centrifuged for 5 min at 27,000$g$ and 4°. The supernatant is aspirated and the pellet gently resuspended in 3 ml of cold STN with 5 m$M$ $MgCl_2$ (STN–$Mg^{2+}$). After centrifugation (as above), the pellet is resuspended in a small volume of cold STN–$Mg^{2+}$ and the chlorophyll concentration determined. Stored covered and on ice, the $\varepsilon$-deficient membranes lost little activity over at least 4 hr.

*Reconstitution of $CF_1(-\varepsilon)$ in Solution with $\varepsilon$*

$CF_1(-\varepsilon)$ is desalted into 20 m$M$ Tricine-NaOH (pH 8.0). $CF_1(-\varepsilon)$ (20 $\mu$g in 320 $\mu$l) is mixed with 160 $\mu$l of recombinant or native $\varepsilon$ at various concentrations. For wild-type recombinant $\varepsilon$, concentrations used range

---

[16] N. Nelson and E. Eytan, in "Cation Fluxes Across Biomembranes" (Y. Mukahata and L. Packer, eds.), p. 409. Academic Press, New York.

from 3.45 to 34.5 $\mu$g/ml, which yields molar ratios of $\varepsilon : CF_1(-\varepsilon)$ of 0.75 : 1 to 7.5 : 1. The $M_r$ of $\varepsilon$ is 14,700. Using more concentrated $\varepsilon$ solutions, molar ratios of $\varepsilon$ to $CF_1$ in excess of 30 to 1 may be readily achieved. As a control, 320 $\mu$l $CF_1(-\varepsilon)$ is mixed with 160 $\mu$l of $\varepsilon$ buffer (as above). ATPase activity in aliquots equivalent to 5 $\mu$g of $CF_1$ is determined. Figure 1 shows ATPase inhibition by recombinant $\varepsilon$.

*Membrane Studies*

NaBr-treated thylakoids reconstituted with $CF_1(-\varepsilon)$ are diluted with STN–$Mg^{2+}$ to a concentration of 5 $\mu$g chlorophyll per 95 $\mu$l. For typical experiments, 570 $\mu$l of diluted membranes are mixed with 30 $\mu$l folded $\varepsilon$ at various concentrations. For wild-type $\varepsilon$ concentrations range from 0.011 to 0.441 mg/ml. On reconstitution, these concentrations yield molar ratios of $\varepsilon : CF_1(-\varepsilon)$ of 0.75 : 1 to 30 : 1. For a control 570 $\mu$l of membranes is mixed with 30 $\mu$l of control buffer.

*ATPase Activity Assay*

In the absence of $Mg^{2+}$, ATPase activity of $CF_1$ can be measured in the presence of $Ca^{2+}$. When $Mg^{2+}$ is present, $Ca^{2+}$ ATPase activity is inhibited. High rates of $Mg^{2+}$ ATP hydrolysis are obtained in the presence of sulfite. For $Ca^{2+}$ ATPase activity NaBr-treated thylakoids (5 $\mu$g of chlorophyll)

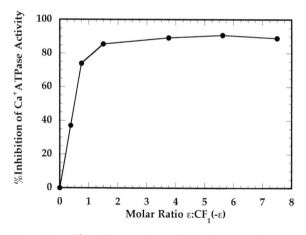

FIG. 1. Inhibition of the $Ca^{2+}$ ATPase activity of $CF_1(-\varepsilon)$ by recombinant $\varepsilon$. By procedures outlined in this article, Cys6 of $\varepsilon$ was mutated to Ser and the recombinant $\varepsilon$ expressed in *E. coli*. A rapid, 10-fold dilution of an 8 *M* urea solution was used to fold the mutant $\varepsilon$. $CF_1(-\varepsilon)$ (20 $\mu$g) was incubated with the recombinant $\varepsilon$ at the amounts shown and $Ca^{2+}$ ATPase activity assayed. (Previously unpublished data of J. A. Cruz (1996).)

are incubated at 37° in 50 m$M$ Tris-HCl, pH 8.0, 5 m$M$ CaCl$_2$, 5 m$M$ ATP, and in a final volume of 1 mL. For Mg$^{2+}$ ATPase activity, samples are incubated at 37° in 50 m$M$ Tris-HCl, pH 8.0, 2 m$M$ MgCl$_2$, 100 m$M$ Na$_2$SO$_3$, and 4 m$M$ ATP in a final volume of 1 mL. After 10 min reactions are stopped by the addition of 1 ml of cold 0.5 $N$ trichloroacetic acid. Phosphate release is measured spectrophotometrically.[17]

*Proton Permeability*

The quenching of the fluorescence of 9-amino-6-chloro-2-methoxy-acridine (ACMA) caused by cyclic photoelectron flow around photosystem I may be used as an indirect measure of the proton permeability of the NaBr-treated membranes. ACMA fluorescence quenching[18] is measured in 50 m$M$ Tricine-NaOH (pH 8.0), 50 m$M$ NaCl, 5 m$M$ $N$-methylphenazonium methosulfate salt, 1 $\mu M$ ACMA, 2 m$M$ ascorbate-NaOH, pH 6.0, and membranes equivalent to 5 $\mu$g of chlorophyll. The reaction volume is 1 ml. Ascorbate is added to the cuvette just before assay. As a positive control for gradient formation, $N,N'$-dicyclohexylcarbodiimide, which blocks proton conductance by CF$_0$, is added to the cuvette to a final concentration of ~50 $\mu M$.

ACMA fluorescence at 475 nm is measured in a RF-5000 Shimadzu (Columbia, MD) spectrofluorimeter with an excitation wavelength of 410 nm. Illumination within the fluorimeter with high intensity red light was achieved as reported.[5] In some cases neutral density filters were used to reduce the intensity of the fluorescence excitation beam. The excitation beam can result in slow, but significant, quenching by driving electron and proton transport. Table I shows reconstitution of ACMA fluorescence quenching by various mutant $\varepsilon$s.

*Mutagenesis*

The strategy used for site-directed mutagenesis (briefly outlined later) was a primer-directed method called unique site elimination (for a review, see Deng and Nickoloff[19]). The protocols used were supplied by a kit from Clonetech. A mutagenized second strand of DNA was synthesized by extension off of two primers. The selection primer (5'-GCTTTACCGCTCGAGCCTCCCGCG-3') was designed to convert a unique *Pvu*II site on pET$\varepsilon$1 into a unique *Xho*I site, while the mutagenic primer (5'-CGGAGTCAGTACACTAAGATTTAAGGTC-3') encoded a

[17] H. H. Taussky and E. Shorr, *J. Biol. Chem.* **202**, 675 (1953).
[18] U. Pick and R. E. McCarty, *Methods Enzymol.* **69**, 538 (1980).
[19] W. P. Deng and J. A. Nickoloff, *Anal. Biochem.* **200**, 81 (1992).

TABLE I
RESTORATION OF ACMA FLUORESCENCE
QUENCHING IN ε-DEPLETED
THYLAKOIDS BY MUTANT $\varepsilon^a$

| ε Added | % Quenching of ACMA fluorescence |
|---------|--------------------------------|
| None | <5 |
| C6S | 75 |
| H37R | 53 |
| H37V | 28 |

[a] NaBr-extracted thylakoids were reconstituted with $CF_1(-\varepsilon)$ as described in the Methods section and reconstituted with the recombinant mutant ε folded by the rapid dilution method at a molar ratio of $\varepsilon:CF_1(-\varepsilon)$ of 30:1. Light-dependent quenching of ACMA fluorescence was then determined. Previously unpublished data of J. A. Cruz (1996).

single base change that would convert cys 6 of the protein primary sequence to ser. After transformation of a *mutS* strain of *E. coli,* a strain defective in mismatch repair, with the hybrid plasmid and subsequent plasmid isolation, a digestion of the preparation with *Pvu*II was used to select against the wild-type plasmid. The digest was then used to transform a standard *E. coli* cloning host (DH5α strain). Mutant plasmids could be screened by restriction digest using the newly introduced *Xho*I site, and mutation of the *atpE* sequence was confirmed by DNA sequencing. The mutated plasmid was denoted pETε2. Like the wild-type recombinant ε, the variant, ε-C6S, is expressed as inclusion bodies. When folded using the protocols described earlier (in the absence of reducing agents), ε-C6S was as active as ε purified from spinach.

Subsequent rounds of mutagenesis off of pETε2 were accomplished using a selection primer (5'-GACTGCTTTACCGCAGCTGCCTCGCGCG-3') that converted the unique *Xho*I site on this plasmid to *Pvu*II.

Conclusions

The mutagenesis/reconstitution approach has yielded interesting information about ε function and the importance of the N-terminal region of the protein. It was, for example, shown[11,20] that significant ATPase inhibition by

[20] J. A. Cruz, C. A. Radkowski, and R. E. McCarty, *Plant Physiol.* **113**, 1185 (1997).

some mutant $\varepsilon$ was not accompanied by the restoration of low proton permeability to thylakoids reconstituted with $CF_1(-\varepsilon)$. Deletion of just five residues from the N terminus abolishes the ability of $\varepsilon$ to inhibit the ATPase activity of $CF_1(-\varepsilon)$ in solution.[20]

The $\delta$ subunit has also been overexpressed[9] in active form. An excellent example of the power of the combination of *in vitro* reconstitution and site-directed mutagenesis is the introduction of Cys residues into $\delta$ at various points along the chain and the search for cross-links and formed by a photoactivable cross-linker attached covalently to these Cys residues.[21]

Recently, the $\gamma$ subunit of $CF_1$ has been overexpressed and obtained in a form that restores very high rates of ATP hydrolysis to $\alpha\beta$ preparations.[22] Thus, the way has been paved for the molecular analysis of the $\gamma$ that is involved in catalysis as well as regulation.

[21] H. Lill, F. Hensel, W. Junge, and S. Englebrecht, *J. Biol. Chem.* **271,** 32737 (1996).
[22] M. L. Richter, personal communication (1997).

# Section III

# Gene Expression of Photosynthetic Components

## [10] Transcriptional Regulation of Photosynthesis Operons in *Rhodobacter sphaeroides* 2.4.1

*By* JILL H. ZEILSTRA-RYALLS, MARK GOMELSKY,
ALEXEI A. YELISEEV, JESUS M. ERASO,
and SAMUEL KAPLAN

## I. Introduction

*Rhodobacter sphaeroides* 2.4.1 is a gram-negative bacterium that, in addition to its ability to carry out anoxygenic photosynthesis (PS), has the capacity to grow by both aerobic and anaerobic respiration, fermentatively, and lithotrophically. This metabolic versatility is accompanied by a complex network of gene regulation designed to effect the smooth transition from one growth mode to another. A reduction in oxygen tension is both necessary and sufficient to induce PS gene expression and synthesis of the *R. sphaeroides* photosystem. In addition to oxygen control, incident light intensity regulates the molecular composition and intracellular abundance of the various photosystems. Our studies of the regulation of PS operons and genes have focused on identifying factors mediating PS gene expression at the transcriptional and posttranscriptional levels.

The photosystem of *R. sphaeroides* 2.4.1 consists of three pigment–protein complexes (Table I), comprising the reaction center (RC), and two light-harvesting (LH) complexes, designated B800-850 (LHII) and B875 (LHI), based on their respective absorption maxima. In this article, we focus on the methodological approaches we have employed to study the transcriptional regulation of those operons encoding the structural polypeptides comprising these complexes, as well as those required for the biosynthesis of the photopigments, carotenoids (Crts), bacteriochlorophyll (Bchl), and bacteriopheophytin.

In the following sections, we describe various approaches that have been successfully utilized to identify and analyze factors involved in regulating transcription of the PS operons. We conclude with a summary of those regulatory genes that have been identified and thus far characterized using these approaches.

TABLE I

MOLECULAR COMPONENTS COMPRISING THE SPECTRAL COMPLEXES OF *Rhodobacter sphaeroides* 2.4.1

| Pigment–protein complex | Polypeptides (genes) | Other components (genes) |
|---|---|---|
| RC | L (*pufL*), M (*pufM*), H (*puhA*) | Bchl (*hem* and *bch* genes) Crt (*crt* genes) Bacteriopheophytin Ubiquinone Non-heme iron |
| B875 | PufB (*pufB*), PufA (*pufA*) | Bchl (*hem* and *bch* genes) Crt (*crt* genes) |
| B800-850 | PucB and PucA (*pucBA* operon) | Bchl (*hem* and *bch* genes) Crt (*crt* genes) |

## II. Approaches to Studying the Regulation of Photosynthesis Gene Expression

### A. Regulatory Mutant Isolation

Transcriptional regulation of the PS operons involves both activation and repression. Thus, it is of use to be able to identify mutations that either increase or decrease gene expression in response to oxygen, light, and metabolic redox. Here, we describe several approaches that have been developed—and demonstrated to work—for isolating such mutants. Details regarding general genetics methods with *Rhodospirillaceae* have been described previously.[1]

Typically, upstream regulatory sequences derived from the gene of interest are positioned between the $\Omega$ transcription–translation termination cassette and the promoterless reporter gene, creating a transcriptional fusion. In many instances detailed transcriptional information is not known about the particular gene(s) under study. Thus, one may choose to utilize variable amounts of upstream sequence such that a set of plasmids is created.[2,3]

To identify *cis* acting mutations among the mutants isolated, the plasmid DNA is isolated from the mutant and subsequently reintroduced into wild-type *R. sphaeroides*.[4,5] In this way, if the mutation leading to altered expression of the reporter gene is present on the plasmid, 100% of the exconjugants

[1] T. J. Donohue and S. Kaplan, *Methods Enzymol.* **204,** 459 (1991).
[2] J. K. Lee and S. Kaplan, *J. Bacteriol.* **174,** 1158 (1992).
[3] J. H. Zeilstra-Ryalls and S. Kaplan, *J. Bacteriol.* **177,** 6422 (1995).
[4] J. K. Lee and S. Kaplan, *J. Bacteriol.* **174,** 1146 (1992).
[5] J. Eraso and S. Kaplan, *J. Bacteriol.* **176,** 32 (1994).

should be able to grow on the selective media. To confirm the presence of a *trans*-acting mutation, the mutant candidates are cured of the transcriptional fusion plasmid by removal of the selective condition, and the parental plasmid is then reintroduced into the plasmid-free segregants.[2] If the mutation leading to higher expression of the reporter gene is *trans* to the plasmid, then the exconjugants of the initial segregants should be able to grow on the selective media in all cases, whereas exconjugants of the wild-type parent strain should not. These approaches have the added benefit of serving as a genetic backcross, which helps to purify the mutation in question.

*1. Isolation of Mutants with Increased Photosynthesis Gene Expression Using Transcriptional Fusions to aph.* To identify sequences that are involved in oxygen control of gene expression, the *aph* gene of Tn*903*, which encodes Km resistance, has been used to provide a positive selection for mutations that increase expression of the gene of interest (see Ref. 4). The transcriptional fusion plasmid(s) are introduced into the wild-type strain, and exconjugants are plated on media containing varying amounts of Km. Since the basal level of Km resistance can vary, depending on the gene under study, this would need to be titrated for any given transcriptional fusion. If expression of the gene of interest is negatively regulated by oxygen, for example, this basal level should be higher when exconjugants are grown under conditions of reduced oxygen tension.[2,4] Both *cis*- and *trans*-acting mutations can be isolated using this approach.

One can utilize a related approach to identify sequences involved in light regulation by plating exconjugants under photosynthetic conditions and varying light intensities. Matings can either be spread on plates or suspended in soft agar media containing selective concentrations of Km. Thus, mutations giving rise to increased resistance to Km, characteristic of cells grown at low light intensity, were isolated at high light intensity.[6] The stringency of these selections can be increased by increasing the concentration of Km.[6,7]

*2. Enrichment for and Isolation of Regulatory Mutants using lacZ Fusions.* We have developed an approach that is based on transcriptional fusion of a PS gene to *lacZ* (encoding β-galactosidase). This method is comprised of an enrichment for spontaneous mutations with altered levels of expression of a particular PS gene, and subsequent screening and isolation of the mutant strains.[8]

Enrichment is based on a selective growth advantage of cells having either a decrease in or loss of photosystem production under dark, semiaero-

[6] J. K. Lee and S. Kaplan, *J. Biol. Chem.* **270**, 20453 (1995).
[7] J. H. Zeilstra-Ryalls and S. Kaplan, *J. Bacteriol.* **177**, 2760 (1995).
[8] M. Gomelsky and S. Kaplan, *Microbiology* **141**, 1805 (1995).

bic or anaerobic conditions,[9] since the presence of the photosystem is gratuitous under these conditions. Photosystem production can decrease through down regulation of PS gene expression, which creates a pool of regulatory "down" mutations. On the other hand, the absence of a requirement for balanced expression of the various PS genes can result in mutations with an abnormally increased expression of one or more of the PS genes, thus creating a pool of putative regulatory "up" mutations.

A sensitive X-Gal (5-bromo-4-chloro-3-indolyl-$\beta$-D-galactoside) overlay technique can be used to monitor expression of the transcriptional fusion. Assays of *lacZ* expression can be performed on colonies of *R. sphaeroides* growing on agar medium under various conditions, e.g., aerobically, anaerobically, in presence or absence of light, or following a shift from one growth mode to another. After overlay with soft agar containing X-Gal, colonies that become blue colored either at a faster ("up" mutations) or slower ("down mutations") rate compared to the parental wild-type population are readily distinguished. High sensitivity of the assay affords identification of both regulatory "up" and "down" mutations within a wide range of $\beta$-galactosidase activities. Expression of the *lacZ* gene can be further used to monitor complementation of the isolated mutations by wild-type gene libraries in *trans*.[8,10]

*3. Isolation of Mutants with Decreased Photosynthesis Gene Expression using Transcriptional Fusions to sacB.* The *sacB* gene of *Bacillus subtilis* encodes levan sucrase, which, when expressed in *R. sphaeroides*, catalyzes the formation of toxic polymers in the presence of sucrose. Thus, the *sacB* gene can be used to select positively for mutants with decreased gene expression.[11,12] The plasmid bearing the transcriptional fusion between the PS gene and the promoterless *sacB* gene is then introduced into the appropriate strain, and exconjugants are subsequently plated on selective media containing sucrose. The basal level of *sacB* expression, as determined by the specific gene fusion under study, determines the concentration of sucrose to be used in the enrichment. From this sucrose challenge, both *cis*- and *trans*-acting mutations can be isolated as the result of survival to the sucrose challenge. For the various kinds of mutations that can be isolated, see Ref. 12.

A variation of the preceding approach can be employed, in which a promoterless *lacZ* gene is inserted between the upstream sequences of the

[9] M. Madigan, J. C. Cox, and H. Gest, *J. Bacteriol.* **150,** 1422 (1982).
[10] M. Gomelsky and S. Kaplan, *in* "Diversity, Genetics and Physiology of Photosynthetic Prokaryotes" (C. Bauer, J. Beatty, M. Madigan, J. Ormerod, and F. Tabita, eds.), p. 38. Bloomington, Indiana, 1996.
[11] J. Lee and S. Kaplan, unpublished results (1995).
[12] M. Sabaty and S. Kaplan, *J. Bacteriol.* **178,** 35 (1996).

gene of interest and the *sacB* gene. In this plasmid, *lacZ* and *sacB* are cotranscribed from the transcription initiation signals of the PS gene, and therefore *lacZ* expression can serve as a nonselective indicator by scoring sucrose-resistant candidates on media containing X-Gal as an indicator of $\beta$-galactosidase activity. This approach provides for a more easily identifiable spectrum of true regulatory mutations.[13] Furthermore, the use of this modification allows for a direct quantitation of the expression of the affected gene by assaying for $\beta$-galactosidase activity (see later discussion).

    *4. Isolation of Spontaneous Mutations by Analysis of Pigmentation and Photosynthetic Growth.* In addition to the use of extrinsic reporter genes, analysis of the intrinsic signals such as pigmentation can be used as a marker to identify photosystem regulators. Under standard conditions, the accumulation of pigments by a colony of *R. sphaeroides* results in a light pink to red pigmentation in wild-type cells grown semiaerobically, or dark red in cells subject to anaerobic growth. In most cases cells affected in pigment biosynthesis or accumulation show different coloration than wild-type cells. A pitfall of using pigmentation as a marker to identify regulators of PS gene expression is that mutations in either pigment biosynthesis genes as well as assembly components can lead to a decrease in pigmentation. Thus, the use of transcriptional fusions combined with an analysis of pigmentation is recommended.

    This approach is best illustrated in experiments that led to the identification of the two-component PRRB/PRRA regulatory system.[2,5,14] A *puc::aph* fusion in wild-type cells was used as previously described. These cells are also highly pigmented and thus the formation of the photosynthetic apparatus under aerobic conditions readily leads to the formation of secondary mutations with decreased pigmentation that are unable to grow photosynthetically. Analysis of PS gene expression in this mutant revealed that the second mutation occurred in a regulatory, and not in a structural gene. Complementation with a cosmid library of *R. sphaeroides* DNA led to the identification of the two-component activation system.

    The loss of pigmentation can additionally be exploited to identify mutations in a regulatory gene resident on a plasmid that complements a chromosomal knockout mutation.[5]

    *5. Mutagenesis.* The preceding protocols describe the isolation of spontaneously arising mutations. A variety of mutagenic approaches, site-directed[15] or randomized, are available and are beyond the scope of this discussion.

[13] M. Sabaty and S. Kaplan, unpublished results (1995).
[14] J. M. Eraso and S. Kaplan, *J. Bacteriol.* **177,** 2695 (1995).
[15] L. Gong, J. K. Lee, and S. Kaplan, *J. Bacteriol.* **176,** 2946 (1994).

Particularly useful here for identifying *trans*-acting factors regulating gene expression is the application of transposon mutagenesis. For *R. sphaeroides*, Tn5 derivatives have been used most frequently. A series of Tn5 derivatives is available that contains different antibiotic markers, such as $Tp^r$, $Km^r$, $Sp^r/Sm^r$, or $Tc^r$ for use in *R. sphaeroides*.[1,16] A plasmid bearing a transcriptional fusion to the promoterless *aph*, *lacZ*, or *sacB* gene is introduced into wild-type 2.4.1. Then, a plasmid-containing exconjugant is used as the recipient strain for the suicide vector bearing the appropriate transposon. Exconjugants from the second mating are plated on media containing the antibiotic resistance marker of the transposon, and expression of the transcriptionally based reporter gene is used in a nonselective screen.[17,18]

Confirmation of the presence of a *trans*-acting mutation requires segregation and reintroduction of the transcription fusion plasmid into the transposon mutant strain, as previously described. The advantage of this approach is that the locus affecting expression of the reporter system is now marked by the transposon.

## B. Regulatory Gene Identification

*1. Complementation of Regulatory Mutants.* An *R. sphaeroides* 2.4.1 cosmid library of approximately 800 clones has been constructed[19] using an IncP1, low copy number vector, pLA2917.[20] This vector is compatible with the *lacZ* transcriptional fusion plasmids, which are based on an IncQ incompatibility system. The cosmid library can be introduced into a presumptive regulatory mutant already carrying a transcriptional *lacZ* fusion to a particular PS gene. This allows one to monitor directly the effect of a cosmid on both *lacZ* expression (using the X-Gal overlay method) and restoration of other mutant phenotypes, e.g., pigmentation and photosynthesis growth.[8,10]

*2. Isolation of Suppressors.* An approach similar to the isolation of multicopy suppressors has been developed in *R. sphaeroides* for PS regulatory gene identification.[8] The rationale for this approach, referred to as *partial complementation*, relies on the ability of certain genes provided in extra copy through the use of a low copy number cosmid to partially restore the phenotype of a given regulatory mutant. Several mechanisms exist by

[16] C. Sasakawa and M. Yoshikawa, *Gene* **56,** 283 (1987).
[17] J. H. Zeilstra-Ryalls and S. Kaplan, *J. Bacteriol.* **178,** 985 (1996).
[18] J. M. Eraso and S. Kaplan, unpublished results (1997).
[19] S. C. Dryden and S. Kaplan, *Nucleic Acids Res.* **18,** 7267 (1990).
[20] L. N. Allen and R. S. Hanson, *J. Bacteriol.* **161,** 955 (1985).

which partial complementation can occur. One involves suppression of a deficiency in the factor acting upstream in a regulatory cascade by increased gene dosage of the downstream regulator.[8,14] Alternatively, extra copies of a regulatory factor acting through another regulatory pathway can produce a discernible effect on the regulatory mutant when compared to wild type. Several regulatory factors were identified using partial complementation.[8,10,12,21]

To identify interacting regulatory loci, one can search for suppressor mutations of a known regulatory mutation, with subsequent suppressor gene isolation. Suppressor mutations are anticipated to overcome, at least partially, the impairment imposed by the available regulatory mutation. Suppression of both regulatory "up"[5] and "down"[22] mutations have been exploited for the study of PS gene expression. This method also permits isolation of a series of suppressor point mutations in the same genetic locus, thus providing a valuable base for studying structure–function relationships.[22]

3. *Use of a Heterologous Host to Unmask Photosynthesis Regulatory Genes.* To identify transcriptional regulators of PS gene expression, we have developed a genetic screen in a nonphotosynthetic organism, *Paracoccus denitrificans*, which belongs to the same α3 subdivision of the *Proteobacteria* as does *R. sphaeroides*. The "relatedness" to *R. sphaeroides* enables the high G + C *P.* denitrificans to express *R. sphaeroides* genetic information, including PS genes. The use of *P. denitrificans* permits monitoring effects of regulatory genes in the absence of multiple and complex secondary effects due to the presence of the PS genetic apparatus. This approach presumes that endogeneous *P. denitrificans* regulatory factors will have a minimal effect on PS gene expression, but caution must be exercised. Note that *E. coli*, a traditional host for foreign gene expression, does not express most *R. sphaeroides* genes.

A transcriptional *lacZ* fusion of a given PS gene is introduced into *P. denitrificans* and basal *lacZ* expression is assessed. The *R. sphaeroides* 2.4.1 cosmid library, or a specific clone, is then introduced. Activators or repressors may increase or decrease *lacZ* expression respectively, while most cosmids have no effect.[23] It is worth mentioning that use of *P. denitrificans* as a heterologous host often provides insights into whether a given regulatory factor is unique to PS gene expression[24] or has a broader range of action.[23]

[21] M. Gomelsky and S. Kaplan, p. 41B. Paper presented at the VIII International Symposium on Phototrophic Prokariotes (G. Tedioli, S. Ventura, and D. Zannoni, eds.), Consiglio Nazionale delle Ricerche, Urbino, Italy, 1994.

[22] M. Gomelsky and S. Kaplan, *J. Bacteriol.* **179**, 128 (1997).

[23] M. Gomelsky and S. Kaplan, *J. Bacteriol.* **178**, 1946 (1996).

[24] M. Gomelsky and S. Kaplan, *J. Bacteriol.* **177**, 1634 (1995).

*4. Mapping and Cloning.* The availability of a physical map of the *R. sphaeroides* 2.4.1 genome greatly facilitates mapping of regulatory mutations. In addition, several cosmid contigs have been defined, covering most of chromosomes I and II,[25,26] and therefore localization of a regulatory gene to a specific cosmid is made easier. As more *R. sphaeroides* genes are identified, it becomes increasingly useful to survey genetic maps of other related organisms for potential similarities between loci and their immediate linkage relationships. This can provide an additional means to identify other useful regulatory factors,[3] or provide an indication of the structural organization of a particular gene and its regulon.

If transposon mutagenesis has been used to identify a regulatory locus, the location of the transposon insertion can be directly determined, as earlier described.[17,27] Additionally, DNA molecules used to inactivate genes for physiologic studies may contain rare restriction endonuclease sites, which can then be used for mapping disrupted genes.[27]

In *R. sphaeroides*, a number of genes and their homologs have been found. Among these are PS genes, including *hemA* and *hemT* encoding isozymes of ALA synthase,[28] and *hemZ*[3] and *hemN*,[29,30] which encode isozymes of (anaerobic) coproporphyrinogen III oxidase. Other duplicate genes include *rdxA*[31] and *rdxB*,[17,31,32] encoding proteins that are putative redox carriers, *groESL(1)* and *groESL(2)* encoding the GroES and GroEL chaperonins,[33] and *rpoN(1)*[34] and *rpoN(2)*,[35] which encode $\sigma^{54}$ transcription factors. Although not all of these genes may be regulatory, the presence of these duplicate genes can cloud interpretation of data. For this reason, it has become almost routine in our laboratories to analyze for the presence of duplicate copies of any gene that is cloned. If identified, one can directly map the duplicated sequences by probing genomic DNA restricted with the rare-cutting enzymes used previously.[27] The duplicated sequences can then be cloned, as described earlier.

[25] M. Choudhary, C. Mackenzie, K. S. Nereng, E. Sodergren, G. M. Weinstock, and S. Kaplan, *J. Bacteriol.* **176,** 7694 (1994).

[26] M. Choudhary, K. S. Nereng, and S. Kaplan, unpublished results (1996).

[27] A. Suwanto and S. Kaplan, *J. Bacteriol.* **171,** 5840 (1989).

[28] T.-N. Tai, M. D. Moore, and S. Kaplan, *Gene* **70,** 139 (1988).

[29] S. A. Coomber, R. M. Jones, P. M. Jordan, and C. N. Hunter, *Mol. Microbiol.* **6,** 3159 (1992).

[30] J. H. Zeilstra-Ryalls, A. Simmons, and S. Kaplan, unpublished results (1996).

[31] E. L. Neidle and S. Kaplan, *J. Bacteriol.* **174,** 6444 (1992).

[32] J. P. O'Gara and S. Kaplan, *J. Bacteriol.* **179,** 1951 (1997).

[33] W. T. Lee, K. C. Terlesky, and F. R. Tabita, *J. Bacteriol.* **179,** 487 (1997).

[34] W. G. Meijer and F. R. Tabita, *J. Bacteriol.* **174,** 3855 (1992).

[35] J. H. Zeilstra-Ryalls, R. C. MacKenzie, G. Weinstock, and S. Kaplan, unpublished results (1997).

5. *Reverse Genetics.* The genetic approaches presented earlier are complemented by biochemical methods when purification of a regulatory protein precedes identification of its coding sequences. These methods are most applicable for the isolation of DNA-binding proteins, which can be extracted from protein mixture using a DNA fragment containing a putative binding site within the upstream sequences of a given PS gene. Several experimental designs are possible but are beyond the scope of this study.[36] Microsequencing of the DNA-binding protein allows for the design of degenerate primers for polymerase drain reaction (PCR) or DNA hybridization, enabling the cloning of the corresponding gene.[37] The expanding database of *R. sphaeroides* gene sequences has provided for the development of a codon usage table, which is useful in designing such primers.[38]

## C. Analysis of the Transcriptional Regulatory Factors

Once a gene involved in the regulation of PS gene expression is identified, elucidation of its function, mechanism of action, and other components of the regulatory pathway with which it interacts becomes the prime objective. Some of these approaches that have been successfully employed for use in *R. sphaeroides* are described next.

*1. Sequence Analysis.* Analysis of the gene of interest may involve computer-assisted database searches for possible relationships. Computer analysis may also help to reveal structurally distinct and functionally important domains in the deduced gene product, thus providing insights for further phenotypic and biochemical characterization of the regulatory factor(s).[5,39]

*2. Phenotypic Characterization.* Functional characterization of the regulatory factor(s) can be performed using either null or missense mutations, or by introduction of an extrachromosomal copy(ies) of the gene of interest into a mutant or wild-type background. The following techniques have proven useful in analysis of the phenotypic effect of regulatory factor(s) so far identified.

1. Growth characteristics of the mutant can be determined and compared under a variety of cultivation conditions: photoheterotrophically, at high (100 W/m²), medium (10 W/m²) or low (3 W/m²) light; anaerobically in the dark in the presence of the alternative electron acceptor DMSO

---

[36] J. P. Jost and H. P. Saluz, eds., "A Laboratory Guide to *in Vitro* Studies of Protein–DNA Interactions," *Biomethods,* Vol. 5, 32b. Birkhauser, Basel, Boston (1991).

[37] H. Shimada, T. Wada, H. Handa, H. Ohta, H. Mizoguchi, K. Nishimura, T. Masuda, Y. Shioi, and K. Takamiya, *Plant Cell Physiol.* **37,** 515 (1996).

[38] J. M. Eraso and S. Kaplan, unpublished results (1994).

[39] R. J. Penfold and J. M. Pemberton, *J. Bacteriol.* **176,** 2869 (1994).

(dimethyl sulfoxide); and aerobically or semiaerobically, in the presence of 20–30% $O_2$ or 2–3% $O_2$, respectively. Adaptation experiments, when chemoheterotrophically grown cells are "shifted" from high to low oxygen tensions,[3,14] or from aerobic to photosynthetic (anaerobic) conditions,[10,40] may also provide important information on the functional activity of the regulatory factor of interest.

2. The qualitative and quantitative analysis of spectral complexes is a routinely employed method to assess the phenotypic effects of PS gene regulators.[41,42] The ability of *R. sphaeroides* to form the various photosystems not only during photoheterotrophic growth, but also during chemoheterotrophic growth under limited aeration, or anaerobically in the dark in the presence of DMSO, allows for the analysis of spectral complexes in those regulatory mutants that are impaired in photosynthetic growth.

3. Bchl and Crt are the two major classes of nonprotein structural components of the *R. sphaeroides* photosystem (see Table I), and their synthesis is known to be tightly regulated in response to oxygen availability and/or light intensity.[43] Analysis of the Bchl and Crt content of *R. sphaeroides* cells is routinely performed.[43] The composition of Crt pigments is assessed using high-performance liquid chromatography (HPLC).[44,45] Analysis of the relative abundance of the two major Crts (spheroidene and spheroidenone) has proven to be useful in our studies of the regulator(s) involved in maintaining cellular redox and redox control of PS gene expression, and therefore affecting biosynthesis of the other structural components of the photosystem in this bacterium.[32]

3. *Quantitation of Photosynthesis Expression.* Quantitation of gene expression is used to analyze the transcriptional effect(s) of the regulatory factors. This can be performed by (1) quantitating relative transcription levels by assaying enzyme activity in cell extracts bearing appropriate transcriptional fusions or (2) quantitating PS gene transcripts directly. In the case of transcriptional fusions, we routinely use the *lacZ*, *aph*, or *xylE* genes as reporters, as described in previous sections or elsewhere.[46]

Quantitation of PS gene transcripts is another approach toward analyzing the expression of PS genes. However, the amount of message reflects

[40] J. H. Zeilstra-Ryalls and S. Kaplan, in preparation (1997).
[41] S. W. Meinhardt, P. J. Kiley, S. Kaplan, A. R. Crofts, and S. Harayama, *Arch. Biochem. Biophys.* **236**, 130 (1985).
[42] P. J. Kiley, A. Varga, and S. Kaplan, *J. Bacteriol.* **170**, 1103 (1988).
[43] G. Cohen-Bazire, W. R. Sistrom, and R. Y. Stanier, *J. Cell. Comp. Physiol.* **49**, 25 (1957).
[44] M. B. Evans, R. J. Cogdell, and G. Britton, *Biochim. Biophys. Acta* **935**, 292 (1988).
[45] A. A. Yeliseev, J. M. Eraso, and S. Kaplan, *J. Bacteriol.* **178**, 5877 (1996).
[46] S. Dryden and S. Kaplan, *J. Bacteriol.* **175**, 6392 (1993).

not only the rate of transcription initiation, but also the turnover of mRNA. We routinely extract RNA according to the method of Zhu and Kaplan,[47] with modifications.[5] To normalize relative RNA concentrations to correct for differences in isolation, loading, etc., we routinely use labeled DNA probes encoding *R. sphaeroides* rRNA.[46] Standard procedures are used for slot-blotting and electrophoresis, as well as for transfer to membranes and DNA or riboprobe synthesis.

4. *Structure–Function Analysis.* Several genetic approaches can be used to identify specific function(s) of regulatory proteins. In general the identification of mutations encoding mutant proteins, and their separation into functionally distinct classes, can be used as a means to characterize the properties of proteins. If DNA sequence analysis of the gene provides information on putative functional domains, standard site-directed mutagenesis can be used to test domain function(s).[18] Alternatively, the isolation of spontaneous suppressor mutations is a useful tool, since the mutations allow mapping of specific domains. Both, extragenic, as in PpsR, AppA,[22] as well as intragenic suppressors,[39] are useful in this respect. The expression of isolated protein domains, which may retain functionality, can also be useful in such studies.[24,48]

5. *Biochemical Characterization of Regulatory Proteins.* A commonly used approach to the biochemical characterization of regulatory protein factors involves purification of the protein of interest after its overexpression in *R. sphaeroides*, or in a heterologous host (*E. coli, P. denitrificans*). Once purified, the protein can be used in a variety of biochemical and molecular biological systems to study the mechanism(s) underlying its functional activity(ies). Although these studies have been initiated only recently, a number of PS regulatory protein factors have been overexpressed and partly purified.[10,49–51] For the expression and purification of proteins in *E. coli*, commercially available vectors have proven to be useful but are beyond the scope of this discussion. The expressed product can then be purified by affinity chromatography, immunoprecipitation, or other conventional techniques, and then analyzed as a fusion, or as an isolated protein, on removal of the tag sequence.[49]

Study of the cellular localization of a given regulatory factor can provide valuable information on its functional activity(ies). This analysis includes immunodetection of the protein of interest and separation of the subcellular

[47] Y. S. Zhu and S. Kaplan, *J. Bacteriol.* **162,** 925 (1985).
[48] M. Gomelsky and S. Kaplan, unpublished results (1997).
[49] A. A. Yeliseev and S. Kaplan, *J. Biol. Chem.* **270,** 21167 (1995).
[50] K. Inoue, K. J.-L. K., C. S. Mosley, and C. E. Bauer, *Biochemistry* **34,** 391 (1995).
[51] S. N. Ponnampalam, J. J. Buggy, and C. E. Bauer, *in* "Diversity, Genetics and Physiology of Photosynthetic Prokaryotes" (C. Bauer, J. Beatty, M. Madigan, J. Ormerod, and F. Tabita, eds.), p. 37. Bloomington, Indiana, 1996.

fractions by standard techniques. Subcellular fractions from null mutant cells can be used as a negative control to verify the specificity of immunodetection.[49]

The topology of membrane-localized regulatory proteins is analyzed by using translational fusions to the *phoA* gene, encoding alkaline phosphatase.[14,31,52] This analysis can be complemented by the use of *lacZ* traditional fusions, since β-galactosidase is active cytoplasmically.[31]

6. *Interactions between Regulatory Factors.* Establishing the hierarchy of factors within a given regulatory pathway, as well as relationships between different regulatory pathways, can be achieved by studying the interaction(s) between identified regulatory factors.

1. One can construct null mutations in two (or more) regulatory genes and study the phenotypes of the double or triple mutant strains, compared to the phenotype of a single null mutant strain. An additive effect of the second mutation may indicate independent action of the two regulatory factors while, e.g., epistasis could indicate participation in the same regulatory pathway.[22,32] Isolation and study of extragenic suppressors represents a variation of this approach (see Section II,B,2).

2. Extra copies of a given regulatory gene can be introduced into a defined genetic background consisting of a regulatory mutant and a wild type. If the effect in either genetic background alone is similar, this could implicate independence of the regulatory pathways. If the effect is significantly different, one might anticipate an interaction between the two regulatory factors.[22]

3. One can use the heterologous host, *P.* denitrificans, to access potential interactions between regulators of PS gene expression.[22]

Studies of such interactions *in vitro*, between purified factors is another important tool. However, most of the protein components of the regulatory network governing PS gene expression have yet to be purified. Some of the interactions between regulatory proteins from *R. sphaeroides*[53] and a related photosynthetic bacterium, *Rhodobacter capsulatus*,[51] have been demonstrated *in vitro*.

7. *Protein–DNA Interactions.* In vitro study of the interactions between a DNA-binding protein and its target sequences provides important insight into the mode of action of a given transcriptional regulator. To test for DNA binding, gel shift assays using both crude cell extracts and purified

[52] A. R. Varga and S. Kaplan, *J. Bacteriol.* **171,** 5830 (1989).
[53] J. Newman and T. J. Donohue, *in* "Diversity, Genetics and Physiology of Photosynthetic Prokaryotes" (C. Bauer, J. Beatty, M. Madigan, J. Ormerod, and F. Tabita, eds.), p. 29. Bloomington, Indiana, 1996.

components are employed.[36,51] Use of crude extracts ensures the presence of necessary cofactors (if any), which may be important for DNA binding. Cell extracts of heterologous hosts, e.g., *P. denitrificans* or *E. coli*, expressing a given *R. sphaeroides* DNA-binding protein, which is specific to PS gene expression, can also prove useful.[48] Use in DNA-binding assays of the purified *E. coli* homolog in place of the *R. sphaeroides* transcriptional factor, when these homologs are anticipated to perform similar functions (e..g, IHF factor), has proven beneficial in assessing the likelihood of involvement of these factors in binding to specific DNA regions.[55]

Purification of a DNA-binding factor(s) and its subsequent use in experiments *in vitro* is the next logical step in analyzing protein–DNA interactions. A variety of methods are applicable for these studies.[37] It is worth mentioning that genetic and physiologic studies often provide insight into important biochemical properties of the DNA binding proteins, e.g., sensitivity of some PS gene transcription factors to redox conditions.[22,51] When basic components of the transcription apparatus are available, and specific regulatory proteins are purified, it may become possible to reconstitute transcription *in vitro* either in part or *en toto*.[53,56–59]

## III. Summary and Future Goals

Our studies, using the methods described, have led to the identification of a number of genes and their products that regulate PS gene expression in *R. sphaeroides* 2.4.1. Some of these factors are clearly DNA-binding proteins, such as IHF, HvrA (SPB), PpsR, and FnrL; some comprise a two-component regulatory system; and some are sensors, as well as signal generators. The identification of these factors has permitted an understanding of some of the individual regulatory pathways in which they are involved. Table II lists factors so far identified and their salient features. The search continues for additional transcriptional factors that are predicted to exist but have not been discovered to date. Additionally, we have studied and continue to study regulation of the regulators,[18,22,60] and we have also begun

[54] D. Ma, D. N. Cook, D. A. O'Brien, and J. E. Hearst, *J. Bacteriol.* **175,** 2037 (1993).
[55] K. J. Lee, S. Wang, J. M. Eraso, J. Gardner, and S. Kaplan, *J. Biol. Chem.* **268,** 24491 (1993).
[56] J. W. Kansy and S. Kaplan, *J. Biol. Chem.* **264,** 13751 (1989).
[57] R. K. Karls, D. J. Jin, and T. J. Donohue, *J. Bacteriol.* **175,** 7629 (1993).
[58] J. Brooks, R. K. Karls, P. Rossmeissl, and T. J. Donohue, *in* "Diversity, Genetics and Physiology of Photosynthetic Prokaryotes" (C. Bauer, J. Beatty, M. Madigan, J. Ormerod, and F. Tabita, eds.), p. 28. Bloomington, Indiana, 1996.
[59] R. Karls, B. MacGregor, and T. J. Donohue, *in* "Diversity, Genetics and Physiology of Photosynthetic Prokaryotes" (C. Bauer, J. Beatty, M. Madigan, J. Ormerod, and F. Tabita, eds.), p. 30. Bloomington, Indiana, 1996.
[60] J. H. Zeilstra-Ryalls and S. Kaplan, unpublished results (1997).

TABLE II
FACTORS REGULATING TRANSCRIPTION OF PS GENES OF *Rhodobacter sphaeroides* 2.4.1

| Name | Relevant characteristics | (Proposed)[a] PS gene(s) target | Reference |
|---|---|---|---|
| CcoNOQP | Terminal cytochrome *c* oxidase; proposed to sense and interpret a redox signal | *hemA, bch, crt, pucBA, puf* | *b, c* |
| RdxBHIS (and RdxA) | Proposed to comprise a membrane-localized redox protein cluster, involved in redox sensing and interacting with CcoNOQP | *bch, crt, pucBA, puf* | *c* |
| PrrA | Response regulator of a two-component regulatory system involved in $O_2$ sensing | *bch, crt, pucBA, puf, puhA, ccoNOQP* | *d* |
| PrrB | Histidine kinase/phosphatase of a two-component regulatory system involved in $O_2$ sensing | *bch, crt, pucBA, puf, puhA* | *e, f* |
| PrrC | Periplasmically localized protein involved in PS gene expression, in conjunction with PrrB/PrrA | (*puf, puhA*) | *e* |
| PpsR | Oxygen- and light-responsive transcriptional repressor; DNA-binding site is TGT-$N_{12}$-ACA | *crt, bch, pucBA, bch* | *g–j* |
| AppA | Flavoprotein involved in the PpsR regulatory pathway | *bch, crt, pucBA, puf* | *j, k* |
| Ppa | Putatively involved in the PpsR regulatory pathway | *bch, crt, pucBA* | *l* |
| FnrL | Anaerobic regulatory protein; putative DNA-binding site is TTGAT-$N_4$-ATCAA | (*ccoNOQP, rdxBHIS, hem*), *hemA, pucBA, bch* | *b, m, n* |
| TspO | Outer membrane-localized sensory protein, may operate through the PpsR pathway | *crt, bch, pucBA* | *o* |
| Spb (HvrA) | Histone-like protein; contains a leucine zipper motif | (*puf*) | *p, q* |

to purify some of the protein components involved in the regulation of gene expression. This and the use of an *R. sphaeroides*-specific *in vitro* transcription–translation system will greatly complement our *in vivo* studies and allow for the more rapid identification and isolation of other components of this system, strengthening the scope of our biochemical approach, through the manipulation of wild-type as well as mutant proteins.

We have also begun to study cross-pathway interactions, to define interrelationships and hierarchies among the so far identified regulators.[22,41,61]

[61] J. O'Gara, J. M. Eraso, and S. Kaplan, in preparation (1997).

TABLE II (*continued*)

| Name | Relevant characteristics | (Proposed)[a] PS gene(s) target | Reference |
|---|---|---|---|
| IHF | Integration host factor, DNA-binding site is TTTCAAGCCGTTA | *pucBA* | *r, s* |
| RpgS1 | Putative DNA binding protein that regulates PS gene expression | *pucBA, puf* | *t* |
| RpgS2 | Putative DNA binding protein that regulates PS gene expression | *pucBA, puf* | *t* |
| MgpS | Activator of PS gene expression | *pucBA, puf* | *u* |

[a] PS genes indicated in parentheses are proposed to be target(s) of regulation by the factors indicated.

[b] J. H. Zeilstra-Ryalls and S. Kaplan, *J. Bacteriol.* **178**, 985 (1996).

[c] J. P. O'Gara and S. Kaplan, *J. Bacteriol.* **179**, 1951 (1997).

[d] J. Eraso and S. Kaplan, *J. Bacteriol.* **176**, 32 (1994).

[e] J. M. Eraso and S. Kaplan, *J. Bacteriol.* **177**, 2695 (1995).

[f] J. M. Eraso and S. Kaplan, *J. Bacteriol.* **178**, 7037 (1996).

[g] R. J. Penfold and J. M. Pemberton, *Curr. Microbiol.* **23**, 259 (1991).

[h] R. J. Penfold and J. M. Pemberton, *J. Bacteriol.* **176**, 2869 (1994).

[i] M. Gomelsky and S. Kaplan, *J. Bacteriol.* **177**, 1634 (1995).

[j] M. Gomelsky and S. Kaplan, *J. Bacteriol.* **179**, 128 (1997).

[k] M. Gomelsky and S. Kaplan, *J. Bacteriol.* **177**, 4609 (1995).

[l] M. Gomelsky and S. Kaplan, paper presented at the VIII International Symposium on Phototrophic Prokaryotes (G. Tedioli, S. Ventura, and D. Zannoni, eds.), Consiglio Nazionale delle Ricerche, Urbino, Italy, 1994, p. 41B.

[m] J. H. Zeilstra-Ryalls and S. Kaplan, *J. Bacteriol.* **177**, 2760 (1995).

[n] J. H. Zeilstra-Ryalls and S. Kaplan, in preparation (1997).

[o] A. A. Yeliseev and S. Kaplan, *J. Biol. Chem.* **270**, 21167 (1995).

[p] H. Shimada, T. Wada, H. Handa, H. Ohta, H. Mizoguchi, K. Nishimura, T. Masuda, Y. Shioi, and K. Takamiya, *Plant Cell Physiol.* **37**, 515 (1996).

[q] P. Sen, J. M. Eraso, and S. Kaplan, in preparation (1997).

[r] J. K. Lee, J. Wang, J. M. Eraso, J. Gardner, and S. Kaplan, *J. Biol. Chem.* **268**, 24491 (1993).

[s] P. Sen and S. Kaplan, in preparation (1997).

[t] P. Sen and S. Kaplan, paper presented at the 95th General Meeting of the American Society for Microbiologists, Washington, D.C., 1995, p. 494.

[u] M. Sabaty and S. Kaplan, *J. Bacteriol.* **178**, 35 (1996).

These analyses should reveal how the transduction of signal(s) emanating from different sensory pathways is achieved, and how this information is coprocessed and integrated by the organism, to allow photosystem development and control.

Finally, although the focus of this review has been on transcriptional regulation, the study of posttranscriptional as well as translational regulation and photosystem assembly is also in progress in our laboratory.[62–64] This should provide a more complete picture of the regulation of photosystem development in *R. sphaeroides*. It remains to be determined whether or not posttranscriptional control processes effect transcriptional control of PS gene expression.

### Acknowledgments

We wish to acknowledge those members of this laboratory, past and present, who have contributed to the development of the approaches, biological tools, and protocols presented here. This work was supported by Public Health Service grant GM1550 from the National Institutes of Health.

[62] A. Varga and S. Kaplan, *J. Biol. Chem.* **268**, 19842 (1993).
[63] L. Gong and S. Kaplan, *Microbiology* **142**, 2057 (1995).
[64] J. M. Eraso and S. Kaplan, *J. Bacteriol.* **178**, 7037 (1996).

## [11] Use of *Synechocystis* 6803 to Study Expression of a *psbA* Gene Family

*By* CHRISTER JANSSON, GAZA SALIH, JAN ERIKSSON, RONNEY WIKLUND, and HAILE GHEBRAMEDHIN

### Introduction

The unicellular cyanobacterium *Synechocystis* 6803 offers an excellent model system for mutagenesis studies of the reaction center proteins in Photosystem II (PSII) and their genes for at least three reasons: (1) The reaction center in *Synechocystis* 6803 is very similar, both at the structural and the functional level, to that of higher plants[1,2]; (2) the *Synechocystis* 6803 cells can be easily transformed and the added DNA is incorporated into the chromosome by homologous recombination; and (3) the cells can grow heterotrophically as well as photoautotrophically. Therefore, mutants that partially or completely lack PSII activity can be cultured and analyzed.

The D1 and D2 reaction center proteins of PSII are encoded by the *psbA* and *psbD* genes, respectively. In plants and algae these genes are located on the plastid genome and are typically present as one unique copy

[1] B. Andersson and S. Styring, *in* "Current Topics in Bioenergetics" (C. P. Lee, ed.), p. 2. Academic Press, San Diego, 1991.
[2] H. B. Pakrasi, *Annu. Rev. Genet.* **29**, 755 (1995).

per genome. In the prokaryotic cyanobacteria, on the other hand, *psbA* and *psbD* exist as small multigene families with between two and more members for *psbA*[3] and two members for *psbD*.[4] In *Synechocystis* 6803, the *psbA* gene family consists of three members, *psbA1*, *psbA2*, and *psbA3*.[5]

Studies on regulation of *psbA* gene expression in *Synechocystis* 6803 displays several interesting features. The *psbA* genes are differentially expressed in a light-dependent manner with approximately 95% of the transcripts being produced by *psbA2* and the rest by *psbA3*.[3,6,7] Inactivation of the highly expressed *psbA2* leads to an eightfold up-regulation of *psbA3*.[3] The transcript stability varies between *psbA2* and *psbA3* and seems to be controlled by the photosynthetic electron transport.[3,7] The *psbA1* gene is silent and its presence in *Synechocystis* 6803 presents an enigma.[3] When the *psbA1* gene was activated by directed mutagenesis it was found to encode a novel but functional D1 protein.[8] Activation of the *psbA1* gene also provided two new mutant strains of *Synechocystis* 6803 with different combinations of active and inactive *psbA* genes; strain A1-K with *psbA1* as the only active *psbA* gene, and strain K with all three *psbA* genes active (see Table II later in this article).

### Growth Conditions

*Synechocystis* 6803-G[9] cells are grown to mid-log phase ($A_{730} \approx 0.6$)[10] at 32° in BG-11 medium,[9,11] supplemented with $NaHCO_3$ to a concentration of 5–10 m$M$,[12] $Na_2S_2O_3$ (solid media only) to a concentration of 0.3% (w/v), and TES, pH 8.0 (solid media only), to a concentration of 20 m$M$, and a constant photon flux density of 40–70 $\mu$E m$^{-1}$ s$^{-1}$. For incubations under low and high light conditions, cells are transferred to 10–30 and 1000–1500 $\mu$E m$^{-1}$ s$^{-1}$, respectively. High-light illumination is provided by a 250-W projector lamp. During this illumination, cells are stirred continuously (a normal stir plate is used) and the temperature is maintained at 25–30°. If required, a fan is installed for air circulation. For dark incubations, cultures

---

[3] A. Mohamed, J. Erikson, H. D. Osiewacz, and C. Jansson, *Mol. Gen. Genet.* **238**, 161 (1993).
[4] S. A. Bustos and S. S. Golden, *J. Bacteriol.* **173**, 7525 (1991) and references therein.
[5] C. Jansson, R. J. Debus, H. D. Osiewacz, M. Gurevitz, and L. McIntosh, *Plant Physiol.* **85**, 1021 (1987).
[6] A. Mohamed and C. Jansson, *Plant Mol. Biol.* **13**, 693 (1989).
[7] A. Mohamed and C. Jansson, *Plant Mol. Biol.* **16**, 891 (1991).
[8] G. Salih and C. Jansson, *Plant Cell* **9**, 1 (1997).
[9] J. G. K. Williams, *Methods Enzymol.* **167**, 766.
[10] $A_{730} \approx 0.25$ corresponds to $\sim 10^8$ cells ml$^{-1}$.
[11] R. Rippka, J. Derulles, J. B. Waterbury, M. Herdman, and R. Y. Stanier, *J. Gen. Microbiol.* **111**, 1 (1979).
[12] It is our experience that $NHCO_3$ gives better results than $NaCO_2$.

TABLE I
SUPPLEMENTS DURING CULTIVATION OF *SYNECHOCYSTIS* 6803

| Supplement | Function | Concentration |
|---|---|---|
| Kanamycin | Translation inhibitor—selection agent | 5 $\mu$g ml$^{-1}$ |
| Streptomycin | Translation inhibitor—selection agent | 10 $\mu$g ml$^{-1}$ |
| Spectinomycin | Translation inhibitor—selection agent | 20 $\mu$g ml$^{-1}$ |
| Chloramphenicol | Translation inhibitor—selection agent | 5 $\mu$g ml$^{-1}$ |
| Lincomycin | Translation inhibitor—to prevent D1 synthesis during photoinhibitory conditions | 400 $\mu$g ml$^{-1}$ |
| Rifampicin[a] | Transcription inhibitor—to prevent *psbA* transcription during assays of transcript stability | 500 mg ml$^{-1}$ |
| DCMU | PSII inhibitor—to prevent selection for revertants or pseudorevertants; to study the influence of PSII electron transport on *psbA* transcription | 10 $\mu M$ |
| Methylviologen | PSI electron acceptor—to study the influence of PSI electron transport on *psbA* transcription | 300 $\mu M$ |
| Glucose | Carbon source—to allow for phenotypic expression of transformants; to prevent selection for revertants or pseudorevertants; to yield higher culture densities for DNA isolation | 6.5 m$M$ |

[a] Because rifampicin is subjected to photodegradation an addition of 100 mg ml$^{-1}$ was repeated every hour during the incubation.

are wrapped in aluminum foil. For liquid cultures, orbital shaking incubators are used, normally without $CO_2$-enriched atmosphere.[13] For isolation of DNA, glucose is added (Table I) because it usually speeds growth and allows higher culture densities.[9] When required, additional supplements are added (Table I).

The growth conditions described here, and variations thereof, represent a fairly common procedure for *Synechocystis* 6803. Other conditions for liquid cultures that are often employed are based on air perfusion in bubbling flasks.[9] For culture sizes of 5 liters and above, we use a similar technique. Large (10-liter) culture flasks containing an ordinary stir bar are sealed with rubber stopper. A bubbling tube is pushed through the stopper and is equipped with a 0.2-$\mu$m bacterial filter (Gelman, Pall Gelman Sciences, Lund, Sweden) on the outside. After autoclaving, the UV-sterilized probe of an IKA-Tron, ETS-D2 thermostat (IKA Labortechnik, Labassco,

[13] We have observed that during summer (in Sweden, mid-May through August) our cultures of *Synechocystis* 6803 grow slower than the rest of the year. The cells also start to degrade their phycobilisomes, manifested by bleaching. This problem can be circumvented by enriching the incubator atmosphere with $CO_2$. This is usually achieved by administering one to two 30-min pulses of 5% $CO_2$ per day.

Stockholm, Sweden) is pushed through the rubber stopper. The thermostat is connected to the outlet of a temperature-controlled hot plate and the temperature is set to 32°. The flask is placed on the hot plate and the stirring is adjusted to a slow agitation. Air enriched with 5% $CO_2$ is led through the filter and the bubbling tube with a pressure of ~2 bar. Illumination, provided by two ramps of light tubes, is similar to that for small-scale cultures.

## Isolation of *Synechocystis* 6803 Genomic DNA

### Rapid Small-Scale DNA Isolation for PCR Analysis

Preparation of genomic DNA for PCR amplification is prepared with either of two methods, each yielding 1–5 $\mu$g of DNA.

1. The isolation is based on the procedure developed by Möller *et al.*[14] Cell cultures (1–2 ml) are pelleted by centrifugation and the pellet suspended in 500 $\mu$l TE buffer [10 m*M* Tris-Cl, pH 8.2, and 1 m*M* ethylenediaminetetraacetic acid (EDTA)]. Alternatively, a single colony is suspended in 500 $\mu$l TE buffer. The cells are disrupted by sonication for 2 × 1.5 min at low effect [15%; this parameter has to be optimized for the sonicator and the size of the DNA fragment to be amplified by polymerase chain reaction (PCR)]. To precipitate the DNA, 225 $\mu$l isopropanol is added and the mixture is placed at −20° for 15–30 min. The DNA is pelleted by centrifugation for 15 min at 14,000*g* at 4°. The pellet is washed once in 70% ethanol and subsequently resuspended in 100 $\mu$l TE buffer. The Wizard DNA Clean Up System (Promega, Madison, WI) is used to purify the DNA. The DNA is eluted with 55 $\mu$l water.

2. This isolation is based on the procedure described by Ohad and Hirschberg.[15] A volume of 1 ml cell suspension, or a single colony suspended in 1 ml of water, is pelleted by centrifugation and the pellet is resuspended in 200 $\mu$l TE buffer (10 m*M* Tris-Cl, pH 8.2, and 1 m*M* EDTA). An equal volume of glass beads (0.2 mm, G1509; Sigma, St. Louis, MO) and 4 $\mu$l of 10% sodium dodecyl sulfate (SDS) solution are added, and the mixture is vortexed vigorously three times for 10 sec at room temperature. The resulting homogenate is centrifuged for 10 min at 14,000*g*. The supernatant is extracted once with phenol and then once with chloroform. The DNA in the aqueouse phase is precipitated with the addition of ammonium acetate to the final concentration of 2 *M* and an equal volume of isopropa-

[14] A. Möller, A.-M. Norrby, K. Gustafsson, and J. Jansson, *FEMS Microbiol. Lett.* **129,** 43 (1995).
[15] N. Ohad and J. Hirschberg, *Plant Cell* **4,** 273 (1992).

nol. The DNA is pelleted by centrifugation for 15 min at 14,000$g$ at 4°. The pellet is washed once in 70% ethanol and resuspended in 20–50 $\mu$l TE buffer.

### Isolation of DNA for DNA Gel Blot Analysis

Isolation of chromosomal DNA from *Synechocystis* 6803 is accomplished according to Salih *et al.*[16] based on the protocol described by Dzelzakalns *et al.*[17]

Cells are harvested at OD$_{730}$ = 0.5–1 by centrifugation at 4000$g$ for 10 min at 4°. The cells are washed once with growth medium (BG-11). The pellet is frozen in liquid nitrogen followed by thawing on ice. The pellet is then resuspended in 2.5 ml TE buffer (10 m$M$ Tris-HCl, pH 8.0, and 50 m$M$ EDTA), 200 $\mu$l of 10% SDS, and 50 $\mu$l of 3 $M$ NaAc, pH 5.0. This mixture is then ground in liquid nitrogen. The powder is allowed to thaw at room temperature and then extracted with phenol/chloroform three times. The DNA is precipitated with 1 volume of isopropanol for 20 min at −70° and then centrifuged for 15 min at 12,000$g$. The pellet is washed with 1 ml of 70% ethanol. The pellet is then suspended in 400 $\mu$l TE buffer. Subsequently, 8 $\mu$l of 10 mg/ml DNase-free RNase A is added and the sample incubated for 30 min at 37°. The mixture is then extracted twice with phenol/chloroform. The DNA is precipitated with 2.5 volumes of ethanol and 0.1 volumes of 3 $M$ NaAc, pH 5.0, for 15 min at −70°, followed by centrifugation for 15 min at 12,000$g$. The pellet is washed with 1 ml of 70% ethanol and then resuspended in 50 $\mu$l of TE buffer.

### Isolation of RNA from *Synechocystis* 6803

RNA is routinely isolated from 50–100 ml cultures according to the protocol described by Mohamed and Jansson.[6] Cultures are harvested at A$_{730}$ ≈ 0.5 by centrifugation at 6000$g$ for 10 min together with 50 ml crushed ice in a 250-ml centrifuge bottle. The pellet is frozen in liquid nitrogen and thawed on ice. This step is repeated once. The cells are suspended in resuspension buffer (0.3 $M$ sucrose, 10 m$M$ NaAc, pH 4.5). The suspension is transferred to an Eppendorf tube and pelleted at 12,000$g$ for 5 min. The pellet is suspended in 250 $\mu$l resuspension buffer supplemented with 75 $\mu$l 250 m$M$ Na$_2$–EDTA, pH 8.0, and the suspension is incubated on ice. After 5 min, 375 $\mu$l lysis buffer [2% (w/v) SDS, 10 m$M$ NaAc, pH 4.5] is added, followed by incubation at 65°. After 3 min, hot (65°) phenol (redistilled,

[16] G. Salih, R. Wiklund, C. Gerez, T. Tyystjärvi, P. Mäenpää, and C. Jansson, *in* "Proceedings of the Phytochemical Society of Europe—38" (I. M. Møller and P. Brodelius, eds.), p. 161. Oxford University Press, Oxford, 1995.
[17] V. A. Dzelzakalns and L. Bogorad, *J. Bacteriol.* **165,** 964 (1986).

molecular biology grade, IBI, KEBO, Stockholm, Sweden) is added, followed by incubation at 65° for 3 min and then at −70° for 15 sec. The suspension is centrifuged at 12,000$g$ for 5 min. The upper phase is collected and subjected to two additional rounds of hot phenol extraction, followed by an extraction with hot phenol:chlorophorm (1:1). To the final upper phase is added 1/10 volume of 3 $M$ NaAc, pH 5.2, and 2.5 volumes of 99% EtOII. The RNA is precipitated at −20° for 30 min. The pellet is washed once with 80% EtOH. Finally, the pellet is suspended in $H_2O$ and stored in aliquots at −70°. If necessary, the RNA preparation can be treated with RNase-free DNase (Promega). The average RNA yield is 150 and 130 $\mu$g for illuminated and dark-incubated cultures, respectively.

Solutions used for RNA isolation are treated with 0.1% (v/v) diethyl pyrocarbonate (Sigma) according to instructions from the manufacturer. Glassware is treated with 0.1% diethyl pyrocarbonate and baked.

Transformation

*Synechocystis* 6803, like several other cyanobacteria, is naturally competent;[9,18,19] i.e., they can be transformed without any specific pretreatment. Cyanobacteria also possess homologous recombination. Thus exogenous DNA taken up by a cyanobacterial cell can integrate into the chromosome(s) by a double crossover event, provided homologous sequences exist between the added and chromosomal DNA.[18,19] In this way, a chromosomal segment can be deleted or interrupted by insertion of a selectable marker gene (usually an antibiotic resistance gene). Likewise, a chromomal segment can be replaced by a modified version.

The protocol outlined here for transformation of *Synechocystis* 6803, which is successfully used in our laboratory, is a modified version of that described by Williams.[9] Cells of *Synechocystis* 6803 from fresh stock BG-11 plates are used for transformation. Cultures in mid- to late-log phase ($A_{730} \approx 0.6$) are diluted 100 times and allowed to grow further for 2–3 days until they retain midexponential growth. Cultures of 1–2 ml are centrifuged at 6000$g$ for 5 min and the cells are washed with 1 ml BG-11 medium to remove cell residues, which can create problems during transformation (e.g., by containing nuclease activities). The washed cells are resuspended in fresh BG-11 media to yield a final concentration of approximately $10^9$ cells ml$^{-1}$.

[18] C. Jansson and A. Lönneborg, *Progr. Botany* **52,** 226 (1991).
[19] P. Nixon and C. Jansson, *in* "Molecular Biology of Photosynthesis" (B. Andersson, A. H. Salter, and J. Barber, eds.), p. 197. IRL Press, Oxford, 1996.

Plasmid DNA[20] (2–10 $\mu$g) is added to 100 $\mu$l of the concentrated cells and the samples are mixed by pipetting several times. The samples are then incubated for 4–6 hr in a growth chamber under optimal growth conditions without shaking. Transformation mixtures are spread on nitrocellulose membranes placed on nonselective BG-11 plates. After incubation under optimal growth conditions for 48 hr, the membranes are transferred to BG-11 plates supplemented with appropriate antibiotics, glucose and, DCMU (Table I) and incubated under optimal growth conditions. After 2–3 weeks transformant colonies appear.

Transformation frequencies vary between different *Synechocystis* 6803 cultures and between wild-type and mutant strains. As a rule, between 0 and 100 colonies appear on the membranes after 2–3 weeks. Selected colonies (4–8 colonies) are restreaked on selective BG-11 plates. From each plate, a well-separated colony is restreaked on new selective BG-11 plates. The colonies are then innoculated in liquid BG-11 media supplemented with appropriate antibiotics, glucose, and DCMU (Table I). The correct genotype of the mutants is confirmed by PCR amplification followed by DNA sequencing. That the mutants are isogenic with respect to the mutated is confirmed by DNA gel blot analysis.

Also the plate transformation protocol developed by Dzelzkalns and Bogorad[21] can be also used for transformation of *Synechocystis* 6803.[22,23]

## Vectors for Transformation of *Synechocystis* 6803

In principle, six different vectors have been employed in the constructions of various *psbA* mutants of *Synechocystis* 6803 (Figs. 1 and 2). Vectors pGS1[24,25] and pGS2[25] were designed for introducing mutations in the 3' and central portion of the *psbA2* gene, respectively, and vectors pGS3, pGS31, and pGS32 for mutagenesis of the 5' portion of the gene. Vector pGR1[26] was designed for mutagenesis of the *psbA1* gene.

[20] Transformation with linear plasmids and chromosomal fragments has also been described by P. J. Nixon, D. Chisholm, and B. Diner, *in* "Plant Protein Engineering" (P. Shewry, and S. Gutteridge, eds.), p. 93. Cambridge University Press, Cambridge, Massachusetts, 1992.
[21] V. A. Dzelzkalns and L. Bogorad, *EMBO J.* **10,** 1619 (1988).
[22] V. K. Shukla, G. E. Stanbekova, S. V. Shetakov, and H. B. Pakrasi, *Mol. Microbiol.* **6,** 947 (1992).
[23] S. V. Shestakov, P. R. Anbudurai, G. E. Stanbekova, A. Gadzhiev, L. K. Lind, and H. B. Pakrasi, *J. Biol. Chem.* **269,** 19354 (1994).
[24] P. Mäenpää, T. Kallio, P. Mulo, G. Salih, E.-M. Aro, E. Tyystjärvi, and C. Jansson, *Plant Mol. Biol.* **22,** 1 (1993).
[25] G. Salih, R. Wiklund, T. Tyystjärvi, P. Mäenpää, C. Gerez, and C. Jansson, *Photosynth. Res.* **49,** 131 (1996).
[26] G. Salih, R. Wiklund, and C. Jansson, *in* "Photosynthesis: From Light to Biosphere" (P. Mathis, ed.), Vol. III, p. 529. Kluwer Academic Publishers, Dordrecht, 1995.

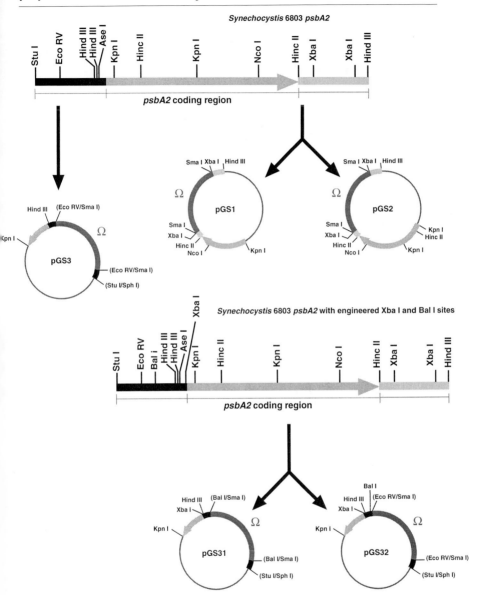

FIG. 1. Transformation vectors designed for mutagenesis of the *psbA2* gene in *Synechocystis* 6803. The resident plasmid in pGS1 and pGS2 is pUC18 (Amersham Pharmacia Biotech, Stockholm, Sweden), and in pGS3, pGS31, and pGS32 pGEM-3Z (Promega). Symbol Ω designates the linked streptomycin and spectinomycin resistance genes from pHP45Ω [P. Prenki and H. M. Krisch, *Gene* **29**, 303 (1984).]

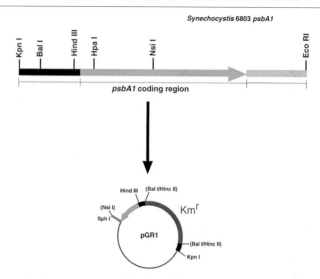

FIG. 2. Transformation vector designed for mutagenesis of the *psbA1* gene in *Synechocystis* 6803. The resident plasmid is pGEM-3Z (Promega). Km$^r$ designates the kanamycin resistance gene from pUC4K [J. Viera and J. Messing, *Gene* **19**, 259 (1982).]

The pGS1 vector was constructed in the following way: The Ω cassette from pHP45Ω[27] was excised as a *Bam*HI–*Bam*HI fragment and cloned into *Bam*HI-digested pRL171.[25,28] The Ω cassette was then re-excised from pRL171-Ω as an *Xba*I–*Xba*I fragment and was substituted for the *Xba*I–*Xba*I segment of pKW1266 (which contains the 3′ half plus flanking region of *psbA2* as a *Kpn*I–*Hind*II fragment[5]). For mutagenesis of the 3′ end of the *psbA2* gene, the *Kpn*I–*Hind*II fragment of the *psbA2* gene in pGS1 was subcloned into *Kpn*I–*Hinc*II-digested M13 mp18, mutagenized, and then substituted for the *Kpn*I–*Hinc*II fragment on pGS1. Alternatively, prior to the substitution, mutations were introduced into the *psbA2* gene during PCR amplification (see "Mutagenesis" section later), followed by excision of the *Kpn*I–*Hinc*II fragment.

The pGS2 vector was constructed by ligating the *Kpn*I–*Kpn*I fragment of a full-length *psbA2* clone[3] to the *Kpn*I site on pGS1. For mutagenesis within the *Kpn*I–*Kpn*I portion of the *psbA2* gene, the *Kpn*I–*Kpn*I fragment of pGS2 was subcloned into the *Kpn*I site on M13 mp18, mutagenized, and then substituted for the corresponding fragment on pGS2. Alternatively,

[27] P. Prenki and H. M. Krisch, *Gene* **29**, 303 (1984).

[28] The pRL vector series was developed by J. Elhaj and P. Wolk, Plant Research Laboratory, Michigan State University.

prior to the substitution, mutations were introduced into the *psbA2* gene during PCR amplification (see "Mutagenesis" section later), followed by excision of the *Kpn*I–*Kpn*I fragment. The orientation of the inserted fragment was determined by *Hinc*II and *Kpn*I digestions.

For mutagenesis of the 5' region of *psbA2*, the pGS3 vector was constructed by first cloning the PCR-amplified 5' *Stu*I–*Kpn*I fragment of *psbA2* into *Kpn*I–*Hind*II-digested M13 mp18. Following mutagenesis, the *Kpn*I–*Sph*I fragment of M13 mp18–*psbA2*, encompassing the cloned and mutated fragment, was cloned into *Kpn*I–*Sph*I-digested pGEM-3Z. Finally, the Ω cassette was ligated as a *Sma*I–*Sma*I fragment into the Eco RV site on pGEM-3Z–*psbA2*, yielding pGS3 with a mutated *psbA2* construct.

The vectors pGS31 and pGS32 were constructed from pGS3 by engineering restriction sites for *Bal*I and *Xba*I in the 5' flanking region of *psbA2*, yielding the *psbA2-A* construct (see "Activation of the Silent *psbA1* Gene in *Synechocystis* 6803" section later). In pGS31, the Ω cassette was inserted as a *Sma*I–*Sma*I fragment into the *Bal*I site, and in pGS32 into the Evo RV site. The pGS31 and pGS32 vectors were designed to allow for exchange of upstream regions between the *psbA2* and *psbA1* genes.

The pGR1 vector was constructed by subcloning the *Kpn*I–*Nsi*I 5' flanking and upstream region of the *psbA1* gene from pKW1214 (which contains the entire *psbA1* gene plus flanking sequences[5]) into *Kpn*I–*Pst*I-digested M13 mp18. Following mutagenesis, the *Kpn*I–*Sph*I fragment pf M13 mp18–*psbA1*, encompassing the cloned and mutated fragment, was cloned into *Kpn*I–*Sph*I-digested pGEM-3Z. Finally, the kanamycin resistance gene from pUC4K[29] was inserted as a *Hinc*II–*Hinc*II fragment into the *Bal*I site of pGEM-3Z–*psbA1*.

For inactivation of one or more *psbA* genes, the protocol and vectors described in the following section were used.

## Mutagenesis

Site-directed mutagenesis was performed either with the M13 method using the Sculptur™ *in vitro* mutagenesis system (RPN 1526, Amersham Pharmacia Biotech, Stockholm, Sweden) according to Olsen and Eckstein[30] or with the overlap extension PCR procedure.[31] Insertional inactivation of individual *psbA* genes in *Synechocystis* 6803 by interposon mutagenesis[18] has been described by Jansson et al.[5] For interruption of the *psbA1* gene, the kanamycin resistance gene was isolated as a *Hinc*II fragment from

[29] J. Viera and J. Messing, *Gene* **19,** 259 (1982).
[30] D. B. Olsen and F. Eckstein, *Proc. Natl. Acad. Sci. U.S.A.* **82,** 488 (1990).
[31] S. N. Ho, H. D. Hunt, R. M. Horton, J. K. Pullen, and L. R. Pease, *Gene* **77,** 51 (1989).

pUC4K[29] and inserted into the *Hinc*II site of *psbA*. For interruption of the *psbA2* gene, the Ω cassette was isolated as a *Sma*I–*Sma*I fragment from pHP45[27] and inserted into the blunt-ended *Nco*I site in *psbA2*. For interruption of the *psbA3* gene, the chloramphenicol resistance gene from pKT210[32] was isolated as a *Hinc*II fragment from pRL171[28] and inserted into the blunt-ended *Nco*I site in *psbA3*. Naturally, any combination of suitable antibiotic resistance cassettes, or other marker genes, can be used for inactivation of the *psbA* genes. Alternatively, the same technique can be used to delete the entire, or major portion, of single *psbA* genes.[33,34]

Apart from the utilization in studies on *psbA* gene expression, a genetic background with either *psbA2* or *psbA3* as the only functional *psbA* gene is necessary for studies directed toward structure–function analyses of the *psbA* gene product D1.[35]

### Studies on the Differential Expression of the *psbA* Genes in *Synechocystis* 6803

#### Analysis of Light-Controlled Transcription

For analysis of the light-dependent transcription of the *psbA* genes in *Synechocystis* 6803, cells were grown to mid-log phase ($A_{730} \approx 0.6$) under low-light or high-light conditions. For dark incubations, cells were kept for 48 hr under otherwise normal conditions. For low-light and high-light incubations, aliquots were taken from dark-adapted cultures and incubated for 20 hr at the respective light regime.

[32] M. Bagdasarian, M. M. Bagdasarian, S. Coleman, and K. N. Timmis, *in* "New Vector Plasmids for Gene Cloning in *Pseudomonas*" (K. N. Timmis and A. Pühler, eds.), p. 411. Elsevier, Amsterdam, 1979.

[33] R. J. Debus, B. A. Barry, G. T. Babcock, and L. McIntosh, *Proc. Natl. Acad. Sci. U.S.A.* **85,** 427 (1988).

[34] J. G. Metz, P. J. Nixon, M. Rögner, G. W. Brudvig, and B. A. Diner, *Biochemistry* **28,** 6960 (1989).

[35] To study structure–function relationships in the D1 protein of *Synechocystis* 6803, mutated constructs can be introduced either into the *psbA2* gene [R. J. Debus, B. A. Barry, I. Sithole, G. T. Babcock, and L. McIntosh, *Biochemistry* **27,** 9071 (1988); P. Mäenpää, T. Kallio, P. Mulo, G. Salih, E.-M. Aro, E. Tyystjärvi, and C. Jansson, *Plant Mol. Biol.* **22,** 1 (1993); N. Ohad and J. Hirschberg, *Plant Cell* **4,** 273 (1992); G. Salih, R. Wiklund, T. Tyystjärvi, P. Mäenpää, C. Gerez, and C. Jansson, *Photosynth. Res.* **49,** 131 (1996)] or into the *psbA3* gene [J. G. Metz, P. J. Nixon, M. Rögner, G. W. Brudvig, and B. A. Diner, *Biochemistry* **28,** 6960 (1989)]. Although perhaps not of primary concern in studies on *psbA* gene regulation, keep in mind that, since *psbA2* is by far the most predominantly expressed *psbA* gene in the wild-type strain of *Synechocystis* 6803 [A. Mohamed, J. Eriksson, H. D. Osiewacz, and C. Jansson, *Mol. Gen. Genet.* **238,** 161 (1993)], a control strain with *psbA3* as the only active *psbA* gene might not necessarily always exhibit the same phenotype as the wild type.

RNA was isolated from dark, low-light and high-light cultures and subjected to RNA gel blot analysis following the protocol described by Mohamed and Jansson.[6] The RNA was challenged with *psbA*-, *psbD*-, *rbcLS*-, and *rrn*-specific probes.[6] The results from these investigations showed that no *psbA*, *psbD*, *psbD-C*, or *rbcL-S* transcripts were produced in dark-incubated cells, whereas cells grown under illuminated conditions accumulated high levels of these transcripts.[6,36] The levels of *psbA*, *psbD*, and *psbD-C* transcripts increased with light intensity while those of *rbcL-S* decreased.[6,36] No light dependency was observed for the *rrn* genes.[6]

With the RNA gel blot assay outlined earlier, no transcription of the *psbA3* gene could be detected. To examine the contribution of individual genes in *psbA* transcript production, primer extension analysis was carried out.[3] These experiments demonstrated that both the *psbA2* and *psbA3* genes responded similarly to light intensity. They also revealed that approximately 95% of the *psbA* transcripts in cells grown under high- or low-light conditions originated from *psbA2* and only ~5% from *psbA3*.[3] No extension product was detected with a *psbA1*-specific primer. Furthermore, an eight-fold increase in *psbA3* transcript levels was observed after inactivation of the *psbA2* gene.[3]

A truncated but stable *psbA2* transcript was constructed[3] by inactivating the *psbA2* gene by insertion of the $\Omega$ fragment[27] (see the preceding section). The same approach has been used to produce a truncated *psbD* transcript in *Anacystis nidulans*.[37] Because accumulation of the truncated *psbA2* transcript followed the same pattern as that of the full-length *psbA2* transcript, one can conclude that the 3' end of the *psbA2* gene is not involved in the light-regulatory process.[3]

Promoter Characterization

The transcription start sites for the *psbA2* and *psbA3* genes were mapped by primer extension analysis to positions $-49$ and $-88$, respectively, relative to the ATG site. Sequence analysis showed that both the *psbA2* and *psbA3* basal promoters conformed fairly well to typical *E. coli* $\sigma^{70}$ promoters with readily recognizable -35 and -10 sequences.[3,19] From the expression patterns it can be concluded that the *psbA2* promoter represents a strong promoter in *Synechocystis* 6803, whereas the *psbA3* promoter is considerably weaker. Because the *psbA2* and *psbA3* genes share identical -35 signals but differ in their -10 signals (50% identity),[3,19] it is possible that the latter is a major determinant for the differential expression of the two genes.

[36] A. Mohamed and C. Jansson, *in* "Current Research in Photosynthesis" (M. Baltscheffsky, ed.), Vol. III, p. 568. Kluwer Academic Publishers, Dordrecht, 1990.
[37] S. S. Golden and G. W. Stearns, *Gene* **67**, 85 (1988).

To identify and characterize putative light-responsive elements and other regulatory motifs of the *psbA* genes in *Synechocystis* 6803, overlapping deletion mutagenesis of the 5' region of the *psbA2* gene has been carried out. Using this technique, it could be shown that a region of 120 nucleotides or less upstream of the ATG site contained all information necessary for both light-dependent and high-light-stimulated transcription.[38]

### Analysis of *psbA* Transcript Stability and Termination in *Synechocystis* 6803

To monitor the decay of *psbA* transcripts, cultures of *Synechocystis* 6803 were incubated either in darkness or under low-light conditions in the presence of the transcription inhibitor rifampicin (Table I). RNA was isolated at different intervals and subjected to RNA gel blot analysis using *psbA*-, *psbD*-, and *rbcL-S*-specific probes.[3] The results demonstrated that the half-life of *psbA* transcripts in low light was ~15 min and in darkness ~7 hr. In contrast, the stability of the *psbD* and *psbD-C* transcripts was essentially the same under both dark and low-light conditions, with half-lives of ~20 min. The stability of the *rbcL-S* transcripts was higher in darkness than in low light.

The influence of photosynthetic electron transport on *psbA* transcript stability was followed by repeating the experiments above with RNA isolated from cultures that had been incubated in the presence or absence of the inhibitors DCMU and methylviologen (Sigma; Table I).[7] DCMU blocks the photosynthetic electron transport by competing with plastoquinones ($Q_B$) at the binding site on the D1 protein, and methylviologen intercepts electrons from photosystem I. The rationale with the use of these inhibitors is that whereas DCMU alone prevents linear electron transport and $NADP^+$ reduction but allows formation of a $H^+$ gradient through cyclic electron transport, and thereby adenosine triphosphate (ATP) production, DCMU in combination with methylviologen prevents the production of both NADPH and ATP. The results from these studies showed that in the presence of both inhibitors, the stability of *psbA* transcripts under illuminated conditions was as high as in darkness, and the conclusion was that the stability of *psbA* transcripts in *Synechocystis* 6803 is under metabolic control.[7]

The decay of the individual *psbA2* and *psbA3* transcripts was examined by employing *Synechocystis* strains A2 and A3, respectively (Table II). RNA gel blot analysis of RNA isolated from these strains suggested that the *psbA2* transcripts are more stable than the *psbA3* transcripts in darkness.[3]

---

[38] G. Salih, J. Eriksson, H. Ghebramedhin, and C. Jansson, 1997.

TABLE II

STRAINS OF *SYNECHOCYSTIS* 6803 WITH DIFFERENT COMBINATIONS OF
ACTIVE OR INACTIVE *PSBA* GENES

| *Synechocystis* 6803 strain | Genotype with active (+) or inactive (−) *psbA* genes | Reference |
|---|---|---|
| Wild type[a] | *psbA1−* *psbA2+* *psbA3+* | b |
| A1 | *psbA1+* *psbA2−* *psbA3−* | b,c |
| A2 | *psbA1−* *psbA2+* *psbA3−* | c |
| A3 | *psbA1−* *psbA2−* *psbA3+* | c,d |
| A1-K | *psbA1+* *psbA2−* *psbA3−* | e |
| A2-K | *psbA1+* *psbA2+* *psbA3−* | f |
| K | *psbA1+* *psbA2+* *psbA3+* | e |

[a] Wild type refers to strain *Synechocystis* 6803-G [J. G. K. Williams, *Meth. Enymol.* **167**, 766 (1988)].
[b] C. Jansson, R. J. Debus, H. D. Osiewacz, M. Gurevitz, and L. McIntosh, *Plant Physiol.* **85**, 1021 (1987).
[c] A. Mohamed and C. Jansson, *Plant Mol. Biol.* **13**, 693 (1989).
[d] C. Jansson, R. J. Debus, H. D. Osiewacz, M. Gurevitz, and L. McIntosh, *Plant Physiol.* **85**, 1021 (1987).
[e] G. Salih and C. Jansson, *Plant Cell* **9**, 1 (1997).
[f] G. Salih and C. Jansson, 1998.

Transcription termination for the *psbA2* and *psbA3* genes in *Synechocystis* 6803 was mapped by S1 nuclease protection analysis to nucleotide 65 downstream of the open reading frame for both genes.[39] The possible formation of a strong stemloop structure followed by an A-T-rich region suggested that termination is rho independent.

[39] J. Eriksson, H. Ghebramedhin, and C. Jansson *in* "Photosynthesis: From Light to Biosphere" (P. Mathis, ed.), Vol. III, p. 533. Kluwer Academic Publishers, Dordrecht, 1995.

Activation of the Silent *psbA1* Gene in *Synechocystis* 6803

A series of 11 different constructs was made to study the influence of proximal and distal sequences on *psbA1* expression[8] (Fig. 3). In *psbA1-A*, the kanamycin resistance gene[29] (Km[r]) was inserted in the *Bal*I site at position -323 relative the ATG site. In *psbA1-B, psbA1-C*, and *psbA-E*, the -35, -10 and Shine–Dalgarno sequences, respectively, of *psbA2* were engineered at the corresponding positions. In *psbA1-D*, both promoter elements are present and *psbA1-F* contains the promoter elements plus the Shine–Dalgarno sequence. A second *Bal*I site was generated in *psbA1-G*. In *psbA1-H*, the region between the two *Bal*I sites was deleted and replaced with the Km[r] gene. The same region is deleted in *psbA1-I*, which contains the promoter and Shine–Dalgarno elements of *psbA1-F*. The *psbA1-J* construct was obtained by generating an *Xba*I site downstream of the Shine–Dalgarno sequence in *psbA1-G*. Replacing the 360-nucleotide-long upstream *Bal*I–*Xba*I fragment of *psbA1-J* with the 160-nucleotide-long upstream *Bal*I–*Xba*I fragment of *psbA2-A* and reinserting the Km[r] gene in the *Bal*I site yielded *psbA1-K*. The *psbA2-A* construct was derived by generating a *Bal*I site at position -170 and a *Xba*I site at position -10 in the *psbA2* gene. Ligation of the Ω fragment[27] into the *Bal*I site generated *psbA2-B*. The *Bal*I–*Xba*I fragment of *psbA1-J* was then exchanged for the corresponding fragment of *psbA2-B* and the Ω fragment inserted in the *Bal*I site, resulting in the *psbA2-C* construct.

It was found that engineering the *psbA1* upstream region so as to contain -35 and -10 sequences identical to those of the highly expressed *psbA2* gene did not suffice to activate the gene. In the active gene, *psbA1-K*, a 160-nucleotide-long upstream fragment of *psbA1* was exchanged for an upstream fragment of *psbA2*.[8] The *Synechocystis* strains A1-K and A2-K harbor, respectively, *psbA1-K* and *psbA1-K* plus *psbA2* as the only active *psbA* gene, whereas in strain K, all three *psbA* genes are active (Table II). The novel D1 protein, D1′, produced in strain A1-K differs significantly from the D1 protein encoded by *psbA2* and *psbA3*. However, D1′ is fully operational and *Synechocystis* strain A1-K grows photoautotrophically with a rate comparable to that of the wild type.[8]

Analysis of Pseudorevertants of *psbA* Mutants in *Synechocystis* 6803

Characterization of pseudorevertants can be a powerful method to identify *cis*-acting elements required for gene expression. This approach has so far been used only to a limited extent in the study of *psbA* gene regulation in *Synechocystis* 6803. In our laboratory, one pseudorevertant of strain A1 is currently being investigated. This strain was obtained by culturing ~2000

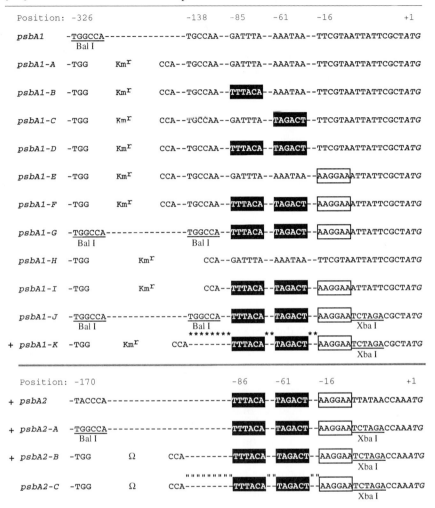

FIG. 3. Construction of mutations in the 5' region of *psbA1* and *psbA2*. Distances are given relative to the ATG start codon. The -35 and -10 promoter elements are indicated by black boxes and Shine–Dalgarno sequences by open boxes. Restriction sites are underlined. A plus sign (+) in front of the constructs denotes an active gene. Asterisks over the active *psbA1-K* gene indicate inserted *psbA2* sequences that are not present in the inactive *psbA1-J* gene. Conversely, quotation marks over the inactive *psbA2-C* gene indicate inserted *psbA1-J* sequences that are not present in the active *psbA2-B* gene. Dashes indicate unspecified nucleotides. Km[r] designates the kanamycin resistance gene from pUC4K [J. Viera and J. Messing, *Gene* **19,** 259 (1982)] and Ω the streptomycin–spectinomycin cassette from pHP45Ω [P. Prenki and H. M. Krisch, *Gene* **29,** 303 (1984)]. [Reprinted after modification with permission from G. Salih and C. Jansson, *Plant Cell* **9,** 1 (1997).]

colonies of A1 on solid media under nonpermissive conditions, i.e., in the absence of glucose, and selecting for photoautotrophic mutants. Preliminary analyses point to major rearrangements of the *psbA* genes and flanking sequences in the pseudorevertant.[40]

[40] G. Salih, J. Eriksson, H. Ghebramedhin, and C. Jansson, 1997.

# [12] Analysis of Light-Regulated Gene Expression

*By* Paul P. Dijkwel, Fred Rook, and Sjef C. M. Smeekens

## Introduction

Light is important for plant development and the expression of a large number of genes is regulated by light. The quality, amount, and duration of the light treatment needed to induce or repress the expression of a gene of interest provide information about the way light controls its expression. The phytochromes are the best studied family of photoreceptors and several excellent reviews about its properties have been published.[1,2] Briefly, phytochromes can exist in two photoconvertible forms: Pr, which absorbs maximally in red light ($\lambda_{max}$ = 666 nm), and Pfr, which absorbs maximally in far-red light ($\lambda_{max}$ = 730 nm). Phytochrome is synthesized in the Pr form, which is biologically inactive in most cases. On irradiation with red light, phytochrome is photoconverted into the active Pfr form, which can be converted back to the Pr form on subsequent absorption of far-red light. The role of phytochrome in light-regulated gene expression can be determined relatively easily. Phytochrome can function as a molecular switch because many responses that are triggered by a pulse of red light can be reduced when the plant is subsequently exposed to far-red light. For example, expression of genes encoding chlorophyll *a/b* binding proteins (CAB) or small subunits of Rubisco (RBCS) are induced by a short red light pulse whereas this induction can be reduced when the red-light is followed by a far-red light pulse.[3]

The small crucifer *Arabidopsis thaliana* is widely used as a model organism in plant molecular biology. The methods described below are optimized

[1] R. E. Kendrick and G. H. M. Kronenberg, eds., "Photomorphogenesis in Plants," 2nd Ed. Kluwer Academic Publishers, Dordrecht, The Netherlands, 1994.
[2] H. Smith, *Annu. Rev. Plant Physiol. Plant Mol. Biol.* **46,** 289 (1995).
[3] J. Silverthorne and E. M. Tobin, *Proc. Natl. Acad. Sci. U.S.A.* **81,** 1112 (1984).

for this weed. However, with minor changes these methods can be adapted to other dicotyledonous plants.

## Plant Growth Conditions

Seeds are surface sterilized by washing for 5 min in 20% (v/v) commercial bleach and rinsing four times with sterile water. When equal distribution of seed on agar plates is required, add 10 ml of water for each 15-cm plate. Swirl the tube to distribute the seeds equally and pour the seeds on a Murashige and Skoog agar plate.[4] The seeds will stick to the agar surface and the excess water can be removed by slightly tilting the plate. It is convenient to cover the agar with a piece of filter paper because the seeds will stick even better and clumps of seeds can be separated manually without destroying the agar surface. Also, the harvesting of intact seedlings is simplified because the roots will not grow into the agar. Alternatively, the surface sterilized seeds can be distributed in a small amount of 0.1% agarose and sown onto the agar using a pipette. Seed germination is promoted by a cold treatment of the plates for several days at 4°. Seedlings that are to be grown in the dark develop elongated hypocotyls and should be grown in plates that allow sufficient space such as the Falcon 1015 (Becton-Dickinson, Lincoln Park, NJ). Treatment of the seeds with 1 hr of light prior to their transfer to darkness promotes uniform germination. After this light treatment, wrap the plates in several layers of aluminum foil and allow for growth in a standard growth chamber. When the seeds are grown on soil, seed surface sterilization is not required. Instead, distribute the dry seeds equally on a piece of thoroughly wetted Miracloth (Calbiochem, La Jolla, CA) placed on moist soil in a translucent container and treat as described above.

## Phytochrome Responses

To determine the role of phytochrome in the light-regulated expression of the gene of interest, a red and far-red light source are required. A red light source can be obtained by filtering the red light portion from an incandescent bulb with filters such as those used in photography. Moreover, several types of red-light bulbs and fluorescent tubes are also commercially available (e.g., Sylvania Biosystems, Wageningen, The Netherlands) and their use is much preferred. Spectra of these light sources are often available from the supplier and any unwanted wavelengths can be filtered. Specific far-red light sources are more difficult to obtain, but Sylvania Biosystems

---

[4] T. Murashige and F. Skoog, *Physiol. Plant.* **15,** 473 (1962).

supplies handmade fluorescent tubes that emit light mainly in the far-red light range of the spectrum. Alternatively, incandescent light bulbs can be used because they emit a significant amount of far-red light. The light of several bulbs should be filtered with red and blue filters to abolish the wavelengths below 700 nm. In addition, these bulbs produce a lot of heat, which should be eliminated by cooling or filtering the light through, for example, a layer of water.

After exposure of etiolated (or dark-adapted) seedlings to short pulses of red and far-red light, the expression of the gene of interest can be analyzed by measuring the mRNA abundance (see below). First, make sure that the plants grow in absolute darkness because some phytochrome responses are very sensitive to light and can be induced by a small amount of green safelight. It is useful to optimize the light treatment by analyzing the expression of a control gene for which phytochrome-dependent expression is well established, such as a *CAB* or *RBCS* gene. The plates or containers with dark-grown seedlings are treated with four different light conditions. One plate should be kept in the dark and another treated with a saturating pulse of red light (1–10 min). A pulse is saturating if an increase of the length of exposure does not result in a higher induction of the expression of the control gene, i.e., the phytochrome pool is fully converted into the Pfr form. A third plate should be treated with the red light pulse followed by a saturating pulse of far-red light (1–10 min). In addition, treat a plate with the far-red light pulse only to determine the effect of far-red light alone. Several phytochrome responses can be induced by far-red light alone, because far-red light itself converts a small amount of phytochrome in the Pfr form. Leave the light-treated and dark control plants for 1–12 hr in the dark and subsequently measure the mRNA abundance of the control gene and the gene of interest. The mRNA levels of the *CAB* and *RBCS* genes should be maximal at around 2 hr after red-light treatment.

## Analysis of mRNA Levels

### Northern Blot Analysis

Genes that encode structural proteins are often expressed at levels that allow detection of mRNA abundance by making use of standard Northern hybridization methods. Several RNA isolation procedures are available and a particularly simple microfuge-tube isolation method for etiolated *Arabidopsis* seedlings has been described by Brusslan and Tobin.[5] This method can be adapted for light-grown plants by increasing the volumes

[5] J. A. Brusslan and E. M. Tobin, *Proc. Natl. Acad. Sci. U.S.A.* **89,** 7791 (1992).

of the first two steps in the procedure. An efficient electrophoretic separation and blotting method was described by Fourney et al.[6] An adaptation of this method is the inclusion of 1 $\mu$l of a 1 mg/ml ethidium bromide solution in the loading buffer. This allows visual monitoring of the separation and transfer of the RNA. A simple Northern blot hybridization method is described below. We found this method to be more sensitive than other methods. After hybridization the membrane can be stripped and hybridized several times using different probes. To strip the signal, pour a boiling 0.1× SSC/0.1% (w/v) SDS solution on the RNA side of the membrane and allow to cool to room temperature. However, some probe may remain and therefore first use the membrane to visualize the least abundant mRNA.

*Northern Blot Hybridization Protocol*

Probe preparation and labeling by random priming:

1. Boil 9 $\mu$l (50 ng) purified DNA fragment for 2 min, transfer to 0°.
2. Add 11.4 $\mu$l labeling solution, 1 $\mu$l bovine serum albumin (BSA) (1 mg/ml), 1 $\mu$l (1 U) Klenow DNA polymerase, and 5 $\mu$l (50 $\mu$Ci) [$\alpha$-$^{32}$P]dCTP.
3. Incubate at room temperature for 1 hr.
4. Boil for 5 min and transfer to 0°.
5. Spin briefly in a microfuge and add to the hybridization solution.

*Hybridization:*

1. Separate total RNA and transfer to a membrane as described by Fourney et al.[6]
2. Prehybridize the membrane for 1 hr with hybridization buffer.
3. Discard the buffer and add fresh hybridization buffer (0.1 ml/cm$^2$).
4. Add labeled probe and hybridize for 16–24 hr at 65°.
5. Wash two times with 0.5× SSC/0.1% (w/v) SDS at room temperature for 15 min and two times with the same solution at 55° for 20 min, or, for stringent hybridization, wash two times with 0.1× SSC/0.1% (w/v) SDS at 65° for 20 min.
6. Briefly dry the membrane, wrap in a plastic foil, and expose to X-ray film.

*Reagents*

Labeling solution: 0.44 $M$ HEPES, pH 6.6; 0.11 $M$ Tris-HCl, pH 8.0; 11 m$M$ MgCl$_2$; 22 m$M$ 2-mercaptoethanol; 44 $\mu M$ dATP/dGTP/dTTP; 0.75 $\mu$g/$\mu$l random hexamer oligo nucleotides

[6] R. M. Fourney, J. Miyakoshi, R. S. Day, III, and M. C. Paterson, *Focus* **10**, 5 (1988).

Hybridization buffer[7]: 0.25 $M$ sodium phosphate buffer, pH 7.2; 1 m$M$ ethylenediaminetetraacetic acid (EDTA); 7% (w/v) SDS; 1% (w/v) BSA. Filter the solution before use.
SSC (1×): 150 m$M$ NaCl; 15 m$M$ sodium citrate.

*RNase Protection*

Northern hybridization is the most commonly used method to detect and quantify the levels of a specific mRNA. However, when transcript levels are low, more sensitive methods are required. To study the expression of genes with low transcript levels such as those encoding transcription factors, we use RNase protection assays.[8,9] RNase protection involves the hybridization in solution of a radiolabeled antisense RNA probe to a RNA sample. The antisense RNA will only hybridize to the mRNA of interest, and the unhybridized single-stranded RNA is removed with RNases. The protected double-stranded radiolabeled hybrid fragments are recovered by ethanol precipitation. The protected fragments are subsequently analyzed by denaturing polyacrylamide gel electrophoresis and autoradiography.

*Templates for RNA Probe Synthesis*

Radiolabeled RNA probes of high specific activity can be efficiently produced *in vitro,* using plasmid vectors containing a polycloning site downstream from promoters derived from the *Salmonella typhimurium* bacteriophage SP6 or the *Escherichia coli* bacteriophages T7 and T3.[10,11] All or part of the gene of interest is cloned into the polycloning site into an orientation that leads to the production of an antisense RNA probe. A large number of plasmids, containing bacteriophage promoters, are available. Several plasmids contain two different bacteriophage promoters flanking the polycloning site, allowing the production of RNA probes in both orientations. For instance, the pBluescript series of plasmids (Stratagene, La Jolla, CA) contain both the T3 and T7 promoters flanking the polycloning site. The recombinant plasmid is linearized by digesting it with a restriction enzyme that cuts it at a suitable site within the gene or at a site in the plasmid

[7] G. M. Church and W. Gilbert, *Proc. Natl. Acad. Sci. U.S.A.* **81,** 1991 (1984).
[8] N. Quaedvlieg, J. Dockx, F. Rook, P. Weisbeek, and S. Smeekens, *Plant Cell* **7,** 117 (1995).
[9] N. Quaedvlieg, J. Dockx, G. Keultjes, P. Kock, J. Wilmering, P. Weisbeek, and S. Smeekens, *Plant Mol. Biol.* **32,** 987 (1996).
[10] D. A. Melton, P. A. Krieg, M. R. Rebagliati, T. Maniatis, K. Zinn, and M. R. Green, *Nucleic Acids Res.* **12,** 7035 (1984).
[11] J. Sambrook, E. F. Fritsch, and T. Maniatis, *in* "Molecular Cloning: A Laboratory Manual," 2nd Ed. Cold Spring Harbor Laboratory Press, New York, 1989.

Fig. 1. Autoradiogram of an RNase protection assay showing the light regulated expression of the *Arabidopsis* basic domain/leucine zipper transcription factor gene *ATB2*. Seedlings were dark adapted by growing them in continuous fluorescent light for 5 days (5dL) and subsequently transferring them to darkness for 1 hr (5dL1hD), 3 hr (5dL3hD), 8 hr (5dL8hD), 1 day (5dL1dD), and 2 days (5dL2dD) and returning them to the light for 1 hr (5dL2dD1hL), 3 hr (5dL2dD3hL), and 8 hr (5dL2dD8hL). All samples contained 10 μg of *Arabidopsis* total RNA and 10 μg of tRNA. The control sample contained 20 μg of tRNA (tRNA), the full-length probe was loaded in lane P. Typically two protected fragments are observed in this assay, which differ by approximately five bases in size. Their relative intensity differs somewhat when using different concentrations of RNase T1 and RNase A. This is probably a result of different processing of the duplex RNA by the two RNases. The RNA probe used is 257 nucleotides long.

downstream of the gene. The linearized plasmid serves as a template for the appropriate bacteriophage RNA polymerase to produce radiolabeled RNA.

The 5′ terminus of the produced RNA is fixed by the bacteriophage promoter. The position of the restriction site used to linearize the template DNA determines the size of the produced RNA probe and the protected fragment. As the protected fragments are analyzed by polyacrylamide gel electrophoresis, an RNA probe with a length of between 100 and 250 nucleotides is most convenient. The presence of some plasmid sequences is required to distinguish between the protected fragment and the presence of residual probe (Fig. 1). These plasmid sequences do not hybridize and are removed during the RNase treatment.

Particular care should be taken with the choice of the restriction enzyme used to linearize the template plasmid. Restriction enzymes that generate blunt ends or protruding 5′ termini should be used. Transcription of templates, prepared by digestion with restriction enzymes that leave 3′ protruding ends, result in the production of significant amounts of long RNA transcripts that are aberrantly initiated at the termini of the template. These transcripts are also copied from the noncoding template strand, which interferes with the RNase protection assay. If the use of restriction enzymes that create 3′ protruding ends cannot be avoided, these ends can be modified by treatment with Klenow DNA polymerase prior to the transcription reaction.[12]

[12] E. T. Schenborn and R. C. Mierendorf, *Nucleic Acids Res.* **13**, 6223 (1985).

*RNase Protection Assay Protocol*

Specific radiolabeled antisense RNA probes are synthesized *in vitro* using the following protocol. Mix in an Eppendorf tube: 4 $\mu$l 5× transcription buffer, 1 $\mu$l 100 m$M$ dithiothreitol (DTT), 0.8 $\mu$l RNasin (placental RNase inhibitor; 25 U/$\mu$l), 4 $\mu$l 2.5 m$M$ GTP/ATP/CTP, 1.2 $\mu$l 100 $\mu M$ UTP, 1 $\mu$l linearized DNA template (1 $\mu$g/$\mu$l), 8.0 $\mu$l (160 $\mu$Ci) [$\alpha$-$^{32}$P]UTP, and 0.7 $\mu$l RNA polymerase (T3, T7, or SP6; 19 U/$\mu$l). After an incubation for 1 hr at 37°, the DNA template is digested by the addition of 1 $\mu$l of RQ1 DNase (RNase-free, Promega, Madison, WI) for 10 min at 37°. This reaction is stopped by the addition of 160 $\mu$l $H_2O$ and by phenol extraction. The supernatant is transferred to a tube containing 20 $\mu$g tRNA (10 mg/ml) and then precipitated with 60 $\mu$l 8 $M$ $NH_4Ac$ and 600 $\mu$l ethanol. After spinning the tube in a microfuge, the supernatant is removed and the pellet resuspended in 180 $\mu$l $H_2O$. The RNA is precipitated a second time with 60 $\mu$l 8 $M$ $NH_4Ac$ and 600 $\mu$l ethanol. After removal of the supernatant, the pellet is washed with 250 $\mu$l 80% ethanol, air dried, and resuspended in 30 $\mu$l of hybridization buffer (see below). Use a small sample to determine the amount of incorporated radiolabel. Usually, total incorporation is above $100 \times 10^6$ cpm.

Isolate total plant RNA as described[5] or by any other suitable method. The radiolabeled RNA probe can now be hybridized with the mRNA of the gene of interest, which is present in the total RNA sample. Typically, 10 $\mu$g of total RNA is mixed with 10 $\mu$g of yeast tRNA (10 mg/ml) in an Eppendorf tube and subsequently dried. A sample containing 20 $\mu$g of tRNA can be used as a negative control. Each sample is resuspended in 30 $\mu$l of hybridization buffer containing $0.5 \times 10^6$ cpm of radiolabeled RNA probe. A drop of mineral oil is added to prevent evaporation. The tubes are subsequently heated to 85° for 5 min to denature the RNAs, and then incubated overnight at 45°.

After hybridization, the samples are thoroughly mixed with 300 $\mu$l of ice-cold RNase digestion buffer containing the appropriate amounts of RNases. The optimal RNase concentrations should be determined empirically, because they can differ between probes. Let the RNA digestion proceed for 1 hr at 30°. After digestion, add 5 $\mu$l of proteinase K (10 mg/ml) and 10 $\mu$l 20% (w/v) SDS, and incubate for 10 min at 37°. Extract the samples with 300 $\mu$l of phenol/chloroform/isoamylalcohol (24:24:1) and centrifuge. The aqueous layer is transferred to a new tube containing 10 $\mu$g tRNA (10 mg/ml) and 900 $\mu$l ethanol is subsequently added. The samples are left at room temperature for at least 15 min, and then centrifuged for 15–30 min at 12,000$g$ in a microfuge. The supernatant is carefully removed using a drawn-out Pasteur pipette. The pellet is allowed to air dry for a few minutes, and resuspended in 3.5 $\mu$l of standard sequencing dye. After heating to 90° for 1.5 min the samples are loaded onto a 6% polyacrylamide/ 7 $M$ urea sequencing gel. After running and drying of the gel, the protected fragments can be visualized by autoradiography (Fig. 1). The bands can be quantified by cutting them out and counting in a scintillation counter, or by use of a Phospho-Imager.

*Reagents*

Transcription buffer (5×): 200 m$M$ Tris-HCl, pH 8.0; 125 m$M$ NaCl; 40 m$M$ MgCl$_2$; 10 m$M$ spermidine-HCl.

Hybridization buffer: 80% (v/v) deionized formamide; 40 m$M$ PIPES buffer, pH 6.7; 400 m$M$ NaCl; 1 m$M$ EDTA.

RNase digestion buffer: 10 m$M$ Tris-HCl, pH 7.5; 5 m$M$ EDTA; 300 m$M$ NaCl; 5 $\mu$g/ml RNase A (Boehringer Mannheim, Mannheim, Germany); 250 units/ml RNase T1 (GIBCO-BRL, Gaithersburg, MD). Chill the digestion buffer on ice before adding it to the samples because this may prevent background hybridization.

Sequencing dye: 80% (v/v) formamide; 10 m$M$ EDTA, pH 8.0; 1 mg/ml xylene cyanol FF; 1 mg/ml bromphenol blue.

## Measurement of Gene Expression using Luciferase as a Reporter Gene

### In vitro Luciferase Activity Measurement

The expression of a light-regulated gene can be efficiently estimated by measuring the activity of a reporter gene in transgenic plants harboring a promoter–reporter gene fusion. The firefly luciferase gene[13] has been widely used as a reporter gene. The activity of this gene product can be detected by a fast and sensitive method as described below. This method makes use of the Promega luciferase assay system. The solutions are available from Promega but can also easily be prepared (see below). The efficiency of the assay buffer varies between different batches, therefore, test the efficiency of the buffer with a standard luciferase protein extract before measurement of unknown samples.

Homogenize approximately 10 etiolated or 2 green seedlings in 50 $\mu$l of extraction buffer at room temperature or on ice. This is best done with a pestle that fits exactly in a microfuge tube. Spin down the debris and

---

[13] D. W. Ow, K. V. Wood, M. Deluca, J. R. De Wet, D. R. Helinski, and S. H. Howell, *Science* **234,** 856 (1986).

keep the supernatant in the dark, on ice, for several minutes up to several hours to reduce the autofluorescence of chlorophyll. Mix 5 $\mu$l extract with 50 $\mu$l of assay buffer and immediately measure the light emission with a luminometer or scintillation counter. When using a scintillation counter, turn the coincidence circuit off. If this is not possible, the sensitivity of the measurement is lower and the luciferase activity is proportional to the square root of the measured counts. The protein content can be measured according to Bradford.[14]

*Reagents*

   Luciferase extraction buffer (Promega): 25 m$M$ Tris-phosphate, pH 7.8; 2 m$M$ DTT, 2 m$M$ 1,2-diaminocyclohexane-$N,N,N',N'$-tetraacetic acid, 10% glycerol, 1% Triton X-100.
   Luciferase assay buffer (Promega): 20 m$M$ Tricine; 1.07 m$M$ Mg(CO$_3$)$_4 \cdot$ Mg(OH)$_2$; 2.67 m$M$ MgSO$_4$; 0.1 m$M$ EDTA; 33.3 m$M$ DTT; 270 $\mu$M ATP; 470 $\mu$M Luciferin; 270 $\mu$M coenzyme A. This solution can be aliquoted and stably stored at $-70°$. Before use, allow the solution to equilibrate to room temperature.

*In vivo Imaging of Luciferase-Induced Luminescence*

   Luciferase has the unique property that one of its reaction products, light, can be detected *in vivo* without destroying the tissue. Advantages are that the luciferase activity of many seedlings can be measured at once. This allows the luciferase gene to be used for genetic screens: from a mutagenized population of transgenic promoter-luciferase plants, mutants can be selected and propagated that show an altered level of luciferase-induced luminescence. In addition, tissue-specific expression can be determined.[15] However, expensive video imaging equipment is required.

   A schematic representation of how video imaging equipment can be set up is shown in Fig. 2. The camera is mounted on a height-adjustable stand and the optics used are similar to those used in photography. It is useful to have a range of optics that allows measurement of a single to several hundreds of seedlings. The system includes an adjustable dim light source to make light images and for focusing purposes. Mounted on the base of the stand is a mold that exactly fits the plate containing the seedlings. For example, make a mold that fits Falcon 1015 plates. These plates have a grid on the bottom and are sufficiently high to allow growth of etiolated

---

[14] M. M. Bradford, *Anal. Biochem.* **72**, 255 (1976).
[15] A. J. Millar, S. R. Short, K. Hiratsuka, N.-H. Chua, and S. A. Kay, *Plant Mol. Biol. Rep.* **10**, 324 (1992).

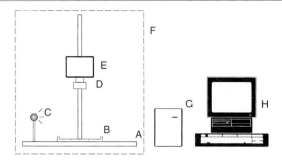

Fig. 2. Schematic representation of video imaging equipment. Video imaging equipment consists of A, height adjustable stand; B, mold that exactly fits a suitable plate; C, dim, adjustable light source; D, optics similar to those used in photography; E, intensified tube or cooled CCD camera; F, light-tight dark box; G, camera controller; H, computer.

seedlings. Alternatively, square plates can be used. The camera is contained in an absolutely dark chamber because the faintest light source can be detected by the camera. The computer is placed close to the camera for focusing purposes.

Measurement of luciferase-induced luminescence from transgenic etiolated or light-treated seedlings requires some practice. The seedlings need to be sprayed with a luciferin solution [5 m$M$ luciferin/0.01% (v/v) Triton X-100, filter sterilized]. The solution is stable for at least a week in the dark at 4°. The luciferin can be sprayed onto the seedlings using a standard sprayer from, e.g., an empty hairspray bottle. The seedlings should be sprayed approximately 24 hr before the actual measurement to bring down background levels of luciferase.[15] Etiolated or dark-adapted seedlings should be sprayed in the dark or under a dim green safelight because even a green safelight can induce expression of some genes. Spray again immediately before measurement, position the plate onto the mold, and count for a few minutes to several hours. The mold is especially convenient when measuring, e.g., a time course of induction since the plate can be exactly repositioned. The luciferase-induced luminescence between identical transgenic seedlings can differ considerably. This is probably due to the many variables that are difficult to control. Luciferase activity depends on its substrates, luciferin and oxygen. The distribution of these compounds is not tightly controlled and stomata of dark-grown plants are often closed as opposed to light-grown seedlings. Also we find that chlorophyll can give some quenching of the light signal. Another issue that should be taken into account when doing experiments with dark-grown seedlings is that the light emitted by the reaction may induce expression of a light-regulated gene by itself.

# [13] Translational Regulation of Chloroplast Gene Expression in *Chlamydomonas reinhardtii*

*By* AMYBETH COHEN, CHRISTOPHER B. YOHN, RICHARD K. BRUICK, and STEPHEN P. MAYFIELD

## Introduction

The photosynthetic apparatus of the chloroplast consists of four unique protein complexes (Photosystem I and II, cytochrome $b_6f$ and the ATPase complex) that collectively carry out the light reactions of photosynthesis. Each of these complexes is comprised of both nuclear- and chloroplast-encoded proteins. In Photosystem II (PSII), the core chloroplast-encoded polypeptides include D1, D2, P5, P6, and cytochrome $b559$, while the nuclear-encoded proteins include those of the oxygen evolving complex, chlorophyll $a/b$ binding proteins, and other less defined polypeptides. Expression of the nuclear-encoded proteins within PSII has been primarily attributed to transcriptional control, while translational control has been identified as a key regulatory step in the expression of the chloroplast-encoded proteins.[1,2] Translational regulation of chloroplast protein expression has been shown to require the interaction of nuclear-encoded proteins with the 5' untranslated regions (UTRs) of specific chloroplast-encoded mRNAs.[2,3] Determining the manner in which these nuclear-encoded factors control the translation of chloroplast-encoded mRNAs is essential to our understanding of translational regulation.

Translational regulation of chloroplast-encoded mRNAs has been studied extensively in the unicellular green alga *Chlamydomonas reinhardtii*. A number of nuclear-encoded proteins have been characterized within this organism that are required for translation of specific chloroplast-encoded, photosynthesis-related mRNAs.[2,3] These proteins have been found to act at different levels, including mRNA stability, processing, translation initiation, and elongation.[2,3] This chapter describes the methods utilized in our laboratory to study translational regulation of chloroplast mRNAs that encode for PSII proteins, with specific emphasis on the D1 protein, encoded by the *psbA* gene. The methods described here can be applied to studying

[1] J. E. Mullet, *Plant Physiol.* **103**, 309 (1993).
[2] S. P. Mayfield, C. B. Yohn, A. Cohen, and A. Danon, *Annu. Rev. Plant Physiol. Plant Molec. Biol.* **46**, 147 (1995).
[3] J.-D. Rochaix, *Plant Mol. Biol.* **32**, 327 (1996).

translation of other chloroplast-encoded mRNAs within *C. reinhardtii*, and to some degree within higher plants.

### Generation and Selection of Photosynthetic Mutants of *C. reinhardtii*

A number of nuclear mutants deficient in PSII have been isolated in *C. reinhardtii*.[2,3] Despite the absence of photosynthetic activity in these mutants, they can be maintained in media (liquid or solid) supplemented with an external carbon source. Within this set of nuclear mutants are those that lack the ability to translate specific chloroplast-encoded mRNAs. Molecular genetic and biochemical analysis has revealed a correlation between the absence of translation in these PSII-deficient mutants and an alteration in the binding activity of proteins to the 5′ UTRs of these chloroplast-encoded mRNAs.[2,3]

One of the simpler methods utilized to generate nuclear mutants of *C. reinhardtii* is insertion of exogenous DNA into the nuclear genome.[4-6] Use of a selectable marker gene allows for the growth of transformants on selectable media. In our laboratory we introduce by glass bead transformation[7] a plasmid containing the *C. reinhardtii* argininosuccinate lyase gene (*arg7*)[8] into a cell wall-deficient strain lacking a functional *arg7* gene (Arg7/cw15)[8] (Fig. 1). The DNA inserts randomly into the genome, often disrupting other functional genes. Transformants that contain the integrated gene can be selected on media lacking exogenous arginine. These transformants can then be screened for the mutation of interest. PSII-deficient transformants are screened for a high fluorescence (hf) phenotype, because they typically exhibit a high level of chlorophyll fluorescence.[9] Approximately 10% of the hf mutants selected are PSII deficient. To screen for those mutants lacking PSII only, immunoblot analysis using antisera raised against an individual protein in the PSII complex is used. A PSII-deficient mutant will not accumulate any of the core chloroplast-encoded polypeptides, because the loss of one protein results in destabilization of the entire complex.[10] Further analysis, as diagrammed in Fig. 1, must be carried out to characterize these mutants in greater detail.

[4] E. J. Smart and B. R. Selman, *Mol. Cell. Biol.* **11,** 5053 (1991).
[5] M. Adam, K. E. Lentz and R. Loppes, *FEMS Microbiol. Lett.* **110,** 265 (1993).
[6] L.-W. Tam and P. A. Lefebvre, *Genetics* **135,** 375 (1993).
[7] K. L. Kindle, *Proc. Natl. Acad. Sci. U.S.A.* **87,** 1228 (1990).
[8] R. Debuchy, S. Purton, and J.-D. Rochaix, *EMBO J.* **8,** 2803 (1989).
[9] P. Bennoun and P. Delepelaire, *in* "Methods in Chloroplast Molecular Biology" (M. Edelman, R. B. Hallick, and N.-H. Chua, eds.), p. 25. Elsevier Biomedical Press, Amsterdam, 1982.
[10] J. M. Erickson, M. Rahire, P. Malnoë, J. Girard-Bascou, Y. Pierre, P. Bennoun, and J.-D. Rochaix, *EMBO J.* **5,** 1745 (1986).

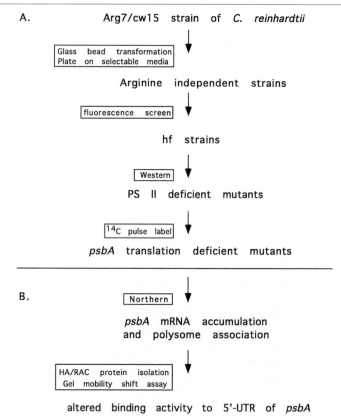

A.          Arg7/cw15  strain  of  *C.  reinhardtii*

Glass  bead  transformation
Plate  on  selectable  media

Arginine  independent  strains

fluorescence  screen

hf  strains

Western

PS  II  deficient  mutants

$^{14}$C  pulse  label

*psbA*  translation  deficient  mutants

B.          Northern

*psbA*  mRNA  accumulation
and  polysome  association

HA/RAC  protein  isolation
Gel  mobility  shift  assay

altered  binding  activity  to  5'-UTR  of  *psbA*

FIG. 1. Flowchart for generating, screening, and characterizing high fluorescence (hf), *psbA* translation-deficient mutants of *C. reinhardtii*. The result of each step is indicated by a description of the expected phenotype. The method of screening (A) and characterization (B) are boxed to the left of each step.

## Glass Bead Nuclear Transformation of C. reinhardtii

1. An *Arg7.8* DNA fragment,[8] subcloned into the pBS-SK plasmid (Stratagene, La Jolla, CA), is linearized using a restriction endonuclease that cleaves a site outside of the *Arg7* gene. The DNA is extracted once with chloroform to remove contaminating bacteria and/or fungus.

2. The *C. reinhardtii* Arg7/cw15 strain is grown in complete liquid media (Tris-acetate-phosphate [TAP])[11] containing 50 mg/l L-arginine to a density

[11] E. H. Harris, "The Chlamydomonas Sourcebook." Academic Press, San Diego, 1989.

of approximately $1 \times 10^6$ cells/ml with agitation under fluorescent lights. The cells are pelleted at 4000g for 5 min at 4°, washed once with TAP, pelleted again, and resuspended in TAP at approximately $5 \times 10^7$ cells/ml.

3. For each glass bead transformation,[7] 400 $\mu$l of cells is placed in a 15-ml sterile conical tube with 300 $\mu$l of sterile, acid-washed glass beads (0.5-mm average diameter). The linearized *Arg7.8*/pBS-SK DNA (0.1–2 $\mu$g in a small volume of TE, pH 8.0) and 5% (w/v) polyethylene glycol (MW 8000) are added to the glass beads. The transformation mixture is vortexed at full speed for 10 sec. After the beads have settled, the cells are carefully removed from the tube and immediately spread on TAP plates containing 1.5% agar. The plates are placed in dim light for 10–14 days until transformants appear.

*Comments.* This procedure is for cell wall-deficient strains of *C. reinhardtii* only. Autolysin treatment must be used on strains containing cell walls to remove the wall prior to addition of the DNA.[11] In addition, the quantity of DNA used should be determined empirically. Linearized *Arg7.8*/pBS-SK (1–2 $\mu$g), transformed into 400 $\mu$l of cells ($5 \times 10^7$ cells/ml), yields approximately 300–500 transformants per plate with mostly single insertion events. To maintain this transformation efficiency, the PEG solution should be prepared fresh every 6–8 weeks.

*Fluorescence Screening of hf Mutants*

1. *Arg7* transformants with a hf phenotype are identified with a video system as described by Bennoun and Delepelaire.[9] Plates containing *arg7* transformants are illuminated with a 300-W quartz-iodine lamp equipped with a blue glass filter (peak transmittance 480 nm; Oriel Corp., Long Beach, CA). A CCD camera (XC-77; Sony, San Jose, CA) with a red filter (Wratten Gelatin Filter 89B, peak transmittance 720 nm; Kodak, Rochester, NY) attached to the lens is used to visualize chlorophyll fluorescence in the transformed colonies via import of the video signal to a Macintosh computer operating NIH-Image 1.57 (public domain program written by Wayne Rasband at the U.S. National Institutes of Health and available from the Internet by anonymous ftp from zippy.nimh.nih.gov or on floppy disk from NTIS, 5285 Port Royal Rd., Springfield, VA 22161, part number PB93-504868) via a framegrabber (LG-3; Scion Corp., Frederick, MD). Integration of four to eight frames allows for easy visualization.

2. Hf mutants are selected by matching the pattern of colonies on the computer screen with those on the TAP plate. Hf mutants will fluoresce brighter than wild-type (wt) cells. The hf colonies are plated individually on fresh TAP plates for further analysis.

*Screening hf Mutants for a PSII Deficiency*

1. The hf mutant strains are grown in liquid TAP to a density of $5 \times 10^6$ cells/ml with agitation under fluorescent lights. Approximately 10 ml of the cell culture is pelleted at 4000g for 5 min at 4°.

2. The cells are resuspended in 0.5 ml of buffer A (750 m$M$ Tris-HCl, pH 8.0, 15% sucrose, 100 m$M$ 2-mercaptoethanol) and disrupted by sonication with a microtip probe (B. Braun, Allentown, PA; Sonic 2000) for 15 sec at 30% maximum power.

3. The lysed cells are pelleted at 10,000g for 15 min at 4°. The supernatant is removed and saved as the soluble protein fraction, while the cell pellet is resuspended in 100 $\mu$l of buffer A to yield the membrane protein fraction.

4. To 5 $\mu$l of protein (membrane or soluble), 5 $\mu$l of gel loading buffer [10% sodium dodecyl sulfate (SDS), 10% 2-mercaptoethanol, 400 m$M$ Tris-HCl, pH 6.8] and 10 $\mu$l of 8 $M$ urea are added. The proteins are heated at 60° for 5 min and separated by electrophoresis in a 12% 1-D SDS polyacrylamide gel.[12]

5. Following electrophoresis, the gels are incubated in 10 m$M$ CAPS (pH 11.0) for 20 min at room temperature, and electroblotted to nitrocellulose (Schleicher and Schuell, Kenne, NH) in 10 m$M$ CAPS (pH 11.0) containing 10% MeOH at 4°.

6. The nitrocellulose filters are blocked in 50 m$M$ Tris (pH 8.0), 150 m$M$ NaCl, 0.1% Tween 20 (TBS-T) containing 5% nonfat dry milk for 30 min at room temperature. Rabbit polyclonal antisera specific for D1 (or other PSII proteins) is added to the blots and incubated at room temperature for 4 hr. The filters are washed in TBS-T at room temperature, and then incubated in TBS-T containing goat anti-rabbit IgG alkaline phosphatase conjugate (Sigma, St. Louis, MO) for 3 hr at room temperature. The filters are washed in TBS-T and AP buffer (100 m$M$ Tris, pH 9.5, 100 m$M$ NaCl, 5 m$M$ MgCl$_2$) for 10 min at room temperature, respectively, prior to the addition of an alkaline phosphatase substrate (nitro blue tetrazolium and 5-bromo-4-chloro-3-indolyl phosphate) to visualize the proteins.

*In vivo* Labeling of Proteins with [$^{14}$C]Acetate

The analysis of translation products in PSII-deficient nuclear transformants of *C. reinhardtii* allows for detection of altered chloroplast protein synthesis in these mutants. Nuclear mutants of *C. reinhardtii* that are deficient in PSII do not synthesize one or more of the core chloroplast-encoded polypeptides (e.g., D1). The other polypeptides are still synthesized at wt or

---

[12] N.-H. Chua, *Methods Enzymol.* **69,** 434 (1980).

greater levels. Immunoblot analysis will not detect any of these chloroplast-encoded polypeptides, because the absence of one PSII protein results in the destabilization and degradation of the entire macromolecular complex.[10] Therefore, *in vivo* pulse labeling is necessary to identify within a particular hf mutant the PSII protein(s) whose synthesis is absent or significantly decreased.

To examine protein synthesis in *C. reinhardtii* cells, proteins can be labeled *in vivo* with either [$^{35}$S]methionine or [$^{14}$C]acetate. Addition of translation inhibitors to the labeling reaction, such as cycloheximide or chloramphenicol, allows for examination of protein synthesis solely in the chloroplast or cytoplasm, respectively. Labeling for short periods of time, as little as 2 min, coupled with SDS–PAGE (polyacrylamide gel electrophoresis) and autoradiography allows one to examine the relative rate of synthesis for individual proteins. These labeled proteins are also available for immunoprecipitation or other biochemical analysis.

We generally use [$^{14}$C]acetate for labeling *C. reinhardtii* cells, because this radioisotope is incorporated into proteins at a more uniform level than [$^{35}$S]methionine. The use of [$^{35}$S]methionine causes proteins with a higher ratio of sulfur to be radiolabeled to a greater extent than proteins with a lower sulfur content. Despite this discrepancy, [$^{35}$S]methionine incorporates into proteins quite well, and can be used in place of [$^{14}$C]acetate.

*In vivo Labeling of PSII-Deficient Nuclear Mutants of C. reinhardtii*

1. The *C. reinhardtii* strains are grown in liquid TAP with agitation under fluorescent lights to a density of $5 \times 10^6$ cells/ml. The cells are pelleted by centrifugation at 4000$g$ for 5 min at 4°, washed once in TAP lacking acetate (TAP-HCl),[11] and pelleted again under the same conditions.

2. The cells are resuspended at a density of $1 \times 10^7$ cells/ml in TAP-HCl. Fifty milliliters of cells is placed within a 125-ml Erlenmeyer flask and agitated under fluorescent lights for 1 hr to deplete any acetate reserves. At the end of the 1-hr incubation, drugs which inhibit translation are added. For labeling of chloroplast proteins, cycloheximide (Sigma) is added to a final concentration of 10 $\mu$g/ml 10 min prior to the addition of [$^{14}$C]acetate.

3. [$^{14}$C]acetate (4.2 mCi/mmol; ICN, Irvine, CA) is added in liquid form to a final concentration of 1 $\mu$Ci/ml and the cells are agitated under fluorescent light for an additional 10 min. Unlabeled NaOAc is then added to a final concentration of 50 m$M$, and the cells are immediately pelleted by centrifugation at 10,000$g$ for 2 min at 4°. The pelleted cells are frozen in liquid nitrogen and stored at $-70°$ indefinitely.

4. For the analysis of protein synthesis, SDS–PAGE of soluble and membrane protein fractions is carried out as described previously for immunoblot analysis of hf mutants (see steps 2–4 in previous section).

5. For autoradiography, gels are stained in 40% MeOH, 10% acetic acid, and 0.25% Coomassie Blue for 30 min and destained in 40% MeOH and 10% acetic acid to confirm that an equal amount of protein is loaded in each lane. The stained gel is then incubated with En³hance (DuPont–New England Nuclear, Boston, MA) for 30 min, placed into water for 30 min, and dried between two sheets of cellophane. The dried gel is then exposed to X-ray film (DuPont–New England Nuclear) at −70° to visualize labeled protein bands as shown in Fig. 2. The absence of D1 protein synthesis in the hf1 and hf2 PSII-deficient mutants is unambiguous when compared to D1 synthesis in wt cells. This pulse-labeling experiment shows that the nuclear mutations within each of these hf mutants affects the synthesis of an individual chloroplast-encoded PSII protein.

Preparation and Characterization of RNA

Message instability and defective translation initiation and elongation have been shown to result in the absence of chloroplast-encoded protein synthesis in several nuclear mutants in *C. reinhardtii* deficient in PSII.[2,3] It is important to characterize the corresponding mRNA to determine which posttranscriptional step of expression is affected. Analysis of the total and polysomal RNA levels, as well as polysomal distribution, can provide a significant amount of information about the status of a particular chloroplast

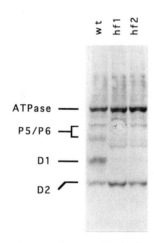

Fig. 2. Synthesis of chloroplast-encoded proteins in *C. reinhardtii*. [¹⁴C]Acetate pulse labeling of wt cells and the PSII-deficient hf1 and hf2 nuclear mutants with the addition of 10 μg/ml of cycloheximide to the labeling reaction. Equal quantities of labeled membrane proteins were loaded in each lane. Positions of the major components of PSII (D1, D2, P5, and P6) and ATPase are indicated.

mRNA within the translation machinery. Analysis of total RNA indicates the steady-state levels of a particular mRNA, while analysis of polysomal RNA identifies RNA actively associated with polysomes. The distribution of polysomes on an mRNA provides additional information that may differentiate between defects in initiation and elongation.

## Isolation of Polysomal RNA from C. reinhardtii Cells

1. The *C. reinhardtii* cells are grown in 250–500 ml of liquid TAP to approximately $5 \times 10^6$ cells/ml under fluorescent lights with agitation. The cells are pelleted at 4000$g$ for 5 min at 4°. The cells are washed with sterile $H_2O$, transferred to a 50-ml conical tube, and pelleted again in a clinical centrifuge at 4000 rpm for 5 min at 4°. The cell pellet is frozen in liquid $N_2$ and stored at −70°.

2. Frozen cells are ground in liquid $N_2$ using a mortar and pestle. The frozen cell powder is transferred to a 15-ml conical tube on ice with a sterile spatula and resuspended in 2.5 volumes of extraction buffer (200 m$M$ Tris HCl, pH 9.0, 200 m$M$ KCl, 35 m$M$ MgCl$_2$, 25 m$M$ EGTA, 200 m$M$ sucrose, 1% Triton-X 100, 2% polyoxyethelene-10-tridecyl-ether) with 0.5 mg/ml heparin, 0.7% 2-mercaptoethanol, and 100 $\mu$g/ml chloramphenicol added prior to extraction.[13,14] The cells are vortexed until the mixture is homogenous.

3. The cell extract is transferred to microcentrifuge tubes and incubated on ice for 10 min in order to solubilize membranes. The cell debris is pelleted by centrifugation at 14,000 rpm for 5 min at 4°.

4. The aqueous cell extract is transferred to new microcentrifuge tubes and 10% sodium deoxycholate is added to a final concentration of 0.05% to solubilize microsomal membranes. All tubes are mixed by inversion several times and then incubated on ice for 5 min. After centrifugation at 14,000 rpm for 15 min at 4°, the soluble cell extract is transferred to new microcentrifuge tubes on ice.

5. The cell extract (400 $\mu$l) is layered over a two-step cushion in a 13- × 51-mm polycarbonate centrifuge tube (Beckman, Fullerton, CA) consisting of 1.5 ml of 1.75 $M$ sucrose in 1× cushion buffer (40 m$M$ Tris HCl, pH 9.0, 200 m$M$ KCl, 30 m$M$ MgCl$_2$, 5 m$M$ EGTA, 0.5 mg/ml heparin, 0.7% 2-mercaptoethanol, 100 $\mu$g/ml chloramphenicol)[13,14] overlaid with 1.0 ml of 0.5 $M$ sucrose in 1× cushion buffer. Centrifugation of the cell extract is carried out in a TLA-100.3 fixed-angle rotor (Beckman) at 103,000$g$ for 3 hr at 4°.

[13] A. O. Jackson and B. A. Larkins, *Plant Physiol.* **57**, 5 (1976).
[14] A. Barkan, *EMBO J.* **7**, 2637 (1988).

6. The polysomal RNA, which has pelleted to the bottom of the tube, is recovered by careful removal of the sucrose layers (the pellet is translucent). The polysome pellet is briefly rinsed with a small volume of polysome pellet resuspension buffer (40 m$M$ Tris-HCl, pH 8.4, 200 m$M$ KCl, 30 m$M$ MgCl$_2$, 5 m$M$ EGTA, 0.5 mg/ml heparin, and 100 $\mu$g/$\mu$l chloramphenicol)[13,14] to remove residual sucrose, and then resuspended in 200 $\mu$l of resuspension buffer. The polysome suspension is transferred to a microcentrifuge tube and clarified by centrifugation at 14,000 rpm for 10 min at 4°. The polysomal suspension is transferred to a new microcentrifuge tube and EtOH precipitated with 2.5 volumes of EtOH and 0.1 volumes of 3 $M$ NaOAc (pH 5.3) at −20°.

7. The polysomal RNA is pelleted by centrifugation at 14,000 rpm for 15 min at 4°. The pellet is washed once with 70% EtOH, and inverted to dry. Each pellet is resuspended in a small volume of TE (pH 8.0) (10–20 $\mu$l for an RNA sample obtained from two sucrose cushions). The optical density of the RNA is measured on a spectrophotometer to determine the concentration. Crude polysomal RNA is stored at −70° after freezing in liquid N$_2$.

8. Northern analysis[15] is performed to characterize the polysomal RNA from a particular cell line. RNA (2 $\mu$g) is fractionated on a formaldehyde denaturing agarose gel, transferred by capillary action to a GeneScreen nylon membrane (DuPont–New England Nuclear) using 25 m$M$ NaPO$_4$ buffer (pH 6.5) as a transfer medium, and cross-linked to the membrane using the GS Gene Linker UV Chamber (Bio-Rad, Hercules, CA). Membranes are prehybridized for 30 min at 42° in 50% formamide, 5× SSPE (pH 7.5), 0.1% SDS, 0.5% nonfat dry milk, and 10 $\mu$g/ml salmon sperm DNA. Hybridizations are carried out in the same solution for 16 hr at 42° with the addition of a random-primed, $^{32}$P-labeled cDNA probe for the mRNA of interest (e.g., *psbA*). Membranes are washed twice at 60° for 20 min in 2× SSPE/0.1% SDS. Hybridization is detected by exposure of the membranes to X-ray film (DuPont–New England Nuclear) at −70°.

*Comments.* All steps should be carried out at 4° or on ice. All solutions, glassware, plasticware, etc., should be RNase free. We find autoclaving is sufficient; however, treatment with a solution of 0.1% diethyl pyrocarbonate or baking at high temperature (250° for 3 hr) may be necessary.[16] The gel apparatus in which the formaldehyde gels are cast is treated with 100 m$M$ NaOH and then washed several times with sterile water.

---

[15] J. Sambrook, E. F. Fritsch, and T. Maniatis, *in* "Molecular Cloning: A Laboratory Manual," Vol. 1, p. 7.37. Cold Spring Harbor Laboratory Press, New York, 1989.
[16] D. D. Blumberg, *Methods Enzymol.* **152**, 20 (1987).

*Isolation of Total RNA from Polysomal Cell Extract*

1. The crude *C. reinhardtii* cell extract used to recover polysomal RNA can be used to obtain high-quality total RNA. Approximately 300–400 $\mu$l of the cell extract is added to a microcentrifuge tube. SDS and EDTA (pH 8.0) are added to a final concentration of 0.5% and 20 m$M$, respectively. The tubes are mixed by inversion, extracted once with phenol/chloroform, and EtOH precipitated with 2.5 volumes of EtOH and 0.1 volumes of 3 $M$ NaOAc (pH 5.3) at $-20°$.

2. The total RNA is pelleted by centrifugation at 14,000 rpm for 15 min at 4°. The pellet is washed once with 70% EtOH and inverted to dry. Each pellet is resuspended in 200 $\mu$l TE (pH 8.0) and treated with 1 $\mu$l of RNase-free DNase (10 U/$\mu$l; Boehringer Mannheim, Indianapolis, IN) for 10 min at 37° in the presence of 6 m$M$ MgCl$_2$ to remove any endogenous DNA. The RNA is extracted once with phenol/chloroform and EtOH precipitated with 2.5 volumes of EtOH and 0.1 volumes of 3 $M$ NaOAc (pH 5.3) at $-20°$.

3. The total RNA is pelleted by centrifugation at 14,000 rpm for 15 min at 4°. The pellet is washed once with 70% EtOH and inverted to dry. Each pellet is resuspended in 40 $\mu$l TE (pH 8.0). The optical density of the RNA is measured on a spectrophotometer to determine the concentration. Total RNA is stored at $-70°$ after freezing in liquid N$_2$.

4. To analyze total RNA, Northern analysis[15] is performed as described in the previous section (see step 8). The total and polysomal RNA can be analyzed in parallel by fractionating 2 $\mu$g of each sample from a particular cell line. As shown in Fig. 3, total and polysomal RNA isolated from wt cells grown in constant light or dark were separated on a formaldehyde denaturing agarose gel and hybridized with a random-primed, $^{32}$P-labeled *psbA* cDNA probe. The steady-state level of *psbA* total RNA is approximately equal for cells grown in the light and in the dark. The amount of polysome-associated *psbA* mRNA isolated from wt cells grown in the dark, however, is considerably lower than the amount isolated from wt cells grown in the light. These results suggest that a lower amount of *psbA* is actively translated in the dark than in the light.

*Fractionation of Polysomes on Continuous Sucrose Gradients*

1. The crude *C. reinhardtii* cell extract used to recover polysomal and total RNA can be directly loaded onto 15–55% continuous sucrose gradients in 1× gradient buffer (40 m$M$ Tris-HCl, pH 8.0, 40 m$M$ KCl, 10 m$M$ MgCl$_2$) to examine the polysome distribution on a particular mRNA.

2. The sucrose gradients are prepared in advance in 13- × 51-mm polyallomer centrifuge tubes (Beckman) and stored at $-70°$ until needed.

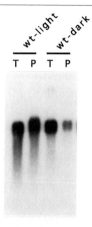

Fig. 3. Northern blot analysis of *psbA* mRNA accumulation wt cells grown in constant light or dark. Total (T) RNA and RNA associated with polysomes (P) were examined for the presence of the *psbA* mRNA.

To prepare gradients, 1.1 ml of 55% sucrose in 1× gradient buffer is carefully added to the bottom of several tubes. The tubes are placed at −70° in a secure, vertical position until the sucrose solution is frozen. The tubes are removed from −70° and 1.1 ml of 40% sucrose in 1× gradient buffer is added immediately. The tubes are again placed at −70° to freeze the next sucrose layer. These steps are repeated with 1.1 ml of 30% sucrose and 1.1 ml of 15% sucrose in 1× gradient buffer, respectively. The number of gradients needed is removed from −70° the night before use, and thawed without agitation at 4° to ensure proper gradient formation.

3. Approximately 100–200 μl of the crude cell extract from 1 g of cells is applied to each gradient. The polysomal RNA is sedimented by centrifugation in an SW-55 Ti rotor (Beckman) at 246,000g for 65 min at 4°.

4. Successive 400-μl aliquots are removed from the top of each gradient and placed in individual microcentrifuge tubes. To each sucrose fraction, SDS and EDTA (pH 8.0) are added to a final concentration of 0.5% and 20 mM, respectively. The tubes are mixed by inversion and then extracted once with phenol/chloroform. The supernatent is transferred to new tubes and 1 ml of 100% EtOH is added (without NaOAc). The RNA is recovered by centrifugation at 14,000 rpm for 15 min at room temperature to exclude precipitation of SDS.

5. The EtOH is removed and the tubes are inverted to dry. Each RNA pellet is resuspended in 40 μl of cold TE (pH 8.0). The RNA samples are frozen in liquid N₂ and stored at −70°.

6. To analyze the distribution of polysomes on a particular mRNA, Northern analysis[15] is performed as previously described, loading successive fractions on a formaldehyde denaturing agarose gel in order of their sedimentation value (from the top of the gradient to the bottom). As shown in Fig. 4, *psbA* mRNA from wt cells grown in constant light is associated with mid- to large-size polysomes (lanes 5–11). In addition, a significant amount of the transcript is found as free RNA or bound to RNP complexes (lanes 1–4), indicating that only a portion of the available *psbA* mRNA pool is actively translated.

*Comments.* To confirm that the observed sedimentation pattern of an mRNA is the result of polysome association, cell extracts can be treated with 20 m$M$ EDTA (pH 8.0), which acts to dissociate polysomes. The protocol described above is carried out, adding EDTA (pH 8.0) to the cell extract to a final concentration of 20 m$M$. EDTA-treated extract is loaded on 15–55% continuous sucrose gradients in 1× EDTA gradient buffer (1× gradient buffer with 10 m$M$ EDTA, pH 8.0, in place of 10 m$M$ MgCl$_2$).[14] Northern analysis of EDTA-treated samples from a linear gradient should reflect a redistribution of the RNA to the first few fractions of the gradient where free- and monosome-associated mRNA is located.

## Affinity Purification of Protein Factors that Specifically Interact with the 5′ Untranslated Region of Chloroplast mRNA

Two levels of purification may be performed in order to isolate protein factors that recognize a specific chloroplast mRNA. A subset of proteins

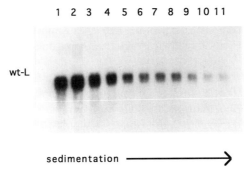

FIG. 4. Distribution of polysomes on the *psbA* RNA in wt cells grown in constant light (wt-L) as determined by a linear sucrose gradient. Gradients were divided into 11 equal fractions and an equal amount of each fraction was analyzed by Northern blot analysis. Sedimentation of the RNA proceeds from left to right.

likely to bind RNA can be obtained by passing a crude cellular lysate over heparin immobilized on an agarose support (heparin-agarose purification).[17] The negative charge of the heparin mimics the negative charge of the RNA phosphodiester backbone, thereby retaining proteins likely to bind to nucleic acids via electrostatic interactions. Proteins in this heparin-agarose (HA) purified fraction that specifically interact with the 5' UTR of an mRNA can then be affinity purified over a column of immobilized 5' UTR RNA.[17]

Once a set of protein factors has been isolated, the specificity for the RNA of interest may be assayed in a number of ways. Gel mobility shift assays[17,18] can provide information regarding the specificity of the protein factors for the RNA as well as highlight the RNA elements that are recognized by the proteins. These elements may be further mapped by footprinting techniques.[19] UV cross-linking experiments can identify which protein(s) from an isolated complex interact directly with the RNA.[17] However, it is the use of biochemical and genetic analysis together that will ultimately confirm the role of the isolated factors in translational regulation.

*Extraction of Crude Cell Lysate from C. reinhardtii Cells*

1. The *C. reinhardtii* cells are grown in 60–70 liters of liquid TAP to approximately $5 \times 10^6$ cells/ml under fluorescent lights with agitation. The cells are pelleted at 4000*g* for 5 min at 4°, washed with double distilled $H_2O$, and pelleted a second time. The cell pellets are frozen in liquid $N_2$ and stored at $-70°$.

2. The frozen cell pellet is resuspended in two volumes (v/w) of freshly prepared extraction buffer (10 m*M* Tris-HCl, pH 7.5, 10 m*M* NaCl, 10 m*M* $MgCl_2$, 2 m*M* DTT) at room temperature.

3. The cells are disrupted by placing the cell suspension in a chilled $N_2$ bomb (Parr Instrument Co., Moline, IL) under approximately 2000 psi of $N_2$ for 5 min. The cells are slowly evacuated into a flask on ice and disrupted a second time.

4. The cell debris is pelleted at 4000*g* for 5 min at 4°. The supernatent is transferred to 25- $\times$ 89-mm quick-seal tubes (Beckman) for centrifugation in a Ti 70 rotor (Beckman) at 208,000*g* for 45 min at 4°. The crude supernatant is transferred to a clean flask on ice. A yield of 200–400 ml is expected from 60–70 liters of cell culture.

[17] A. Danon and S. P. Mayfield, *EMBO J.* **10**, 3993 (1991).
[18] E. A. Leibold and H. N. Munro, *Proc. Natl. Acad. Sci. U.S.A.* **85**, 2171 (1988).
[19] C. Ehresmann, F. Baudin, M. Mougel, P. Romby, J. Ebel, and B. Ehresmann, *Nucleic Acids Res.* **15**, 9109 (1987).

*Comments.* Smaller samples can be prepared in which a liter of cells is grown for extraction. The cells are disrupted in an $N_2$ bomb as described above. The cell extract is transferred to 11- $\times$ 34-mm polyallomar tubes (Beckman) and the cell debris is pelleted at 160,000g in a TLS-55 rotor (Beckman) for 30 min at 4°.

## *Heparin-Agarose Affinity Chromatography of C. reinhardtii Crude Cell Extract*

1. The fast protein liquid chromatography [FPLC] (500 series; Pharmacia, Piscataway, NJ) pumps are washed with either 50 ml of low salt buffer (LS, 20 mM Tris-HCl, pH 7.5, 3 mM $MgCl_2$, 0.1 mM EDTA, pH 8.0, 2 mM DTT) or 50 ml of high salt buffer (HS, 2 M KoAc, 20 mM Tris-HCl, pH 7.5, 3 mM $MgCl_2$, 0.1 mM EDTA, pH 8.0, 2 mM DTT). A 50-ml heparin actigel (Sterogene, Carlsbad, CA) column is washed with 100 ml HS buffer for approximately 30 min, followed by 100–200 ml LS buffer for 30–60 min until a stable baseline is established using a UV detector with a 280-nm filter.

2. The crude *C. reinhardtii* supernatent is placed on ice and loaded onto the column via a P-1 peristaltic pump (Pharmacia) at a flow rate of 2 ml/min. The UV detector will record the flow-through of proteins that do not bind to the heparin matrix as the column is washed with LS buffer.

3. Proteins are eluted from the column in 5-ml fractions with a 0–8% HS gradient of 100 ml. The protein fractions are placed into sterilized[15] Spectra/Por dialysis membrane tubing (Baxter, McGaw Park, IL) with a MW cutoff of 8000–10,000. Dialysis of the proteins is carried out in a large volume (2–4 liters) of dialysis buffer (DB, 20 mM Tris-HCl pH 7.5, 100 mM KoAc, 0.2 mM EDTA, pH 8.0, 2 mM DTT, 20% glycerol) at 4° overnight. The HA fractions are removed from the dialysis tubing and stored at −70°.

4. The HA fractions are assayed for RNA-binding activity using a gel mobility shift assay coupled with RNase protection.[17,18] For a typical gel-shift assay, 1–2 $\mu$l of a HA fraction is preincubated for 15 min at room temperature with 0.5 U Prime RNase Inhibitor (5 Prime $\rightarrow$ 3 Prime, Inc., Boulder, CO) in a final volume of 15 $\mu$l DB supplemented with 4 mM $MgCl_2$. To each sample, 0.065 pmol ($6 \times 10^5$ cpm) of *in vitro* transcribed $^{32}$P-labeled RNA and 20 $\mu$g of wheat germ tRNA (Sigma) are added and the reaction is incubated at room temperature for 5 min. RNase T1 (0.25 U; GIBCO-BRL, Grand Island, NY) is then added to the reaction mixture for 5 min at room temperature to digest RNA at sites that are not protected by protein binding. The samples are loaded onto a 5% nondenaturing polyacrylamide/1$\times$ TBE gel for electrophoresis at 130 V. A marker lane of DB containing bromphenol blue and Orange G is run in parallel with

the samples. The free probe will migrate just below the bromphenol blue marker. The gel is fixed in 40% MeOH, 10% acetic acid, and 0.025% Coomassie Blue for 20 min at room temperature with shaking. The gel is dried between cellophane and exposed to X-ray film (DuPont–New England Nuclear). RNA-binding activity is indicated by a high mobility protein–RNA complex in the gel relative to the free probe as shown in Fig. 5. HA-purified proteins isolated from wt *C. reinhardtii* cells and the hf3 PSII-deficient mutant that lacks D1 synthesis were incubated with an *in vitro* transcribed *psbA* 5′ UTR probe. The RNA-binding activity of hf3 HA-purified proteins to the radiolabeled *psbA* RNA is clearly lower than the RNA-binding activity of wt HA-purified proteins. This difference in RNA-binding activity indicates that the hf3 strain contains an altered profile of *psbA* RNA-binding protein(s) relative to wt.

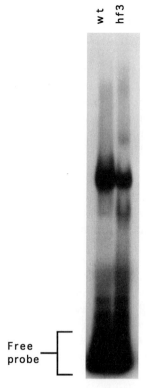

Fɪɢ. 5. *psbA* RNA-binding activity in wt cells and the PSII-deficient hf3 strain. Binding activity from 6 μg of protein was measured by RNase T1 protection coupled with a gel mobility shift assay. The unbound, labeled RNA probes digested by the RNase are indicated (Free Probe).

*Comments.* A 5-ml Econo-Pac heparin column (Bio-Rad) can be used when HA samples are prepared from 1 liter of cells. The proteins are eluted from the column in 1-ml fractions over a 0–50% HS gradient of 15 ml. All FPLC buffers must be passed through a 0.45-$\mu$m filter to prevent debris from entering the FPLC.

### RNA-Affinity Chromatography (RAC) of Heparin Agarose-Purified Proteins

1. An RNA affinity column[17] is prepared by first washing Amino gel 1702 (Sterogene) 4× in 10 ml of 12 m$M$ NaBH$_4$.

2. An equal volume of 20 m$M$ NaIO$_4$ is added to *in vitro* transcribed RNA, and the mixture is incubated in the dark on ice for 30 min.

3. The RNA mixture is added to 1 ml of the washed amino gel slurry. The final NaBH$_4$ concentration is adjusted to 10 m$M$. The mixture is agitated at room temperature for 2 hr as the 3′-terminal ribose residue of the RNA is coupled to the amino gel. An optimal ratio of RNA : amino gel matrix is 0.5 mg RNA : 0.4 ml amino gel slurry.

4. The RNA–amino gel matrix is transferred to a 20-ml poly prep chromatography column (Bio-Rad) and washed with 10 ml H$_2$O, followed by 15 ml high salt buffer (HS, 20 m$M$ Tris-HCl, pH 7.5, 1 $M$ KoAc, 3 m$M$ MgCl$_2$, 0.2 m$M$ EDTA, pH 8.0, 2 m$M$ DTT) and 5 ml low salt buffer (LS, 20 m$M$ Tris-HCl, pH 7.5, 100 m$M$ KoAc, 3 m$M$ MgCl$_2$, 0.2 m$M$ EDTA, pH 8.0, 2 m$M$ DTT).

5. For each 100 $\mu$l of RNA–amino gel column matrix, 1 ml of HA-purified protein is loaded. Prior to loading the HA proteins onto the RNA–amino gel column, Prime RNase Inhibitor (5 Prime → 3 Prime, Inc.), wheat germ tRNA (Sigma), and MgCl$_2$ are added to a final concentration of 5.2 U/ml, 1.3 mg/ml, and 3 m$M$, respectively. The HA protein sample is incubated at room temperature for 20 min.

6. The HA protein sample is applied to the column. The flow-through is collected and reapplied. The column is washed with 5 ml LS buffer, followed by 5 ml LS buffer + RNA (LS buffer containing 0.6 mg/ml wheat germ tRNA and 17 $\mu$g/ml of *C. reinhardtii* total RNA from a strain lacking the RNA of interest) and 5 ml LS buffer.

7. To collect the proteins bound to the RNA matrix, 1.5 column volumes of elution buffer (1 : 1 LS buffer : HS buffer) are passed through the column. The eluted proteins are diluted with 4 volumes of no salt buffer (NS, 20 m$M$ Tris-HCl, pH 7.0, 3 m$M$ MgCl$_2$, 0.2 m$M$ EDTA, pH 8.0, 2 m$M$ DTT). The column is washed with 5 ml HS buffer and 5 ml LS buffer. The diluted protein is reapplied to the column, washed, and eluted a second time.

8. The RAC-purified proteins can be stored under two different conditions. First, glycerol is added to a final concentration of 20% prior to storage at $-70°$. This concentrated protein is appropriate for analysis by 1-D and 2-D PAGE to examine the profile of those proteins that bound to the 5' UTR of an RNA. Alternatively, the proteins are diluted with NS buffer and glycerol so that the final concentration of KoAc is 150 m$M$ and the final concentration of glycerol is 20%. Again these proteins are stored at $-70°$. This diluted RAC-purified protein is appropriate for use in assays that require RNA-binding activity.

*Comments.* The efficiency of RNA coupling to the amino gel matrix can be followed by adding a small amount of [32]P-labeled RNA to the coupling reaction. Isolation of proteins specific for the RNA of interest is further enhanced by the use of competitor RNAs that sequester general RNA-binding proteins. Typically, tRNA is used as a competing RNA. However, we find that it is not necessarily a good mimic of cellular mRNAs. We prefer to use total cellular RNA from a *C. reinhardtii* mutant lacking the mRNA of interest. However, while it is likely that the natural targets for general RNA-binding proteins are represented in total cellular RNA, it is also possible that protein factors specific for a small subset of RNAs, including the one of interest, may also be competed.

Future Prospects

Use of the methods described here will allow for detection of PSII-deficient mutants with specific characteristics of interest. Additionally, an *arg7*-tagged gene within a particular PSII-deficient mutant, as well as the genes encoding the proteins that bind to the 5' UTR of the mRNA, can be isolated using standard cloning techniques. Characterization of these genes may determine the role of the corresponding proteins in translational regulation. The use of molecular, genetic, and biochemical techniques will enable greater insight into the manner in which nuclear-encoded proteins control chloroplast gene expression.

# [14] Use of Antisense Mutants to Study Regulation of Photosynthesis during Leaf Development

*By* STEVEN RODERMEL

## Antisense RNA Technology

Antisense RNA is an efficient means of down-regulating gene expression, and is currently the only effective means of inhibiting the expression of nuclear genes that are members of multigene families. The pioneering experiments of Ecker and Davis[1] paved the way for the use of antisense RNA technology in plants, and inhibition of target gene expression has been observed in transient expression studies, as well as in transgenic plants that have been stably transformed with a variety of constructs.[2–4] Plants also contain naturally occurring antisense mRNAs, which raises the intriguing possibility that antisense RNA may be a normal means of regulating plant gene expression.[4,5]

Despite the importance of artificial and natural antisense RNAs in plants, the mechanism(s) by which antisense RNAs exert their effects are largely unknown. In animals, inhibitory effects have been demonstrated at the levels of target gene transcription, target mRNA processing, transport from the nucleus, and translation.[5] In some cases, target : antisense duplex RNAs have been observed, but in most cases their existence has only been inferred; such structures may be short lived because of rapid turnover. In plants there are few complementary studies concerning the mechanism of antisense RNA inhibition.[4,6] However, because decreased target mRNA levels are generally (though not universally) observed in successful antisense experiments in plants, it is likely that gene expression is down-regulated, at least in part, as a consequence of posttranscriptional degradation of a duplex RNA species.

Antisense RNA technology is gaining widespread currency in plants as a means to dissect the regulation of complicated developmental and

[1] J. Ecker and R. W. Davis, *Proc. Natl. Acad. Sci. U.S.A.* **83,** 5372 (1986).
[2] A. R. van der Krol, J. N. Mol, and A. R. Stuitje, *Gene* **72,** 45 (1988).
[3] S. Rodermel and L. Bogorad, *in* "Antisense RNA and DNA" (J. A. H. Murray, ed.), p. 121. Wiley-Liss, New York, 1992.
[4] J. E. Bourque, *Plant Science* **105,** 125 (1995).
[5] J. A. H. Murray and N. Crockett, *in* "Antisense RNA and DNA" (J. A. H. Murray, ed.), p. 1. Wiley-Liss, New York, 1992.
[6] C.-Z. Jiang, D. Kliebenstein, N. Ke, and S. Rodermel, *Plant Mol. Biol.* **25,** 569 (1994).

biochemical pathways. This chapter describes the use of this technology to understand the impact of photosynthesis on leaf development, using as a paradigm antisense DNA mutants of tobacco with reduced amounts of ribulose bisphosphate carboxylase (Rubisco)—the key regulatory enzyme of photosynthetic carbon assimilation.[7]

## Rubisco Antisense DNA Mutants

### Rubisco

Rubisco catalyzes the first step of both the photosynthetic carbon assimilation and photorespiratory pathways.[7] The holoenzyme is localized in the chloroplast stroma of photosynthetic eukaryotes and is the most abundant protein in the world.[8] Rubisco is typical of most multimeric protein complexes in chloroplasts and is composed of subunits coded for by genes in the nucleus and the organelle.[9] The holoenzyme is composed of eight small subunit (SS) proteins and eight large subunit (LS) proteins.[7] The SS are coded for in the nuclear DNA by a small multigene (rbcS) family, translated as precursors on cytoplasmic 80S ribosomes, and transported into the plastid posttranslationally.[10] The LS, on the other hand, are coded for by a single gene (rbcL) on the multicopy plastid genome, and translated on plastid 70S ribosomes.[10] Assembly of the SS and LS into the holoenzyme is mediated by chaperonin 60.[7]

### Mutant Generation

Antisense RNA technology offers an alternative to classical mutant selection procedures to generate mutations in genes that are members of multigene families (such as rbcS). In designing an antisense experiment, several considerations must be addressed.

*Target Gene Considerations.* It is generally thought that the antisense construct should have high homology to the target gene.[2-4] Consequently, if the target gene is a member of a multigene family, sequences should be used in the antisense construct that are complementary to RNAs from the most highly expressed member of that gene family.

In the case of the Rubisco antisense mutants, Southern analyses revealed that there are at least five rbcS gene family members in *Nicotiana tabacum*.[11]

---

[7] S. Gutteridge and A. A. Gatenby, *Plant Cell* **7,** 809 (1995).
[8] R. J. Ellis, *Trends Biochem. Sci.* **4,** 241 (1979).
[9] L. Bogorad, *Cell Cult. Somat. Cell Genet. Plants* **7B,** 447 (1991).
[10] K. Keegstra, L. J. Olsen, and S. M. Theg, *Annu. Rev. Plant Physiol. Plant Mol. Biol.* **40,** 471 (1989).
[11] B. J. Mazur and C.-F. Chui, *Nucleic Acids Res.* **13,** 2373 (1985).

Two members of this family, which share nearly 100% nucleotide sequence identity, account for the vast bulk of *rbcS* mRNA production in the leaves of this species. The sequences of these genes are nearly 100% identical to cloned *rbcS* sequences in pSEM1, a *Nicotiana sylvestris rbcS* cDNA.[12] Because of its ready availability, pSEM1 was used in generating the Rubisco antisense DNA mutants.

*Choice of Target Sequences to use in the Antisense Construct.* Systematic studies in a handful of plant systems have demonstrated that there are no general "rules" in choosing the region and length of a target gene to use in the antisense construct.[2–5] In some cases, 5' regions are the most effective, while in other cases 3' regions are the most effective. This variability may be due to sequence-specific affects that influence not only the inherent stability of the antisense RNA, but also its capacity to give rise to secondary structures that might interfere with target: antisense RNA duplex formation.

In the case of the Rubisco antisense mutants, a 322-bp fragment of DNA from the 5' region of pSEM1 was used in the antisense construct. This fragment includes *rbcS* sequences from 22-bp upstream to 300-bp downstream of the initiation codon of the gene. We have not targeted other regions of the *rbcS* gene for antisense disruption. However, we have had success using near full-length cDNAs in generating antisense mutants for Rubisco activase, the enzyme that activates Rubisco.[13] Therefore, our current strategy is to engineer several constructs, including one that contains a near full-length cDNA, one that contains sequences from the 5' region of the gene, and one that contains sequences from the 3' region of the gene.

*Promoter Considerations.* Conventional wisdom suggests that antisense RNAs must be present in high quantities to exert an inhibitory effect.[2–5] Antisense RNA abundance is a function of its rate of synthesis and destruction. The latter may be influenced by one's choice of target sequences, while the former is influenced by one's choice of promoter to drive transcription of the antisense sequence. The cauliflower mosaic virus (CaMV) 35S promoter is often used for this purpose because it gives high-level "constitutive" transcription in many tissues.[2–4] However, successful antisense experiments have been reported using various other promoters (e.g., the *Lhcb* promoter, for the chlorophyll *a/b*-binding proteins of PSII, to give light-regulated, tissue-specific expression).[4]

In the case of the Rubisco antisense construct, high levels of antisense expression were deemed crucial because *rbcS* is one of the most highly

[12] M. Pinck, E. Guilley, A. Durr, M. Hoff, L. Pinck, and J. Fleck, *Biochimie* **66,** 539 (1984).
[13] C.-Z. Jiang, W. P. Quick, R. Alred, D. Kliebenstein, and S. R. Rodermel, *Plant J.* **5,** 787 (1994).

transcribed and expressed genes in photosynthetic cells. We thus chose the 35S promoter to drive transcription of the *rbcS* antisense DNA fragment. Our antisense construct was engineered by cloning the 322-bp *rbcS* fragment from pSEM1 in reverse orientation into a 35S promoter/nopaline synthase (*nos*) terminator cassette (pAC1352) in such an orientation that transcription from the 35S promoter would result in the production of antisense *rbcS* mRNAs in transgenic plants.[14] This vector also contains between its left and right T-DNA borders a copy of the *NPTII* gene, conferring kanamycin resistance on transgenic plants. The final construct was named "pTASS" (Tobacco Anti-Small Subunit).

*Growth Medium.* Because Rubisco is the key regulatory enzyme in photosynthetic carbon assimilation, it was thought that any reductions in the amount of this enzyme might be lethal for plant growth under photoautotrophic conditions. Consequently, sucrose was added to the selection medium.

pTASS was introduced into tobacco (*N. tabacum,* cv. SR1) leaf disks by *Agrobacterium*-mediated DNA transfer methods. Kanamycin-resistant plants growing on B5 tissue culture medium (supplemented with 1% sucrose) were randomly selected for further study; nontransformed SR1 tobacco plants maintained under identical conditions served as controls. Genomic Southern hybridization experiments revealed that most of the transformed plants contained a single copy of the *rbcS* antisense DNA, but that one transformant (transformant "5") contained at least four copies of this sequence.

*Nuclear-Chloroplast Interactions in the Rubisco Antisense Transformants: Regulation of LS Protein Accumulation*

*rbcS* and *rbcL* transcript levels were assessed in the transformed and control plants by RNA gel blot experiments. Total cell RNA was isolated from the top three expanding leaves of plants growing on tissue culture medium (supplemented with sucrose), and equal amounts of RNA were electrophoresed through 1.2% MOPS-formaldehyde gels and transferred to nylon filters. The filters were probed with DNAs specific for the tobacco *rbcS* or *rbcL* genes. Densitometric analyses of autoradiographs of the filters revealed that steady-state *rbcS* mRNA levels were depressed up to 90% in the transformed plants. In contrast, *rbcL* mRNA levels were normal in the mutants.

To assess SS and LS protein levels, soluble protein fractions were isolated from the top three expanding leaves of each plant, and equal amounts

---

[14] S. R. Rodermel, M. S. Abbott, and L. Bogorad, *Cell* **55,** 673 (1988).

of protein from each sample were subjected to 15% discontinuous sodium dodecyl sulfate–polyacrylamide gel electrophoresis (SDS–PAGE). The SS and LS bands on these gels were identified by Western immunoblot analysis using antibodies generated against the tobacco SS and LS proteins. Densitometric analyses revealed that the SS and LS proteins were coordinately reduced in amount up to 90% in the mutants. This indicates that the stoichiometry of the two proteins was the same in the mutant and wild-type plants—i.e., excess LS or SS pools did not exist. Because the reductions in LS protein amounts were not matched by corresponding reduction in *rbcL* mRNA amounts, LS protein accumulation is regulated posttranscriptionally in the antisense plants.

*In vivo* pulse-labeling experiments were conducted as a first approach to ascertain the nature of this posttranscriptional control.[15] In these experiments, the top three expanding leaves from plants growing on sucrose-containing tissue culture medium were supplied with [$^{35}$S]methionine for 10 min. Soluble proteins were then isolated and aliquots containing equal cpm were electrophoresed through discontinuous SDS–polyacrylamide gels. Label incorporation into the LS was monitored by PhosphorImage analysis. These experiments revealed that much less label was incorporated into the LS during the pulse in the antisense plants. Although this suggests that *rbcL* mRNA translation is reduced in the mutants, it is also possible that LS turnover is markedly enhanced in these plants.

To investigate directly whether *rbcL* mRNA translation might be impeded in the antisense plants, the polysomal distribution of *rbcL* mRNAs was examined on 12.5–55% analytical sucrose gradients. Because polysomes have higher sedimentation constants than monosomes or free RNAs, the migration of a given mRNA species on these gradients reveals the proportion of an mRNA species in the cell that is associated with polysomes, and thus likely being translated.[16–19] These analyses showed that *rbcL* mRNAs from the antisense plants sediment significantly higher than normal in the gradient, indicating that *rbcL* mRNAs associate with fewer polysomes in these plants. This suggests that *rbcL* mRNA translation initiation is defective in the mutants. This defect appears to be specific for *rbcL* mRNAs, because the polysomal distributions of mRNAs from other representative nuclear and plastid photosynthesis genes (as well as their respective protein levels) are normal.

[15] S. Rodermel, J. Haley, C.-Z. Jiang, C.-H. Tsai, and L. Bogorad, *Proc. Natl. Acad. Sci. U.S.A.* **93**, 3881 (1996).

[16] J. O. Berry, D. E. Breiding, and D. F. Klessig, *Plant Cell* **2**, 795 (1990).

[17] J. O. Berry, J. P. Carr, and D. F. Klessig, *Proc. Natl. Acad. Sci. U.S.A.* **85**, 4190 (1988).

[18] P. Klaff and W. Gruissem, *Plant Cell* **3**, 517 (1991).

[19] A. Barkan, *Plant Cell* **5**, 389 (1993).

The factors that regulate chloroplast translation initiation are poorly understood, but nuclear factors that bind to the 5' untranslated region (UTR) sequences of several chloroplast mRNAs have been identified.[20-24] These factors may activate translation, perhaps by facilitating ribosome binding. The fact that the SS, either directly or indirectly, specifically affects recruitment of ribosomes to *rbcL* mRNA suggests that control of *rbcL* mRNA translation initiation is elicited by one or more gene-specific factors. For example, the SS could be acting as a translational activator, or perhaps as a cofactor or coderepressor. With less SS, less cofactor (or coderepressor) would be present, resulting in less *rbcL* mRNA translation initiation. Alternatively, excess LS or its degradation products could serve to repress *rbcL* mRNA translation initiation when the SS is limiting. The latter mechanism would resemble end-product inhibition at the translational level, as observed for some bacterial genes (e.g., r-protein genes),[25] and may thus be a relic of the prokaryotic nature of the chloroplast and its endosymbiont origins.[26]

## Regulation of Rubisco and Photosynthesis during Leaf Development

### *Patterns of Photosynthesis during Leaf Development*

Photosynthetic rates change during the ontogeny of dicot leaves: they increase during leaf expansion, attain a maximum at full leaf expansion, then undergo by a prolonged senescence decline in the fully expanded leaf.[27] The elements that control these alterations in photosynthetic rates have been broadly defined in a handful of species, and these studies have suggested that photosynthetic rates are largely controlled by Rubisco content (and activity) in some cases, but by the contents (and activities) of other components of the photosynthetic apparatus in other cases.[27-29] It is

[20] J. D. Rochaix, *Annu. Rev. Cell Biol.* **8,** 1 (1992).
[21] N. W. Gillham, J. E. Boynton, and C. R. Hauser, *Annu. Rev. Genet.* **28,** 71 (1994).
[22] S. P. Mayfield, C. B. Yohn, A. Cohen, and A. Danon, *Annu. Rev. Plant Physiol. Plant Mol. Biol.* **46,** 147 (1995).
[23] A. Danon and S. P. Mayfield, *EMBO J.* **10,** 3993 (1991).
[24] A. Danon and S. P. Mayfield, *Science* **266,** 1717 (1994).
[25] B. Lewin, "Genes V." Oxford University Press, New York, 1994.
[26] L. Bogorad, *in* "On the Origins of Chloroplasts" (J. A. Schiff, ed.), p. 277. Elsevier/North Holland, Amsterdam, 1982.
[27] S. Gepstein, *in* "Senescence and Aging in Plants" (L. D. Noodén and A. C. Leopold, eds.), p. 85. Academic Press, San Diego, 1988.
[28] C. J. Brady, *in* "Senescence and Aging in Plants" (L. D. Noodén and A. C. Leopold, eds.), p. 147. Academic Press, San Diego, 1988.
[29] C.-Z. Jiang, S. R. Rodermel, and R. M. Shibles, *Plant Physiol.* **101,** 105 (1993).

likely that these alterations are part of a leaf developmental program that specifies changes in the expression of nuclear and chloroplast genes for photosynthetic proteins. We have shown in soybean, for example, that photosynthetic rates correspond very closely throughout leaf development to Rubisco activity and content, and that coordinate changes in SS and LS protein abundance during this process closely parallel changes in *rbcS* and *rbcL* steady-state mRNA levels.[29] The elements that integrate *rbcS* and *rbcL* transcription are not understood.

The Rubisco antisense DNA mutants are an ideal tool to assess the impact of decreases in Rubisco on leaf developmental processes. One general question is whether leaf developmental programming is sensitive to feedback regulation by the concentrations of various photosynthetic proteins. For example, are there elements of tobacco leaf development that depend on Rubisco expression for their execution?

*Experimental System*

Leaf development is commonly studied by examining a leaf at a given nodal position as a function of leaf age (i.e., during expansion and after full expansion). Because invasive procedures that destroy the leaf are often needed, a large number of replicate isogenic plants are required for these types of analyses.[28–30] One of the original primary *rbcS* antisense transformants (line "5") had at least four copies of the antisense DNA, and segregation of the various copies of this sequence in progeny plants gave rise to a spectrum of plants with Rubisco levels ranging from 10 to 100% of normal.

Because of their wide spectrum of Rubisco concentration, the antisense "5" plants have been notably useful for detailed "flux-control" analyses to study the impact of Rubisco on photosynthesis and plant growth.[31–36] These studies quantified the control that Rubisco exerts on photosynthesis

[30] L. L. Hensel, V. Grbic, D. A. Baumgarten, and A. B. Bleecker, *Plant Cell* **5,** 553 (1993).
[31] W. P. Quick, U. Schurr, R. Scheibe, E.-D. Schulze, S. R. Rodermel, L. Bogorad, and M. Stitt, *Planta* **183,** 542 (1991).
[32] W. P. Quick, U. Schurr, K. Fichtner, E.-D. Schulze, S. R. Rodermel, L. Bogorad, and M. Stitt, *Plant J.* **1,** 51 (1991).
[33] M. Stitt, W. P. Quick, U. Schurr, E.-D. Schulze, S. R. Rodermel, and L. Bogorad, *Planta* **183,** 555 (1991).
[34] W. P. Quick, K. Fichtner, E.-D. Schulze, R. Wendler, R. C. Leegood, H. Mooney, S. R. Rodermel, L. Bogorad, and M. Stitt, *Planta* **188,** 522 (1992).
[35] K. Fichtner, W. P. Quick, E.-D. Schulze, H. A. Mooney, S. R. Rodermel, L. Bogorad, and M. Stitt, *Planta* **190,** 1 (1993).
[36] M. Lauerer, D. Saftic, W. P. Quick, C. Labate, K. Fichtner, E.-D. Schulze, S. R. Rodermel, L. Bogorad, and M. Stitt, *Planta* **190,** 332 (1993).

in plants growing under varying conditions of light, humidity, $CO_2$, or nutrient supply. These experiments required measurements on a large number of first fully expanded leaves of individual plants, each of which had to be tested for its Rubisco content. Physiologic studies, on the other hand, were conducted on whole plants, each of which also had to be tested for its Rubisco concentration.

Because isogenic lines of transformant "5" are not available, we chose to examine the developmental gradient of fully expanded leaves at progressive nodes on wild-type and antisense plants with reductions of either 60 or 80% in Rubisco amount.[37] The growth of these plants was greatly retarded, but they ultimately attained the same height and formed about the same number of leaves as wild-type plants. Direct comparisons could thus be made between developmentally similar leaves (at approximately the same nodal position) on these plants. The measurements were made just prior to flowering.

We have recently found that plant growth is suppressed during a specific phase of shoot morphogenesis in the mutants. In particular, the juvenile phase of shoot development is delayed in the mutant, while the adult and reproductive phases are normal. This suggests that the antisense plants are limiting in some substance required for normal early growth. One possibility is that there is a threshold source strength that is required for the antisense plants to undergo the juvenile-to-adult phase transition. Consistent with this interpretation, all the leaves of the antisense plants, regardless of nodal position, are much longer lived than those from the wild-type plants.

*Control of Photosynthetic Rates by Rubisco*

To examine the factors that influence photosynthetic rates during leaf development in the antisense and wild-type plants, we measured $CO_2$ exchange rates (CER), stomatal conductances, intercellular $CO_2$ concentrations (Ci), chlorophyll contents, chlorophyll *a/b* ratios, total protein contents, Rubisco concentrations, Rubisco "initial" and "total" activities (determined *in vitro*) and Rubisco activation states (the ratio of "initial" to "total" activities). "Initial" Rubisco activities provide an estimate of the amount of activated (carbamylated) enzyme in the leaf sample at the time of harvest, whereas "total" activities provide a measure of the amount of Rubisco that is capable of being activated in the sample. "Total" activities should reflect Rubisco concentrations.

Our experiments revealed, first, that maximal CERs were attained in fully expanded leaves located several nodes from the top of the plant;

---

[37] C.-Z. Jiang and S. R. Rodermel, *Plant Physiol.* **107,** 215 (1995).

CERs progressively declined in subsequent fully expanded leaves—i.e., as one moved down the canopy. This pattern was evident for both the wild-type and antisense plants, and resembled what is normally observed during the expansion of an individual dicot leaf[27]—i.e., after attaining a maximum in the expanding leaf, photosynthetic rates undergo a senescence decline in the fully expanded leaf. Although they displayed similar profiles of change in photosynthetic rates, CERs were markedly depressed in the antisense plants.

The alterations in CER in the mutant and wild-type plants were accompanied by very similar changes in Rubisco initial and total activities. The coincident patterns of change in initial and total activities suggest that the activation state of the enzyme was constant throughout leaf development in both sets of plants. The alterations in CER and Rubisco activities were also accompanied by corresponding alterations in Rubisco concentrations, as determined by densitometry of LS and SS bands on stained SDS–PAGE gels. In contrast, neither chlorophyll content, Cs, nor Ci closely paralleled the changes in CER.

Taken together, the photosynthetic measurements indicate that Rubisco is a primary determinant regulating photosynthetic rates during leaf development in both wild-type and antisense plants. This is consistent with previous measurements on first fully expanded leaves of the antisense and wild-type tobacco plants, which showed that Rubisco is strongly limiting for photosynthesis under moderate to high light intensities.[31,32] For example, the flux control coefficient of Rubisco on photosynthesis is 0.7 under high light, meaning that Rubisco activity can explain 70% of the control on photosynthetic rates under these conditions.

*Control of Rubisco Biosynthesis*

As a first approach to examine the factors that control Rubisco accumulation during tobacco leaf development, we assessed *rbcL* and *rbcS* mRNA abundances by RNA gel blot analysis.[37] These studies revealed that the coordinate alterations in LS and SS amounts during wild-type leaf development are primarily a function of coordinate changes in *rbcS* and *rbcL* transcript abundance. This is consistent with previous studies in a handful of species, including ones we have conducted in soybean,[29] *Phaseolus,*[28] and wheat.[38] This is not a general phenomenon, however, because posttranscriptional controls regulate Rubisco accumulation during at least part of the leaf developmental process in some species.[28,39,40]

[38] N. J. Bate, S. J. Rothstein, and J. E. Thompson, *J. Exp. Bot.* **42,** 801 (1991).
[39] J. E. Mullet, *Annu. Rev. Plant Physiol. Plant Mol. Biol.* **39,** 475 (1988).
[40] R. C. Huffaker, *New Phytol.* **116,** 199 (1990).

As in the wild type, alterations in Rubisco activities during antisense leaf development were accompanied by stoichiometric alterations in LS and SS protein abundance, and by coordinate changes in *rbcS* and *rbcL* mRNA levels. While the overall profiles resembled those in the wild type, protein amounts were much lower. Although *rbcS* mRNA levels were also markedly reduced in amount in the mutants (due to antisense inhibition), *rbcL* mRNA amounts were normal. This indicates that LS protein accumulation is regulated posttranscriptionally during antisense leaf development. We have not tested whether this regulation occurs primarily at the *rbcL* mRNA translational initiation level, as it does in the tissue-culture grown plants.

Alterations in mRNA abundance during leaf development could be a function of altered DNA template amount, altered rates of transcription initiation, and/or altered transcript stability. To estimate *rbcL* DNA template availability, we hybridized Southern filters containing *Bam*HI-digested total cell DNA from wild-type and mutant leaves with tobacco plastid *rbcL* DNA sequences. We found that *rbcL* DNA levels were similar in the wild-type and mutant leaves and that, after attaining a maximum early in development, these DNA sequences declined ~50% during the course of senescence.

One interpretation of our data is that plastid DNA template availability may contribute to alterations in *rbcL* mRNA abundance, especially during the senescence phase of both wild-type and antisense leaf development. This assumes that all plastid DNAs are capable of serving as templates for *rbcL* mRNA transcription—an untested hypothesis. These studies raise the question of why *rbcL* templates decrease during senescence. It is assumed that plastid DNAs are no longer replicating during this process, so that what is being observed is a catabolic event. Is this catabolism due to the destruction of entire plastids? Or is there a selective catabolism of plastid DNAs in each plastid? Little information is available concerning these questions,[41-46] but in rice it has been demonstrated that plastid DNAs are lost during leaf senescence on a per plastid basis.[47] This may also be the case in soybean,[29] where we obtained results similar to those in tobacco. In soybean, whole plastids are lost only late in the leaf developmental

[41] G. K. Lamppa, L. V. Elliot, and A. J. Bendich, *Planta* **148,** 437 (1980).
[42] V. A. Wittenbach, W. Lin, and R. R. Heber, *Plant Physiol.* **69,** 98 (1982).
[43] P. J. Camp, S. C. Huber, J. J. Burke, and D. E. Moreland, *Plant Physiol.* **70,** 1641 (1982).
[44] T. Mae, N. Kai, A. Makino, and K. Ohira, *Plant Cell Physiol.* **25,** 333 (1984).
[45] D. M. Ford and R. Shibles, *Plant Physiol.* **86,** 108 (1988).
[46] T. M. Wardley, P. L. Bhalla, and M. J. Dalling, *Plant Physiol.* **75,** 421 (1984).
[47] Sodmergen, S. Kawano, S. Tano, and T. Kuroiwa, *Protoplasma* **160,** 89 (1991).

process—i.e., in yellow leaves.[45] In the tobacco studies, yellowing had not yet occurred in leaves that had measurable reductions in *rbcL* DNA.

*Feedback Regulation by Rubisco?*

One issue that can be addressed in the antisense mutants is the extent to which developmental programming is sensitive to Rubisco concentrations. One might suppose that some development events are dependent on Rubisco (linear pathways), while other events are not (parallel pathways). Studies with "stay-green" senescence mutants, for example, have emphasized the notion that leaf development is a coordinated series of events that are not obligatorily connected (i.e., parallel pathways).[48–51] These mutants generally have repressed photosynthetic rates and Rubisco accumulation, but the loss of chlorophyll and components of the thylakoid membrane is delayed until very late in the leaf senescence process.

We and others have observed that decreases in Rubisco have little impact on total protein accumulation and/or on the activities of some enzymes, at least in first fully expanded leaves of antisense plants with reductions of up to 80% in Rubisco content.[31,52,53] We also found that this was true for the antisense and wild-type leaves during their ontogeny.[37] Western immunoblot analyses of representative photosynthetic proteins (e.g., CAB, cytochrome *f*, and Rubisco activase) further showed that the profiles of change in these proteins generally resembled those of Rubisco in both sets of plants, and that their accumulation was not altered in the antisense plants. One notable difference was that CAB levels did not decrease as much as wild-type Rubisco levels in senescing leaves of either the wild-type or antisense plants (i.e., overall reductions of 40–50% for CAB versus 70–90% for Rubisco activase, and cytochrome *f*).

The protein accumulation data suggest that the antisense plants do not adapt to genetic alterations in Rubisco content by significantly altering nuclear and chloroplast gene expression—i.e., leaf developmental programs appear to be relatively unaffected by drastic changes in Rubisco content. Consistent with this notion is the finding that chloroplast DNA levels fall in concert throughout leaf development in the wild-type and antisense plants. We have also found that leaf expansion rates are nearly the same

[48] P. Hilditch, H. Thomas, B. J. Thomas, and L. J. Rogers, *Planta* **177,** 265 (1989).
[49] H. Thomas, H. J. Ougham, and T. G. E. Davies, *J. Plant Physiol.* **139,** 403 (1992).
[50] C. M. Ronning, J. C. Bounwkamp, and T. Solomos, *J. Exp. Bot.* **42,** 235 (1991).
[51] J. J. Guiamét, E. Schwartz, E. Pichersky, and L. D. Noodén, *Plant Physiol.* **96,** 227 (1991).
[52] G. S. Hudson, J. R. Evans, S. von Caemmerer, Y. B. C. Arvidsson, and T. J. Andrews, *Plant Physiol.* **98,** 294 (1992).
[53] J. Masle, G. S. Hudson, and M. R. Badger, *Plant Physiol.* **103,** 1075 (1993).

for developmentally similar leaves of the antisense and wild-type plants, even though the leaves of the antisense plants are much longer lived.

It is generally thought that photosynthetic rates are dependent on gene expression as a course control to set enzyme amounts, and consequently maximal potential enzyme activities, whereas photosynthetic rates are fine-tuned by adjustments in substrate and effector levels.[54,55] In this context, our findings on the Rubisco antisense DNA mutants emphasize the metabolic flexibility of plant cells, and their remarkable ability, in the absence of marked changes in gene expression, to adapt to a wide range of concentrations of photosynthetic components to achieve optimal photosynthetic rates.

[54] J. C. Servaites, W. J. Shieh, and D. R. Geiger, *Plant Physiol.* **97**, 1115 (1991).
[55] D. R. Geiger and J. C. Servaites, *Plant Physiol. Biochem.* **32**, 173 (1994).

# [15] Assessing the Potential for Chloroplast Redox Regulation of Nuclear Gene Expression

*By* DION G. DURNFORD, ONDREJ PRASIL, JEAN-MICHEL ESCOUBAS, and PAUL G. FALKOWSKI

## I. Introduction

Photosynthetic organisms respond to light quantity, or photon flux densities (PFDs), by reversibly altering the activities or levels of photosynthetic and metabolic enzymes and biosynthetic molecules to compensate for the changes in electron transport rates.[1-3] Photoacclimation requires a coordinated regulation of both nuclear and organelle gene expression and enzyme activation that enables adjustments in the organism's capacity to utilize light energy. The first step in this response is an ability to sense changes in the light environment. Phytochrome[4] and cryptochrome[5] are examples of sensor molecules that respond to changes in spectral quality; however, a role for these receptors as light quantity sensors in fully differentiated cells has not been demonstrated. Activation of the sensor converts the environmental cue to a biochemical signal. This signal requires a biochemi-

[1] J. M. Anderson and B. Anderson, *Trends Biochem. Sci.* **13**, 351 (1988).
[2] P. G. Falkowski and J. LaRoche, *J. Phycol.* **27**, 8 (1991).
[3] D. G. Durnford and P. G. Falkowski, *Photosyn. Res.* **53**, 229 (1997).
[4] H. Smith, *Annu. Rev. Plant Physiol. Plant Mol. Biol.* **46**, 289 (1995).
[5] T. W. Short and W. R. Briggs, *Annu. Rev. Plant Physiol. Plant Mol. Biol.* **45**, 143 (1994).

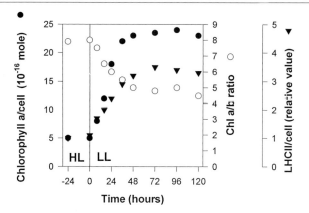

FIG. 1. Kinetics of photoacclimation in *D. tertiolecta* following a high-light (700 μmol quanta·m$^{-2}$·s$^{-1}$) to low-light (70 μmol quanta·m$^{-2}$·s$^{-1}$) transition. Changes in chlorophyll *a* per cell (closed circles), the Chl *a/b* ratio (open circles), and in the levels of LHCII per cell (closed triangles) over a 5-day period. [Figure is modified from A. Sukenik *et al.*, *Plant Physiol.* **92**, 891 (1990).]

cal transduction pathway to initiate changes in gene expression and enzyme activity observed during photoacclimation.

The redox status of the plastoquinone (PQ) pool has been implicated as a sensor, regulating nuclear gene expression during long-term acclimation to light intensity.[6] This hypothesis proposes that the signal transduction pathway is initiated via the PQ pool redox status[6] or the "excitation pressure" on Photosystem II (PSII),[7] thereby coupling cellular regulatory pathways controlling gene expression and enzyme activation to the cell's capacity to utilize light energy. It is our hope that these ideas will stimulate further research into the potential for redox-mediated regulation of nuclear gene expression in other systems.

## II. Selection of Research Organisms

Our primary research organism is the unicellular green alga, *Dunaliella tertiolecta. Dunaliella,* and many other unicellular algae, have a high degree of physiologic plasticity in response to changes in PFDs (Fig. 1). This includes the capacity to change rapidly the amount of chlorophyll per cell,

[6] J.-M. Escoubas, M. Lomas, J. LaRoche, and P. G. Falkowski, *Proc. Natl. Acad. Sci. U.S.A.* **92**, 10237 (1995).

[7] D. P. Maxwell, D. E. Laudenbach, and N. P. A. Huner, *Plant Physiol.* **109**, 787 (1995).

photosynthetic protein levels, cell ultrastructure, and *Lhc* mRNA abundance.[8,9] These large photoacclimation responses are easily measured and provide ideal markers for the regulatory process.[3]

Higher plants, to varying degrees, are able to photoacclimate[10–13]; however, many are genetically adapted to either a sun or shade environment[12,14] and do not display rapid and reversible physiologic responses to changes in PFD observed in many algal taxa.[2] Distinguishing between development and acclimation[3] and the existence of complex optics as a result of leaf morphology[15,16] are significant challenges to the examination of photoacclimation in higher plants.

## III. Algal Cultures

*Dunaliella tertiolecta* Butcher (clone DUN, Provasoli-Guillard Culture Collection, Maine) is grown in a batch culture using artificial seawater (ASW) with f/2 nutrients.[17,18] Cultures are grown under 40-W Cool White fluorescent lights (Sylvania) at 18–25° and bubbled with air passed through sterile bacterial air vents (Gelman, Ann Arbor, MI). For most photoacclimation studies, we maintain *Dunaliella* under a continuous light regime and avoid the use of a light : dark cycle that may introduce circadian rhythm-related cellular changes. *Dunaliella* cells used in experiments are initiated from starter cultures grown under the experimental light regime. These experimental cultures are grown for a minimum of 3 days before harvesting to allow time for complete photoacclimation[8] and we monitor the amount of chlorophyll per cell (Chl/cell) to ensure that cells are acclimated to the experimental light regime. It is crucial to harvest the cells while they are optically thin (ca. $5 \times 10^5$ cells/ml) to minimize self-shading and secondary responses that may arise from nutrient limitation.

We routinely compare cultures of *Dunaliella* grown at PFDs with at

[8] A. Sukenik, J. Bennett, A. Mortain-Bertrand, and P. G. Falkowski, *Plant Physiol.* **92,** 891 (1990).

[9] J. LaRoche, A. Mortain-Bertrand, and P. G. Falkowski, *Plant Physiol.* **97,** 147 (1991).

[10] J. M. Anderson, W. S. Chow, and D. J. Goodchild, *Aust. J. Plant Physiol.* **15,** 11 (1988).

[11] J. M. Anderson, W. S. Chow, and D. J. Goodchild, *Photosyn. Res.* **46,** 129 (1995).

[12] N. K. Boardman, *Ann. Rev. Plant Physiol.* **28,** 355 (1977).

[13] W. S. Chow, A. Melis, and J. M. Anderson, *Proc. Natl. Acad. Sci. U.S.A.* **87,** 7502 (1990).

[14] J. Bennett, J. R. Schwender, E. K. Shaw, N. Tempel, M. Ledbetter, and R. S. Williams, *Biochim. Biophys. Acta* **892,** 118 (1987).

[15] J. Nisho, J. Sun, and T. C. Vogelmann, *in* "Photoinhibition of Photosynthesis" (N. R. Baker and J. R. Bowyer, eds.), p. 407. Bios Scientific Publishers, Oxford, 1994.

[16] U. Schreiber, M. Kuhl, I. Klimant, and H. Reising, *Photosyn. Res.* **47,** 103 (1996).

[17] J. C. Goldman and J. J. McCarthy, *Limnol. Oceanogr.* **23,** 695 (1978).

[18] R. R. L. Guillard and J. Ryther, *Can. J. Microbiol.* **8,** 229 (1962).

least a 10-fold difference to maximize the cellular response to light. With *Dunaliella,* high-light typically exceeds 800 $\mu$mol quanta $\cdot$ m$^{-2}$ $\cdot$ s$^{-1}$ (up to 2000) and low-light is ca. 80 $\mu$mol quanta $\cdot$ m$^{-2}$ $\cdot$ s$^{-1}$. We measure light in the culture vessel with a QSL100 $4\pi$ quantum sensor (Biospherical Instruments, San Diego, CA). Under these conditions the PQ pool is almost entirely reduced under high-light and oxidized under low-light (see Section V).

## IV. Modulation of Plastoquinone Redox State

The photoacclimation-like cellular responses resulting from an *in vivo* manipulation of the PQ redox status define the foundation on which the hypothesis of chloroplast redox control of nuclear gene expression is built. There are several approaches to modifying the PQ redox state. First, direct alteration of the redox status can be accomplished through the use of various inhibitors that prevent the reduction of PQ by electron donation from PSII or the oxidation of PQH$_2$ at the cytochrome (cyt) $b_6/f$ complex. When inhibitors are used at sublethal concentrations, it is possible to inhibit a fraction of donors to, or acceptors from, the PQ pool, thereby altering its redox status under a particular light regime (Section IV,A). A related strategy alters the PQ redox state by using different light qualities to excite preferentially one of the photosystems.[19] This causes imbalances in the rate of electron transfer between the photosystems and consequently alters the PQ redox state. An alternative approach for manipulating PQ redox status involves examining mutants with specific, well-defined lesions in the linear portion of the photosynthetic electron transport chain. These mutations should lead to an overreduction, or oxidation, of the PQ pool, depending on the location of the defect (Section IV,B).

### A. *Inhibitors of Electron Transport*

*Reduction of Plastoquinone.* The reduction of plastoquinone occurs primarily by transferring electrons to plastoquinone at the $Q_B$-binding site of D1. Many urea-substituted herbicides, including DCMU [3-(3',4'-dichlorophenyl)-1,1'-dimethylurea], competitively block the transport of electrons from $Q_A$ to plastoquinone,[20] displacing it from its position within the $Q_B$-binding pocket.[21,22] In the presence of DCMU, plastoquinone reduction at D1 is prevented, increasing the oxidation state of the PQ pool.

[19] Y. Fujita, A. Murakami, and K. Ohki, *Plant Cell Physiol.* **28,** 283 (1987).
[20] W. Tischer and H. Strotmann, *Biochim. Biophys. Acta* **460,** 113 (1977).
[21] B. R. Velthuys, *FEBS Lett.* **126,** 277 (1981).
[22] A. Trebst, *Methods Enzymol.* **69,** 675 (1980).

Beale and Appleman[23] first observed an induction of chlorophyll synthesis on the addition of submicromolar amounts of DCMU to green algal cultures. Several other studies have confirmed this observation in other green algae[6,24] and cyanobacteria.[25] These photoacclimation-like responses are induced under sublethal concentrations of DCMU that result in an inhibition of a fraction of the PSII reaction center population, increasing the PQ pool oxidation state. Consequently, the cells survive and an acclimation response is induced.

The initial PFD is important in determining the response amplitude. For example, addition of DCMU to a low-light acclimated culture produces little observable change in the amount of Chl/cell, but these changes are obvious when DCMU is added to high-light acclimated cultures.[6] Presumably the interaction between light intensity and DCMU reflects the PQ redox state. At low-light, the PQ pool is predominantly oxidized, so the addition of submicromolar concentrations of DCMU would have little additional effect on the oxidation state of the PQ pool and, consequently, little effect on Chl/cell. It is also important to realize that this inhibitor-stimulated response in high-light has a smaller amplitude compared to cells shifted to a lower light intensity.

DCMU (Sigma, St. Louis, MO) is hydrophobic and stock solutions are usually made up in 95% ethanol. When DCMU is added to cultures, the final solvent concentration should be <0.5% (v/v) and equal concentrations of solvent should be added to the control culture. After inhibitor addition, control and trial cultures can be sampled at regular time points, up to 24–48 hr, to assay for the desired cellular response (Section VI). It is critical that the amount of DCMU be standardized to the number of reaction centers per cell, not the final concentration. Because the number of reaction centers/cell in different organisms is variable, the proportion of inhibited PSII centers at various DCMU concentrations and known cell densities must be determined by titration (Fig. 2), whereby increasing amounts of DCMU are added to the cell culture and its effects determined through changes in the rate of $Q_A$ reoxidation as inferred from variable fluorescence kinetics (discussed in Section V). From this curve the amount of inhibitor required to inhibit 50% of the reaction centers can be determined and used to induce a sufficient cellular response.

*Oxidation of Plastoquinol.* The oxidation of plastoquinol occurs at the cyt $b_6/f$ complex.[26] Several inhibitors act at the cyt $b_6/f$ complex, including

[23] S. I. Beale and D. Appleman, *Plant Physiol.* **47**, 230 (1971).
[24] J. Naus and A. Melis, *Photosynthetica* **26**, 67 (1992).
[25] F. Koenig, *Photosyn. Res.* **26**, 29 (1990).
[26] R. Malkin, *Photosyn. Res.* **33**, 121 (1992).

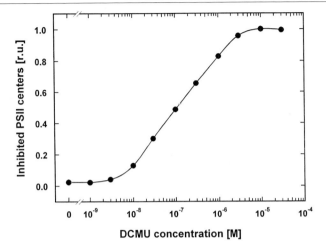

Fig. 2. DCMU titration curve for high-light acclimated *D. tertiolecta* at a cell density of $4.3 \times 10^5$ cells/ml. The fraction of bound $Q_B$ centers was determined by following the kinetics of $Q_A$ reoxidation in cells incubated for 180 sec in the presence of DCMU (see Fig. 3). At this cell density (corresponding to ca. 0.2 $\mu M$ Chl $a + b$), the presence of 100 n$M$ DCMU is sufficient to inhibit half of the PSII reaction centers in high-light acclimated cells. Use of DCMU concentrations that inhibit approximately 50% of the reaction centers should generate a sufficient cellular response in other organisms.

NQNO (2-nonyl-4-hydroxyquinoline *N*-oxide) and DBMIB (2,5-dibromo-3-methyl-6-isopropyl-*p*-benzoquinone).[22,27] Both DBMIB and NQNO bind to cyt $b_6$, thus competitively preventing the oxidation of plastoquinol.[22,28] One DBMIB molecule per cyt $b_6/f$ complex is required for full inhibition of electron transfer through the cyt $b_6/f$ complex.[27]

We use DBMIB (Sigma, listed as dibromothymoquinone) to block cyt $b_6/f$ turnover and maintain the PQ pool in a reduced state. DBMIB is lipid soluble and stock solutions (100 and 10 m$M$) are made in 10% dimethyl sulfoxide (DMSO) (in ethanol) to ensure solubility and to enhance its permeability across cellular membranes. Stock solutions are stored at $-20°$ in a covered bottle. The final DMSO concentration in the culture should be maintained at or below 100 $\mu M$ and equal amounts added to the control.

Appropriate precautions have to be taken when utilizing DBMIB because at higher concentrations ($>1$ $\mu M$) it can inhibit both PQ reduction (at D1) and mitochondrial electron transport.[22,27] Dark respiration rates (as $O_2$ consumption) should be measured to ensure that mitochondrial

[27] P. R. Rich, S. A. Madgwick, and D. A. Moss, *Biochim. Biophys. Acta* **1058**, 312 (1991).
[28] R. W. Jones and J. Whitmarsh, *Biochim. Biophys. Acta* **933**, 258 (1988).

electron transport inhibition is not responsible for any observed effects. DBMIB is also photoactive and exposure to light can cause degradation, affecting its final concentration.[27] In our experience, DBMIB is degraded within 6–24 hr of addition to an illuminated culture, after which the cultures rebound and continue logarithmic growth.

In isolated spinach thylakoids, steady-state electron transport rates are reduced by 50% in the presence of 25 n$M$ DBMIB[28]; however, the optimal DBMIB concentration for use *in vivo* must be determined empirically; a suggested starting range is 10 n$M$ to 1 $\mu M$ keeping in mind the amount of inhibitor per cyt $b_6/f$ complex. In *Dunaliella,* there are approximately 3.4 cyt $b_6$ molecules per PSII reaction center.[29] In low-light-grown *Dunaliella* cultures, 100–700 n$M$ DBMIB is sufficient to induce changes in *Lhc* mRNA steady-state levels, chlorophyll per cell,[6] and fluorescence yields (see Section V). Greater than 1 $\mu M$ DBMIB causes complete inhibition of cell growth.

### B. Mutations Affecting Photosynthetic Electron Transport

To gain more insight into the role of PQ redox status in the regulation of nuclear gene transcription, our laboratory studies mutants with specific lesions in photosynthetic electron transport that alter the rate of PQ reduction at the PSII reaction center, or PQH$_2$ oxidation at the cyt $b_6/f$ complex. These defects are expected to have an altered PQ redox state under a given set of environmental conditions without the use of artificial inhibitors. *Dunaliella* lacks a sexual cycle, is an obligate photoautotroph, and cannot be easily transformed; therefore, it is not possible to apply standard genetic approaches to examine PQ redox control in this organism. *Chlamydomonas,* on the other hand, is an attractive algal system for the examination of this phenomenon because it is amenable to genetic approaches and there are well-developed nuclear and organelle transformation systems.[30] *Chlamydomonas* photoacclimates by altering the total content of cellular chlorophyll, though the magnitude of change in the Chl $a/b$ ratio is relatively small.[31] It is important to compare the mutant phenotype to that of the isogenic wild-type strain because there is considerable variation in the amount of Chl/cell in different "wild-type" *Chlamyodomonas* strains.

Mutants affecting the efficiency of electron transfer within the PSII reaction center are expected to have altered PQ redox states. Site-directed mutation within the region between helices IV and V of the D1 protein

[29] A. Sukenik, J. Bennett, and P. G. Falkowski, *Biochim. Biophys. Acta* **891,** 205 (1987).
[30] J. D. Rochaix, *Annu. Rev. Genet.* **29,** 209 (1995).
[31] D. G. Durnford and P. G. Falkowski, unpublished data (1996).

often have adverse affects on the rate of $Q_A^-$ reoxidation,[32] and are expected to result in a lower PQ reduction state at a particular light intensity. Mutants with impaired PQ reduction rates are hypothesized to have a low-light-grown phenotype, compared to the wild-type strain, when grown under high-light.

Strains with a defect in the rate of plastoquinol oxidation may have mutations in the cyt $b_6/f$ complex, plastocyanin, or any other downstream factor that could potentially cause an electron backup into the PQ pool. Site-directed mutants of the cyt $f$ subunit in *Chlamydomonas*, with reduced abundance of the cyt $b_6/f$ complex, often have alterations in the accumulation of some nuclear and chloroplast encoded gene products.[33] In one cyt $f$-deficient mutant (CC2964, *Chlamydomonas* culture collection, Duke University), $PQH_2$ oxidation is substantially slower than in the wild type, as evidenced by alterations in fluorescence (see Fig. 3B). Our preliminary evidence indicates that this and other *Chlamydomonas* cyt $b_6/f$ mutants tend to have reduced amounts of Chl/cell, resembling cultures grown under high-light.[34] The lower chlorophyll content is hypothesized to be a result of the reduced $PQH_2$ oxidation.

To explain redox control, it is important that the mutants be able to support photoautotrophic growth and not be complete gene knockouts. Null mutants require heterotrophic growth and it is our preference to avoid the exogenous application of acetate or other carbon sources as they repress photosynthetic gene expression.[35,36]

## V. Using Fluorescence to Assess PSII Inhibition and the PQ Redox State

Variable chlorophyll fluorescence can be used to determine the effects of mutation or herbicide presence on the rate of $Q_A$ to $Q_B$ transfer, and to estimate the PQ pool redox state. The fraction of functional PSII reaction centers can be quantified from measurements of changes in Chl $a$ fluorescence that result from the block in reoxidation of primary acceptor $Q_A^-$ with DCMU or similar herbicides. The method we use determines the fraction of the inhibited centers from an analysis of the kinetics of variable fluorescence decay following excitation by short, single turnover flashes (STFs). In the absence of an inhibitor, the major pathway for $Q_A^-$ reoxidation

[32] A. Lardans, B. Förster, O. Prasil, P. G. Falkowski, V. Sobokev, M. Edelman, C. B. Osmond, N. W. Gillham, and J. E. Boynton, *J. Biol. Chem.* **273**, 11082 (1998).
[33] T. A. Smith and B. D. Kohorn, *J. Cell Biol.* **126**, 365 (1994).
[34] D. G. Durnford, O. Prasil, and P. G. Falkowski, unpublished data (1997).
[35] K. L. Kindle, *Plant Mol. Biol.* **9**, 547 (1987).
[36] K. E. Koch, *Annu. Rev. Plant Physiol. Plant Mol. Biol.* **47**, 509 (1996).

is electron transfer to $Q_B$, which occurs with a halftime of 150–600 $\mu s$[37] if the $Q_B$ site is occupied by plastoquinone, or up to 3 ms, if plastoquinone has yet to be bound. When PSII centers are blocked by an herbicide, charge recombination with the donor side is the only pathway for $Q_A^-$ reoxidation, with a room temperature half-time of 0.5–2 sec.[38]

The method for quantitative detection of centers with herbicides bound to the $Q_B$ pocket is based on determination of the amplitude of the slow component of $Q_A^-$ reoxidation kinetics. (1) The kinetics of variable fluorescence decay should be measured for at least 2–5 sec after a STF or until fluorescence decays to the $F_o$ level. Several fluorometers capable of these measurements are commercially available (e.g., PAM system made by Walz, Effeltrich, Germany; PSI fluorometer made by P.S. Instruments, Brno, Czech Republic). Care should be taken that the measuring light flashes have minimal actinic effect on the fluorescence decay kinetics. Samples should be kept in the dark for at least 30 sec before the measurement. (2) The relationship between yields of variable fluorescence and fraction of reduced $Q_A$ in green algae and higher plants is nonlinear,[39] hence the kinetics of $Q_A^-$ reoxidation does not directly follow the decay of variable fluorescence (see Fig. 3). The nonlinearity is related to the connectivity parameter ($p$) which can be determined directly from measurements of PSII apparent cross sections using the fast repetition rate (FRR) fluorometer[40] or a typical value of $p \sim 0.5$ can be assumed. The kinetics of $Q_A^-$ reoxidation can then be calculated from $F_V$ using Eq. (1).[39]

$$[Q_A^-] = F_V/(1 + p(F_V - 1)) \tag{1}$$

If the connectivity is not taken into account, the error in determining $[Q_A^-]$ can be up to 15% for $p = 0.5$. However, the final error in determining the fraction of centers blocked by herbicide $[RC_{herb}]$ would be smaller (see Fig. 3). (3) The $Q_A^-$ reoxidation curve can be fitted to a multiexponential decay function, with either three exponential components (half-time $\sim 500$ $\mu s$, $\sim 2$ ms, $\sim$ hundreds of milliseconds) or two exponentials and a residual.[41] (4) The fraction of $Q_A^-$ at 20 ms after the flash ($[Q_A^-]_{20}$) is calculated from the fitted curve, or it can be determined graphically. (5) Finally, the fraction of centers blocked with herbicide $[RC_{herb}] = ([Q_A^-]_{20} - [Q_A^-]'_{20})/(1 - [Q_A^-]_{20})$, where $[Q_A^-]'_{20}$ is the concentration of long-lived $Q_A^-$ determined in the control samples, without herbicide addition (see Fig. 3).

[37] H. H. Robinson and A. R. Crofts, *FEBS Lett.* **153**, 221 (1983).
[38] P. Bennoun, *Biochim. Biophys. Acta* **216**, 357 (1970).
[39] A. Joliot and P. Joliot, *Comp. Rend. Acade. Sci. Paris* **258**, 4622 (1964).
[40] P. G. Falkowski and Z. Kolber, *Aust. J. Plant Physiol.* **22**, 341 (1995).
[41] Z. Kolber, J. Zehr, and P. G. Falkowski, *Plant Physiol.* **88**, 923 (1988).

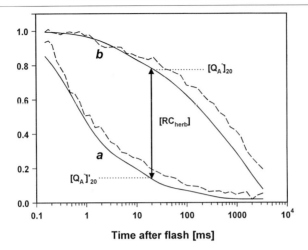

Fig. 3. Determination of the fraction of closed reaction centers from the kinetics of variable fluorescence decay in *Dunaliella* (same conditions as in Fig. 2). Variable fluorescence (dashed line) was measured with FRR fluorometer for 3 sec following the single turnover actinic flash. The connectivity parameter $p$ was calculated from the shape of the variable fluorescence rise during the FRR actinic flash (not shown). The reoxidation kinetics of $Q_A^-$ (solid line) was then calculated using Eq. (1). The $Q_A^-$ reoxidation is much faster in control cells (line $a$) than after addition of 300 n$M$ DCMU (line $b$). The fraction of reaction centers blocked by the herbicide was then calculated as described in the text. In this given example, 70% of centers were blocked.

An estimate of PQ redox state can also be determined using chlorophyll fluorescence. While short ($<100$ $\mu$s) STFs result in $Q_A$ reduction only, longer excitation of milliseconds to seconds duration gradually reduces the entire PQ pool. The actual degree of reduction is highly variable and is a result of competing rates of PQ pool reduction and oxidation. Detecting the actual degree of PQ pool reduction is based on the changes in variable fluorescence yield, which is dependent on the PQ pool redox state: fluorescence yield is low if the PQ pool is oxidized and it increases by up to 50% when the pool is reduced.

We have used this approach to monitor the effects of different DBMIB concentrations on the steady-state redox level of the PQ pool. Figure 4 shows that while the STF yield is the same in control cells and after the addition of 1 $\mu M$ DBMIB to dark-adapted *Dunaliella* cells, the steady-state redox level of the PQ pool differs. In the controls, after an initial transient increase, the fluorescence yield decreases, indicating that the pool is mostly oxidized. In the presence of DBMIB the fluorescence yield remains high, an indication of the highly reduced state of the PQ pool.

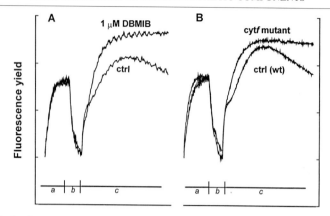

Fɪɢ. 4. Qualitative assessment of the decrease in the reoxidation rate of plastoquinol pool using variable fluorescence. (A) Low-light-grown *Dunaliella* cells were incubated in the presence of 1 $\mu M$ DBMIB for 1 hr. Variable fluorescence was induced by a protocol consisting of the short (100-$\mu$s) single turnover excitation (*a*) that was followed by 3-sec measurement of decay of variable fluorescence (*b*) and by long (100-ms) excitation that caused multiple turnovers of PSII and reduction of PQ pool (*c*). The changes of the fluorescence yield induced by single-turnover excitation were identical indicating that 1 $\mu M$ DBMIB had no effect on the function of PSII. In the control sample, the fluorescence yield induced by long, multiple turnover excitation has initially risen as the PQ pool was transiently reduced, but then a decrease is observed as the dark reactions are activated and the PQ pool gets oxidized faster than it is reduced. On the other hand, in the presence of 1 $\mu M$ DBMIB, fluorescence yield remains high, indicating that the reoxidation of the PQ pool is inhibited. (B) A similar result is seen with cyt *f*-deficient *Chlamydomonas* mutant as compared to the wild-type strain. This illustrates the potential of using variable chlorophyll fluorescence to determine PQ redox states and confirms the effects of the specific inhibitors and mutants used.

## VI. Examining Cellular Changes Related to PQ Redox State

Growth in high and low PFDs results in vast changes in the levels of Chl/cell (both Chl *a* and *b*).[8] Measurements of Chl/cell are an indicator of LHCII abundance, providing an easily measured parameter for the photoacclimation state. Culture samples for chlorophyll measurements are harvested by filtering onto Whatman GF/C glass microfiber filters. If not used immediately, the filters can be folded in half, plunged into liquid nitrogen and frozen at $-20°$. Chlorophyll and other pigments are extracted by homogenizing the filters in 90% acetone. Absorbance spectra are recorded on a spectrophotometer, and the Chl *a* and *b* values ($\mu$g/ml) calculated using the equations of Jeffrey and Humphrey.[42] Cells are fixed in Lugols iodine (0.5% v/v) and counted using a hemocytometer. Eight 1-mm$^3$

[42] S. W. Jeffrey and G. F. Humphrey, *Biochem. Physiol. Pflanzen* **167**, 191 (1975).

blocks are averaged to calculate the number of cells present in 1 ml of culture. The standard error for the determination of Chl/cell is ca. 10% of the mean; therefore, a number of independent replicates are required for each experiment.

A 10-fold decrease in the light intensity induces a 4-fold increase in the steady-state levels of *Lhc* mRNAs.[9] Escoubas *et al.*[6] reported alterations in the steady-state level of *Lhc* mRNA resulting from the addition of specific photosynthetic electron transport chain inhibitors, correlating *Lhc* levels with the redox state of the PQ pool. Oxidation of the PQ pool through DCMU addition (to a high-light culture) or by transfer to low-light, increased steady-state levels of *Lhc* mRNA. Reduction of the PQ pool by the addition of DBMIB or a transfer to high-light caused the inverse effect: a decrease in the steady-state levels of *Lhc* mRNA.

The extraction of RNA from *Dunaliella* and other unicellular algae is routine. Briefly, ca. 150–200 ml of culture is harvested by centrifugation. The culture medium is decanted and the cells immediately resuspended in a 4 $M$ guanidinium thiocyanate/phenol mixture and extracted as described in Ref. 43. We have found the use of Trizol reagent (GIBCO-BRL, Gaithersburg, MD) to be a convenient modification of this procedure. To prevent RNA degradation and rRNA clipping, it is critical to minimize the time between harvesting the cells and guanidinium thiocyanate/phenol extraction. It is also for this reason that we process RNA samples immediately, without freezing the cells. Denaturing gels are loaded with an equal amount of RNA, and message levels detected by Northern blotting are standardized to rRNA levels.[6]

Measurable changes in mRNA abundance occur rapidly (hours), and sampling every 2–4 hr over a 12- to 24-hr time span is necessary to narrow the effective time range for inhibitor-induced changes. For example, with DBMIB addition, we have determined an exposure of 4–8 hr is sufficient to cause a 50% reduction in the *Lhc* mRNA message levels.[31] Escoubas *et al.*[6] reported that 100 n$M$ DCMU elevated the steady-state *Lhc* mRNA levels, resembling a shift toward low-light. Under these conditions, *Dunaliella* cells can survive and continue to support growth. DCMU levels above 1 $\mu M$ prevent cell growth and, in *Dunaliella,* lead to decreases in *Lhc* mRNA levels.[31] The effective concentration will have to be determined empirically for each organism by testing a range of concentrations (10 n$M$ to 10 m$M$), with the goal of inhibiting 50% of the reaction centers (Fig. 2, Section V).

[43] P. Chromczynski and N. Sacchi, *Anal. Biochem.* **162,** 156 (1987).

## VII. Examination of the Signal Transduction Pathway

The specificity of DCMU and DBMIB indicates that the sensor/signal location is within the chloroplast. Because *Lhc* is a nuclear-encoded gene, a chloroplast-to-nucleus signal transduction pathway must exist. The *Lhc* promoter region of *Dunaliella* contains several putative light-responsive elements that resemble those found in the light-regulated promoters of higher plants.[6,44] One strategy for deciphering the signal transduction pathway is to search for DNA-binding proteins that interact with these regulatory elements and whose presence is correlated with the changes in *Lhc* gene expression. We use gel retardation assays[45,46] to detect DNA-binding protein(s) in total protein extracts from high-light and low-light acclimated cells. These growth conditions induce the largest difference in *Lhc* mRNA levels, and maximize the probability of detecting novel regulatory complexes responsible for the PFD-dependent regulation of transcription.

Cells from an optically thin 8 liter culture are harvested by centrifugation and resuspended in a high salt extraction buffer [0.6 $M$ KCl, 25 m$M$ HEPES-KOH, pH 7.9, 12 m$M$ 2-mercaptoethanol, 0.12 m$M$ ethylenediaminetetraacetic acid (EDTA), 0.24 $M$ sucrose, magnesium 0.84 m$M$ Mg acetate] to dissociate the DNA-binding proteins from the chromosomal DNA.[6] The cells are lysed in a mini bead-beater (Biospec Products; Bartlesville, OK) with 0.5-mm glass beads at 4°. The cellular homogenate is clarified by ultracentrifugation (35,000 rpm, SW 50.1 at 4° for 1 hr) to remove membranes and cell fragments, and the protein samples are precipitated on ice with 30% ammonium sulfate (0.166 g/ml). The precipitated proteins are collected by centrifugation at 8000$g$ at 4° for 20 min. The pellet is resuspended in cold storage buffer [40 m$M$ KCl, 25 m$M$ HEPES-KOH, pH 7.9, 5 m$M$ 2-mercaptoethanol, 0.12 m$M$ EDTA, 10% glycerol (v/v)] and dialyzed at 4° for 3–4 hr against 2000 volumes of the same solution. The protein samples are concentrated on Centricon 10 microconcentrator units (Amicon, Beverly, MA) and the concentration determined in triplicate with a BCA assay (Pierce, Rockford, IL). A protease inhibitor cocktail (1 m$M$ phenylmethylsulfonyl fluoride; 5 m$M$ ε-amino-*n*-caproic acid, 1 m$M$ benzamidine HCl, 1 mg/ml leupeptin, and 1 mg/ml antipain) is added to all solutions.

The gel retardation assay can rapidly detect DNA-binding proteins that specifically interact with promoter fragments containing suspected light-responsive elements. These promoter fragments are amplified via polymer-

---

[44] W. B. Terzaghi and A. R. Cashmore, *Annu. Rev. Plant Physiol. Plant Mol. Biol.* **46,** 445 (1995).
[45] J. Cary, *Methods Enzymol* **208,** 103 (1991).
[46] D. Lane, P. Prentki, and M. Chandler, *Microbiol. Rev.* **56,** 509 (1992).

ase chain reaction, purified from 3% agarose gels, and end-labeled with [$\gamma$-$^{32}$P]ATP. The labeled probe (30,000 cpm) is incubated with 10 $\mu$g of the protein extract for 20 min in a 1$\times$ binding mixture [45 m$M$ KCl, 25 m$M$ HEPES, pH7.9, 1 m$M$ EDTA, 0.5 m$M$ dithiothreitol, and 5% glycerol (v/v)] and with poly[dI$\cdot$dC]$\cdot$[dI$\cdot$dC] (Sigma). The addition of polymer DNA to the binding reaction is required to eliminate the interaction of nonspecific DNA-binding proteins with the probe. A variety of concentrations have to be tried (0.5–5 $\mu$g per reaction) to optimize the binding reaction. The binding reactions are fractionated on 4% acrylamide–1$\times$ TBE (Tris-borate-EDTA) gels at 200 V, 4° for 2–3 hr. Gels are fixed in 20% methanol/10% acetic acid, dried on Whatman 3MM and autoradiographed.[6]

Because gel retardation assays are prone to artifacts, it is critical to run experiments that demonstrate promoter-binding specificity by eliminating any detected complexes through the addition of excess amounts of the unlabeled fragment. These competition experiments should be done with 1- to 50-fold amounts of the competitor. Our experiments detect a DNA–protein complex present primarily in high-light-grown cells and absent in protein extracts from cells grown in low-light (Fig. 5). As *Lhc* levels are reduced in high-light, it is hypothesized that this factor is a repressor complex whose binding activity is somehow controlled by the redox state of the PQ pool.

FIG. 5. Gel retardation assay using the 30% ammonium sulfate fraction from total protein extracts of *Dunaliella* high-light (HL) and low-light (LL) acclimated cells. The binding reaction includes a 102-bp end-labeled probe from a *Dunaliella Lhc* promoter region containing a putative light-regulatory motif. A DNA–protein complex is present predominantly in the HL extract and absent in protein extracts from LL. A control lane containing the 102-bp probe without added protein extract (free probe) is shown for comparison.

The present evidence indicates that the PQ pool is a light intensity sensor controlling the expression of *Lhc*,[6] leading to a self-regulation of photosynthesis. One of the first described self-regulatory mechanisms controlling light-harvesting capacity involves a reorganization of the light-harvesting antennas (state transitions) to compensate for unequal excitation of the photosystem.[47] State transitions are mediated through the reversible phosphorylation of LHCII by a thylakoid kinase(s) whose activation is keyed to the redox status of the PQ pool.[48,49] Because of their PQ redox state-dependent activation, it is hypothesized that the LHCII (or similar) kinase(s) activate intermediates in the signal transduction pathway by a specific phosphorylation event when the PQ pool is reduced.[6,49–51] This hypothesis can be tested by adding phosphatase inhibitors that should prevent the dephosphorylation of any phospho intermediates in the signal transduction pathway in the absence of a light cue and retain their activity. This will sustain the cellular signal indicating a reduced PQ pool; therefore, the cell should fail to respond to the oxidation of the PQ pool brought about by a lowering of the light intensity. This is examined by adding phosphatase inhibitors (okadaic acid, microcystin-LR, and tautomycin) in submicromolar amounts (100 n$M$) to high-light acclimated cells and immediately transferring the culture to a lower light intensity. In each case, the presence of the phosphatase inhibitors partially prevents the normal increase in the amount of Chl/cell, suggesting the involvement of a phosphorylation cascade in the redox-regulated photoacclimation response.[6] Although this is intriguing, the intermediates transducing this signal from the chloroplast to the nucleus are unknown, and defining these components will be one of the more challenging aspects to further research in understanding the redox regulation of photosynthetic gene expression.

[47] C. Bonaventura and J. Myers, *Biochim. Biophys. Acta* **189,** 366 (1969).
[48] J. Bennett, *Biochem. J.* **212,** 1 (1983).
[49] J. F. Allen, *Biochim. Biophys. Acta* **1098,** 275 (1992).
[50] J. F. Allen, *Physiol. Plantarum* **93,** 196 (1995).
[51] A. V. Verner, P. J. M. Vankan, P. R. Rich, I. Ohad, and B. Andersson, *Proc. Natl. Acad. Sci. U.S.A.* **94,** 1585 (1997).

# Section IV

## Biogenesis and Adaptation of Photosynthetic Components

## [16] Molecular and Genetic Analysis of Light-Dependent Chlorophyll Biosynthesis

By GREGORY ARMSTRONG and KLAUS APEL

### Introduction

Chlorophylls (Chls) and bacteriochlorophylls (Bchls) play a fundamental role in the energy absorption and transduction activities of photosynthetic organisms. These organisms have evolved two genetically distinct strategies to synthesize Chl/Bchl.[1] The more primitive strategy and the exclusive means of Bchl biosynthesis in anoxygenic photosynthetic bacteria involves the light-independent reduction of Pchlide to Chlide, a precursor of Chl/Bchl. This reaction, which also occurs in all oxygenic photosynthetic organisms except angiosperms, requires at least three gene products that are plastid encoded in eukaryotes and which display sequence homologies to subunits of the nitrogenase enzyme. The second, more highly regulated strategy to control Chl biosynthesis is mediated by the light-dependent plastid localized NADPH-protochlorophyllide oxidoreductase (POR).[2,3] POR, which has no sequence similarity to any of the subunits of the light-independent enzyme, catalyzes the photoreduction of Pchlide to Chlide in all oxygenic photosynthetic organisms including cyanobacteria, algae, liverworts, mosses, ferns, gymnosperms, and angiosperms. Among higher plants angiosperms absolutely require light for Chl biosynthesis and stable accumulation of both nuclear- and plastid-encoded Chl-binding proteins in chloroplasts.

In addition to the light-dependent reduction of Pchlide to Chlide, Chl biosynthesis in higher plants is also controlled at the level of $\delta$-aminolevulinic acid (ALA) synthesis.[1] When seedlings of higher plants are grown in darkness, plastid development is arrested at an early stage, termed the *etioplast*. On illumination the first light-dependent step toward the formation of a chloroplast is the photoreduction of Pchlide to Chlide. At the same time, an earlier block in Chl synthesis is lifted, leading to a light-induced increase in the rate of ALA synthesis that parallels the accumulation of Chl.

Both the POR enzyme and its substrate Pchlide accumulate in etiolated leaves, but their amounts decrease dramatically on exposure to light, due

---

[1] D. von Wettstein, S. Gough, and C. G. Kannangara, *Plant Cell* **7**, 1039 (1995).
[2] W. T. Griffiths, *Biochem. J.* **174**, 681 (1978).
[3] K. Apel, H. J. Santel, T. E. Redlinger, and H. Falk, *Eur. J. Biochem.* **111**, 251 (1980).

0076-6879/98 $25.00

in part to rapid proteolytic turnover of the enzyme.[4] When Chl accumulation in illuminated, previously dark-grown seedlings has reached its maximum rate, POR activity has dropped beyond the limit of detection and only trace amounts of the enzyme protein are measurable. Only recently has it been possible to solve this apparent paradox. In addition to the light-sensitive and highly expressed POR (POR A) originally isolated from dark-grown seedlings of monocotyledonous plants, a second light-dependent and light-sensitive POR (POR B) has been discovered that, in contrast to POR A, is constitutively expressed in dark-grown and light-adapted green plants.[5–6] Although POR A and POR B exhibit biochemically similar light-dependent activities *in vitro,* the different *in vivo* expression patterns of the corresponding genes suggest that the two enzymes may perform unique rather than redundant biological functions during plant development. POR B mRNA and POR B are present throughout the angiosperm life cycle and indeed more or less constitutively during the greening of etiolated seedlings. In contrast, illumination of such seedlings leads to the complete loss of detectable POR A mRNA and POR A within a few hours, before appreciable amounts of Chl have accumulated. These data have led to the proposal that POR B sustains light-dependent Chl biosynthesis in fully greened plants and that POR A performs a specialized function during the early stages of greening in seedlings.[6]

Many aspects of the light-induced changes during the transformation of etioplasts to chloroplasts that lead to the activation of the biosynthetic pathway of Chl are only poorly understood and the roles of POR A and POR B during this developmental step still remain obscure. Moreover, most of the steps that are involved in the delivery of freshly formed Chl to the various parts of the growing photosynthetic membrane have not yet been identified. We have used a combination of molecular and genetic techniques to analyze these problems. Most of our experiments have been carried out with *Arabidopsis thaliana.* This crucifer offers a number of advantages for genetic manipulations including its short generation time, self-fertilization, small size, and generous seed set. Furthermore, *Arabidopsis* has an advantage over many other plants in that it is readily and routinely transformable, and protocols for mutagenesis and mutant screening procedures have been developed. *Arabidopsis* therefore offers an attractive model system to investigate light-dependent Chl biosynthesis.

[4] S. Reinbothe, C. Reinbothe, N. Lebedev, and K. Apel, *Plant Cell* **8,** 763 (1996).

[5] H. Holtorf, S. Reinbothe, C. Reinbothe, B. Bereza, and K. Apel, *Proc. Natl. Acad. Sci. U.S.A.* **92,** 3254 (1995).

[6] G. A. Armstrong, S. Runge, G. Frick, U. Sperling, and K. Apel, *Plant Physiol.* **108,** 1505 (1995).

Analysis of POR A-Deficient Plants

Although it would be desirable to study POR function using a genetic approach, neither POR A nor POR B mutants are currently available. In the absence of POR mutants we have used an alternative approach to deplete seedlings of POR A.[7] This strategy exploits the differential regulation of POR gene expression by light, mediated by the phytochrome photoreceptor system, to abolish the expression of POR A. Seeds of *Arabidopsis* were surface sterilized and distributed on petri plates containing Murashige–Skoog agar.[8] After a 36-hr cold treatment plated seeds were exposed to 1 hr of white light (WL) and were then either placed in darkness at 22° for 4 days or, after 1 day of darkness, were transferred to continuous far-red light (cFR) (16 W m$^{-2}$) for 3 days. Standard cFR with a maximum at 740 nm and a half bandwidth of 123 nm was obtained by filtering WL provided by three 60-W Osram Linestra incandescent lamps through a Plexiglas combination consisting of two Röhm and Haas No. 627/3-mm filters and one No. 501/3-mm filter. The far-red wavelengths obtained with these particular filters permit very little accumulation of Chl while activating phytochrome A-mediated photomorphogenic responses through the far-red high irradiance reaction (FR-HIR). The cFR-grown seedlings of *Arabidopsis* display a morphology typical of wild-type seedlings grown in WL, with open hooks, expanded cotyledons, and short hypocotyls. Such seedlings are yellow, however, because they contain little Chl. Strikingly, cFR-grown seedlings do not green normally on subsequent transfer to WL. This greening defect correlates with the complete absence of POR A mRNA and POR A protein, the disappearance of phototransformable Pchlide-F655 and its replacement with large amounts of nonphototransformable Pchlide-F632 and a strong reduction in the total POR enzymatic activity in extracts from cFR-grown wild-type seedlings.[7] In addition, the amounts of POR B mRNA and POR B protein are also substantially reduced. In contrast to the wt, a cFR-grown phytochrome A-deficient mutant continues to express the POR A gene, accumulates Chl, and visibly greens in WL. All of these data taken together establish a correlation between the complete absence of POR A and abnormal WL-induced greening.[7]

To examine directly the roles of POR A and POR B in greening it was necessary to generate transgenic plants that constitutively overexpress POR. To this end, the POR cDNAs were placed under the control of a modified cauliflower mosaic virus (CAMV) 35 S promoter, CAMV-35 S

[7] S. Runge, U. Sperling, G. Frick, K. Apel, and G. A. Armstrong, *Plant J.* **9,** 513 (1996).
[8] T. Murashige and F. Skoog, *Physiol. Plant* **15,** 473 (1962).

omega, within the context of a binary plant transformation vector that confers resistance to the antibiotic kanamycin (Km) in plant cells. This vector was introduced into *Arabidopsis* by *Agrobacterium*-mediated root or *in planta* transformation. Homozygous transformed lines carrying independent single-copy insertions of the POR transgene were identified on the basis of Km resistance segregation ratios and DNA blot analysis, and were used for subsequent physiological studies.[7,10] Wild-type and transgenic seedlings were initially grown in cFR for 3 days before they were transferred to WL and kept under different light intensities with fluence rates ranging from 0.4 to 500 $\mu$E m$^2$ s$^{-1}$. Overexpression of not only POR A but also POR B increased the fluence rate at which maximal Chl accumulation occurred from 8 to 125 $\mu$E m$^{-2}$ s$^{-1}$ relative to the wild type.[10] Whereas the low Chl content of the wild type decreased steadily with increasing fluence rate from 8 to 250 $\mu$E m$^{-2}$ s$^{-1}$, overexpression of POR allowed substantial amounts of Chl to accumulate at all fluence rates tested from 20 to 500 $\mu$E m$^{-2}$ s$^{-1}$. The relative increase in Chl accumulation mediated by POR overexpression was much smaller at WL fluence rates of 8 $\mu$E m$^{-2}$ s$^{-1}$ or less, indicating that POR-mediated photoprotection rather than a simple increase in the efficiency of light utilization for Pchlide reduction dominated at high fluence rates. Furthermore, POR overexpression not only increased the WL fluence rate at which Chl accumulated maximally but also decreased drastically the steady-state content of Pchlide in cFR-grown seedlings.[10]

## Analysis of Photooxidative Damage during Light-Induced Chloroplast Formation

Chlorophyll and its biosynthetic intermediates, including Pchlide, are one of the few major classes of substances that is capable of absorbing light energy in living organisms. If exposed to light these pigments have a strong tendency to photosensitize the formation of singlet oxygen and oxygen radicals that can cause various types of cellular damage, including lipid peroxidation, membrane lysis, DNA strand breakage, and protein inactivation.[9] In particular, angiosperms accumulate higher levels of Pchlide during seedling development in the dark than in the light and had to evolve strategies to avoid photooxidative damage on illumination. As shown by our studies of POR-overexpressing *Arabidopsis* seedlings, one way of coping with the problem of photooxidation is to synthesize larger amounts of Pchlide-binding POR proteins that form together with Pchlide and NADPH

[9] K. Azada, *Methods Enzymol.* **105,** 422 (1984).
[10] U. Sperling, B. van Cleve, G. Frick, K. Apel, and G. A. Armstrong, *Plant J.* **12,** 649 (1997).

photoactive ternary complexes. Such complexes use the light energy absorbed by Pchlide to catalyze the energy-requiring reduction of Pchlide to Chlide. Once the majority of Pchlide has been photoreduced, the excessive amounts of POR proteins become dispensable, which may explain the rapid decline both of POR A mRNA and POR A protein shortly after the beginning of illumination of etiolated seedlings.[6] Another way of avoiding photooxidative damage is to control tightly the biosynthesis of Pchlide. An as yet undetermined control mechanism restricts the formation of ALA, the first committed precursor of all porphyrins, in etiolated seedlings. This limitation can be overcome by light. Metabolic feedback control of ALA biosynthesis could be exerted by an intermediate in the pathway that accumulates in the dark and disappears in the light. Pchlide has been regarded as a candidate for such a regulatory substance, but also earlier intermediates of Chl biosynthesis have been implicated as possible regulators of ALA synthesis. We have used a genetic approach to identify cellular components that may either act as protectants against photooxidative damage or that may form part of inhibitory feedback loops that help to control tightly the biosynthesis of Mg-porphyrins in dark-grown seedlings.

Three different screening procedures were used to identify mutants of interest. Because many of these mutants were expected to be nonviable, screening was performed with seed families of individual mutagenized $M_1$ plants. In this way a given mutation could be rescued through the identification of a heterozygous plant among the progeny of an $M_1$ plant that also gave rise to the mutant.

Chemical mutagenesis of *Arabidopsis* seeds was performed according to Estelle and Somerville.[11] Approximately 10,000 seeds were resuspended in 100 ml of 0.3% ethyl methane sulfonate (EMS) and incubated at room temperature. Afterward the seeds were washed 10–15 times with distilled water, sown on soil, and kept for 24 hr at 4° before they were transferred into the growth chamber. Individual seedlings were isolated and replanted on soil. After 5–7 weeks, seeds from each $M_1$ plant were collected separately. From each of the $M_1$ seed families 20–30 seeds were used for the subsequent screening steps.

Each seed sample was surfaced sterilized separately for 5 min in 5% (v/v) bleach containing 0.02% Triton X-100 and washed several times with distilled water at room temperature. Finally each seed sample was resuspended in 40° prewarmed 0.5% low-melting-temperature agarose and placed on plates of solidified 0.7% Murashige–Skoog agar. Six to seven seed families were analyzed on a single dish. After preincubation for 24 hr in the dark at 4° the plates were adapted to room temperature and

[11] M. A. Estelle and C. Somerville, *Mol. Gen. Genet.* **206**, 200 (1987).

illuminated for 30 min to induce seed germination. The subsequent handling of the plates varied depending on the screening procedure for which they were used.

### Isolation and Classification of Chlorophyll-Deficient Xantha Mutants

The first screening was aimed at identifying mutants that are either blocked at various steps of the biosynthetic pathway of Chl or that are disturbed in one of the subsequent steps leading to the assembly of an active photosynthetic membrane.[12] Plated seeds were either kept under continuous dim white light (10 $\mu$E m$^{-2}$ s$^{-1}$) or were returned to the dark for another 3 days. Mutants that did not accumulate Chl in the light (xantha mutants) were scored visually. Seed families in which xantha mutants segregated at a ratio of approximately 7 : 1 (wild-type : mutant) were selected for subsequent testing. Two different screening tests were used. First, acetone extracts of three to five xantha mutants of each family were prepared and absorption spectra between 400 and 700 nm measured. Only xantha mutants that showed the same carotenoid spectrum as that of etiolated wild-type seedlings and that were devoid of detectable Chl were used for further studies. In a second screening test, single xantha mutants were transferred to Murashige–Skoog agar plates that contained 0.5% sucrose and were kept under dim light for 5–7 days. Only mutants that did not accumulate detectable amounts of Chl under these conditions were selected for a final ALA feeding test.

In higher plants the accumulation of Pchlide in the dark leads in an unknown manner to the inhibition of ALA synthesis. This physiological block can be bypassed by feeding ALA to dark-grown wild-type seedlings, resulting in a massive accumulation of Pchlide.[13] Hence, none of the enzymatic steps that lead from ALA to Pchlide is rate limiting in these plants. In mutant seedlings with lesions at steps prior to ALA formation or following the synthesis of Pchlide, feeding of exogenous ALA in the dark should lead to the accumulation of high amounts of Pchlide as in the ALA-fed wild-type seedlings, while in mutants with biosynthetic blocks between ALA and Pchlide, no Pchlide accumulation should occur. Using this approach three major classes of Chl-deficient xantha mutants can be defined. Class I mutants have lesions in ALA synthesis: they would be expected to accumulate Pchlide only when the plants are treated with ALA. Class II mutants fail to accumulate Pchlide when fed with ALA, suggesting that the mutation is at a point beyond ALA formation but before the step leading

---

[12] S. Runge, B. van Cleve, N. Lebedev, G. A. Armstrong, and K. Apel, *Planta* **197,** 490 (1995).
[13] S. Granick, *Plant Physiol.* **34,** 18 (1959).

to Pchlide. Class III mutants are blocked beyond the Pchlide synthesis step but before pigment–protein complex assembly as they accumulate this metabolite regardless of whether or not they are fed with ALA.[12]

For the ALA feeding test 4-day-old light-grown xantha mutants were vacuum infiltrated with 10 mM ALA-KOH (pH 7.0) for 1 min and were incubated for 12 hr in the dark. Acetone extracts of seedlings were prepared under green safelight and fluorescence emission spectra recorded at room temperature with an excitation wavelength of 440 nm. Spectra were also recorded from acetone extracts of control seedlings that had been kept in complete darkness. The two cotyledons were harvested under green safelight from individually labeled etiolated seedlings and one each was placed on agar plates maintained either in the dark or in dim white light (10 $\mu$E m$^{-2}$ s$^{-1}$) for 24 hr. Dark-incubated cotyledons of seedlings whose light-incubated cotyledons did not green were extracted and their Pchlide contents determined by fluorescence emission spectroscopy. Thus far, only mutants of classes II and III but no mutants of class I have been identified.[12] Similar results have been reported for Chl-deficient mutants of barley and wheat.[1] This inability to obtain plants with a lesion in ALA formation suggests that such mutations may be lethal even in the heterozygous state.

### Isolation of Mutants with Lesions in the Metabolic Feedback Loop that Controls ALA Synthesis

As mentioned earlier, an as yet undetermined control mechanism restricts the production of ALA in dark-grown angiosperm plants. The only ALA-derived intermediate of Chl synthesis that accumulates to detectable levels is Pchlide. However, this physiological block can be bypassed by feeding ALA to dark-grown wild-type seedlings, resulting in a massive accumulation of Pchlide.[13] While most of the Pchlide that accumulates in untreated control seedlings forms together with POR and NADPH a ternary photoactive complex, the excess Pchlide in ALA-fed seedlings accumulates as nonphotoreducible pigment molecules. When exposed to long-wavelength UV light most of the energy absorbed by Pchlide in control seedlings is quenched and used for the photoreduction of Pchlide, while in ALA-fed seedlings the excited free Pchlide molecules emit a bright red fluorescence. This difference in emitted fluorescence light can be used to screen for mutants with lesions in the metabolic feedback loop that normally prevents the accumulation of excess amounts of free Pchlide. Seeds of mutagenized plants were plated on Murashige–Skoog agar plates as described and germinated in the dark at 22° for 3–4 days.

Etiolated seedlings were exposed to a bright, long-wavelength UV/blue light beam transmitted through a glass fiber conductor (Intralux

100 UV VOL DI). Seedlings that emitted a bright red fluorescence in their cotyledons were identified. Heterozygous plants that segregated for this mutant phenotype were isolated from the $M_1$ seed family and used for the production of larger amounts of seeds. Fluorescence emission spectra were recorded from acetone extracts of the mutant seedlings. These spectra confirmed that the mutants had accumulated excess amounts of Pchlide.[14] Apparently in the mutants the metabolic feedback loop through which wild-type seedlings normally control the synthesis of ALA and the accumulation of Pchlide no longer operates properly.

### Isolation of Mutants Impaired in the Protection Against Photooxidative Damage

POR is an enzyme that is not only necessary for the light-dependent synthesis of chlorophyll but at the same time is required for the protection of the plastid compartment against photooxidative damage during the initial steps of light-induced chloroplast development.[10] A simple screening procedure was used to identify mutants with an impaired protection against photooxidation. Duplicate seed samples from each $M_1$ family were plated on two different petri dishes and, following seed germination in the dark for 24 hr, were placed under light intensities of 10 and 200 $\mu E$ $m^{-2}$ $s^{-1}$, respectively. $M_1$ families yielding mutants that were bleached under high light but not under low light were identified and heterozygous plants isolated that segregated mutants at a $3:1$ ratio.[14] These mutants will be used for the identification of cellular components that are required for protection against light stress during the initial steps of chloroplast development.

[14] M. Nater and K. Apel, unpublished results (1996).

## [17] Molecular Biology of Carotenoid Biosynthesis in Photosynthetic Organisms

By Mark Harker and Joseph Hirschberg

### Introduction

Carotenoids are $C_{40}$ isoprenoids that consist of eight isoprene units joined such that the linking of the units is reversed at the center of the molecule.[1] Carotenes are either linear or cyclized at one or both ends, and xanthophylls are formed from the hydrocarbon carotenes by the introduction of oxygen functions. Carotenoids serve two major functions in photosynthesis: as accessory pigments for light harvesting and in the prevention of photooxidative damage.

The most prominent chemical feature of the carotenoids is the polyene chain, which may extend from 3 to 15 conjugated double bonds. The polyene chain is responsible for the characteristic absorption spectrum and, therefore, color of the carotenoid, and is also responsible for the photochemical properties of the molecule.[2] Due to these photochemical properties carotenoids are essential components for all photosynthetic organisms, where they participate in a number of photosynthetic processes.[3] These include light harvesting resulting in energy transfer to the chlorophylls[4]; photoprotection by quenching triplet-state chlorophyll molecules[5]; and scavenging singlet oxygen and dissipating excess energy.

It has been demonstrated in plants that zeaxanthin is required to dissipate, in a nonradiative manner, the excess excitation energy of the antenna chlorophyll (reviewed in Ref. 3). In algae and plants a light-induced deepoxidation of violaxanthin to yield zeaxanthin, is related to photoprotection processes. The light-induced deepoxidation of violaxanthin and the reverse reaction that takes place in the dark, are known as the "xanthophyll cycle" (reviewed in Ref. 6).

Carotenoids also serve structural functions in the photosynthetic pigment–protein complexes of the reaction centers and the light-harvesting antennae, where they are bound to specific chlorophyll/carotenoid-binding proteins. Carotenoids provide yellow, orange, or red coloring to many flowers, fruits, mushrooms, and animals.

## Cloning Genes of the Carotenoid Biosynthesis Pathway

The carotenoid biosynthesis pathway was postulated over three decades ago by standard biochemical analyses using labeled precursors, specific inhibitors, and characterization of mutants. However, lack of *in vitro* assays

---

[1] T. W. Goodwin, *in* "The Biochemistry of the Carotenoids," Vol. 1. Chapman and Hall, London and New York, 1980.

[2] G. Britton, *FASEB J.* **9**, 1551 (1995).

[3] H. A. Frank and R. J. Cogdell, *Photochem. Photobiol.* **63**, 257 (1996).

[4] H. A. Frank and R. L. Christensen, *in* "Advances in Photosynthesis: Anoxygenic Photosynthetic Bacteria" (R. E. Blankenship, M. T. Madigan, and C. E. Bauer, eds.), p. 373. Kluwer Academic Publishing, 1995.

[5] B. Demmig-Adams, A. M. Gilmore, and W. W. Adams, *FASEB J.* **10**, 403 (1996).

[6] B. Demmig-Adams and W. W. Adams, *Trends Plant Sci.* **1**, 21 (1996).

for most of the enzymes that are involved in the conversion of phytoene to the predominant carotenes and xanthophylls had previously hindered a more detailed elucidation of this pathway. In recent years the molecular-genetic approach to studying carotenogenesis has provided a wealth of information and new perspectives of both the enzyme activities and the regulation of the pathway.

Many genes encoding enzymes that catalyze specific steps in the biosynthesis of carotenoids have been cloned recently from photosynthetic bacteria, cyanobacteria (blue-green algae), green algae, and higher plants.[7] Because of the difficulty isolating the enzymes, most genes were cloned by various genetic methods. In the photosynthetic bacterium *Rhodobacter capsulatus,* a cluster of carotenoid biosynthesis genes (*crt*) was identified and characterized by *in vivo* conplementation of mutants.[8] In *Erwinia* species, which are nonphotosynthetic bacteria, other *crt* gene clusters were cloned by heterologous gene expression in *Escherichia coli,* which subsequently produced carotenoids in the transgenic host cells.[9,10] Carotenoid biosynthesis genes from other bacteria have been identified by nucleotide sequence similarity to the *Rhodobacter* and *Erwinia* genes (reviewed in Ref. 8).

Carotenoid biosynthesis genes in plants, algae, and cyanobacteria are generally not conserved with those of bacteria and, therefore, the latter could not serve as molecular probes for cloning the plant-type genes. In addition, the genes in eukaryotes are dispersed throughout the genome. In cyanobacteria, only two genes, for phytoene desaturase and phytoene synthase, are in the same operon, while all other genes are dispersed.

Cloning of the first plant-type genes took advantage of the fact that the pathway in plants is similar to that of cyanobacteria, which are prokaryotes and are amenable to microbial genetic manipulations. In this manner the plant-type genes for phytoene desaturase (*crtP, Pds*)[11,12] and lycopene β-cyclase (*crtL-b*)[13,14] were first isolated from the cyanobacterium *Synecho-*

[7] G. A. Armstrong and J. E. Hearst, *FASEB J.* **10,** 228 (1996).
[8] G. A. Armstrong, *J. Bacteriol.* **176,** 4795 (1994).
[9] K. L. Perry, T. A. Simonitch, K. J. Harrison-Lavoie, and S. T. Liu, *J. Bacteriol.* **168,** 607 (1986).
[10] N. Misawa, M. Nakagawa, K. Kobayashi, S. Yamano, I. Izawa, K. Nakamura, and K. Harashima, *J. Bacteriol.* **172,** 6704 (1990).
[11] D. Chamovitz, I. Pecker, G. Sandmann, P. Böger, and J. Hirschberg, *Z. Naturforsch.* **45c,** 482 (1990).
[12] D. Chamovitz, I. Pecker, and J. Hirschberg, *Plant Mol. Biol.* **16,** 967 (1991).
[13] F. X. Cunningham, Jr., D. Chamovitz, N. Misawa, E. Gantt, and J. Hirschberg, *FEBS Lett.* **328,** 130 (1993).
[14] F. X. Cunningham, Jr., Z. R. Sun, D. Chamovitz, J. Hirschberg, and E. Gantt, *Plant Cell* **6,** 1107 (1994).

*cystis* sp. PCC7942. *Synechococcus* PCC7942 mutants that are resistant to a specific inhibitor, norflurazon in the case of phytoene desaturase and MPTA in the case of lycopene cyclase, were isolated. Once it was determined that the mutation is related to the target site of the inhibitor, the resistance gene was obtained by complementing the resistance trait in a wild-type strain of cyanobacteria with a genomic library from the mutant.[11] The cyanobacterial genes were successfully used as molecular probes to clone the homologous genes from algae and higher plants.[15-17]

The cyanobacterial gene for phytoene synthase was discovered due to its localization downstream of *crtP* on the same operon, and its ability to be functionally expressed in *E. coli.*[18] This sequence in cyanobacteria (*crtB*) was found to share some sequence similarity with the tomato cDNA pTOM5 that was originally cloned randomly as a gene whose expression is specific to fruit ripening.[19,20] Confirmation that pTOM5 encodes phytoene synthase was obtained by anti-sense silencing of the gene in transgenic tomato plants[21] and by functional complementation in bacteria.[22] A second *Psy* cDNA, which is constitutively expressed in the leaves, was also found in tomato.[23]

A unique method to clone genes for carotenoid biosynthesis enzymes, termed *color complementation,* has been developed. It takes advantage of the ability of the carotenogenic enzymes to function in cells of *E. coli.* In this method a recombinant plasmid is constructed on a pACYC184 vector, with previously cloned genes from different species in such a way that they are functionally expressed in *E. coli.* Cells of *E. coli* that carry this plasmid produce a specific carotenoid that can serve as a precursor for enzymes that can convert it to other carotenoids. These carotenoids accumulate in the bacterial cells and impart to the colonies a typical color that is visible when grown on petri plates. Identification of carotenoid biosynthesis gene is done by screening for change in colors of colonies that express a gene

[15] G. E. Bartley, P. V. Viitanen, I. Pecker, D. Chamovitz, J. Hirschberg, and P. A. Scolnik, *Proc. Natl. Acad. Sci. U.S.A.* **88,** 6532 (1991).

[16] I. Pecker, D. Chamovitz, H. Linden, G. Sandmann, and J. Hirschberg, *Proc. Natl. Acad. Sci. U.S.A.* **89,** 4962 (1992).

[17] I. Pecker, R. Gabbay, F. X. Cunningham, and J. Hirschberg, *Plant Mol. Biol.* **30,** 807 (1996).

[18] D. Chamovitz, N. Misawa, G. Sandmann, and J. Hirschberg, *FEBS Lett.* **296,** 305 (1992).

[19] J. A. Ray, C. R. Bird, M. Maunders, D. Grierson, and W. Schuch, *Nucleic Acids Res.* **15,** 10587 (1987).

[20] J. A. Ray, P. Moureau, A. S. Bird, D. Grierson, M. Maunders, M. Truesdale, P. M. Bramley, and W. Schuch, *Plant Mol. Biol.* **19,** 401 (1992).

[21] C. R. Bird, J. A. Ray, J. D. Fletcher, J. M. Boniwell, A. S. Bird, C. Teulieres, I. Blain, P. M. Bramley, and W. Schuch, *Bio/Technology* **9,** 635 (1991).

[22] G. E. Bartley, P. V. Viitanen, K. O. Bacot, and P. A. Scolnik, *J. Biol. Chem.* **267,** 5036 (1992).

[23] G. E. Bartley and P. A. Scolnik, *J. Biol. Chem.* **268,** 25718 (1993).

(or cDNA) library. The technique of color complementation in *E. coli* has been applied to clone a number of *crt* genes such as *zds* (*crtQ*)[24] from cyanobacteria, *crtO* from the green alga *Haematococcus pluvialis*,[25] and the cDNAs for lycopene ε-cyclase (*CrtL-e*)[26] and β-carotene hydroxylase (*CrtR-b1*) from *Arabidopsis*.[27]

Transposon tagging was used to clone phytoene synthase from maize[28] and zeaxanthin epoxidase from tobacco[29] with subsequent rescue of the transposon DNA.

Three genes from plants were cloned by screening cDNA libraries following purification of the enzymes and raising antibodies. The cDNAs of GGDP synthase (*Ggpps*), phytoene synthase (*Psy*), and capsanthin capsorubin synthase (*CCS*) were cloned from a fruit cDNA library of pepper (*Capsicum annuum*),[30–32] and the cDNA of violaxanthin deepoxidase was cloned from a lettuce leaf cDNA library.[33]

The gene for ζ-carotene desaturase was identified among the randomly sequenced cDNA clones (ESTs) of *Arabidopsis* (Accession No. T46272) due to its sequence similarity to *Pds* and then cloned from pepper.[34] Functional expression of this cDNA in *E. coli* has confirmed its identity. The homologous gene from the cyanobacterium *Synechocystis* PCC6803, *crtQ*, has been identified during random sequencing of the whole genome (gene slr0940 in Accession No. D90914). It is interesting to note that a gene termed *zds*, coding for an enzyme with ζ-carotene desaturase activity, was cloned from the cyanobacterium *Anabaena* PCC7120.[24] However, its amino acid sequence is similar to the bacterial-type phytoene desaturase (*crtI*) and not to the plant-type ζ-carotene desaturase (*crtQ*).

All carotenoid biosynthesis genes in plants and algae are nuclear encoded and their polypeptide products are all imported to the plastids.

[24] H. Linden, A. Vioque, and G. Sandmann, *FEMS Microbiol. Lett.* **106**, 99 (1993).
[25] T. Lotan and J. Hirschberg, *FEBS Lett.* **364**, 125 (1995).
[26] F. X. Cunningham, B. Pogson, Z. R. Sun, K. A. Mcdonald, D. Dellapenna, and E. Gantt, *Plant Cell* **8**, 1613 (1996).
[27] Z. R. Sun, E. Gantt, and F. X. Cunningham, *J. Biol. Chem.* **271**, 24349 (1996).
[28] B. Buckner, P. S. Miguel, D. Janickbuckner, and J. L. Bennetzen, *Genetics* **143**, 479 (1996).
[29] E. Marin, L. Nussaume, A. Quesada, M. Gonneau, B. Sotta, P. Hugueney, A. Frey, and A. Marionpoll, *EMBO J.* **15**, 2331 (1996).
[30] M. Kuntz, S. Römer, C. Suire, P. Hugueney, J. H. Weil, R. Schantz, and B. Camara, *Plant J.* **2**, 25 (1992).
[31] P. Hugueney, S. Römer, M. Kuntz, and B. Camara, *Eur. J. Biochem.* **209**, 399 (1992).
[32] F. Bouvier, P. Hugueney, A. Dharlingue, M. Kuntz, and B. Camara, *Plant J.* **6**, 45 (1994).
[33] M. Nishikitani, K. Kubota, A. Kobayashi, and F. Sugawara, *Biosci. Biotechnol. Biochem.* **60**, 929 (1996)
[34] M. Albrecht, A. Klein, P. Hugueney, G. Sandmann, and M. Kuntz, *FEBS Lett.* **372**, 199 (1995).

Early Reactions of Carotenoid Biosynthesis

All carotenoids are related biosynthetically and share a common early pathway with other isoprenoids. These reactions, in which mevalonate is an important intermediate, yield the monomer building block for all products in the isoprenoid pathway, the $C_5$ molecule isopentyl diphosphate (IDP). For a detailed review of the biochemistry and molecular biology of the isoprenoid pathway see Ref. 35. An alternative route for the biosynthesis of IDP in eubacteria, higher plants, and green algal chloroplasts has recently been identified.[6,36] In this nonmevanolate pathway IDP is formed by the condensation of pyruvate and glyceraldehyde 3-phosphate followed by a transposition reaction to yield IDP. Isomerization of IDP to the homoallylic dimethylallyl diphosphate (DMADP) is catalyzed by IDP isomerase, whose gene has been cloned from a number of plants and bacteria (e.g., Ref. 37). Geranylgeranyl diphosphate (GGDP) $C_{20}$ is produced by GGDP synthase, which catalyzes the sequential $1'$-4 condensation of three molecules of IDP with DMADP to produce GGDP (Fig. 1). Geranyl diphosphate $C_{10}$ and farnesyl diphosphate $C_{15}$ are intermediates in this stepwise reaction and act as substrates for condensation with IDP to produce GGDP. The GGDP synthase gene products cloned from higher plants[30,38] show significant homologies to one another with over 70% identity. They also show a high degree of similarity with GGDP synthases cloned from archaeabacteria and eubacteria, indicating that they belong to a large family of structurally related prenyltransferases.[39,40] However, individual GGDP synthases exhibit different affinities with respect to the chain length of the allylic substrate.[41]

Phytoene Synthesis

The formation of phytoene is the first specific step of the carotenoid biosynthesis pathway. Phytoene is synthesized by the tail-to-tail dimerization of two molecules of GGDP into prephytoene diphosphate, which is then converted into phytoene, in the 15-*cis* configuration in most organisms[42] (Fig. 1). A single gene product, phytoene synthase (*PSY*, *CRTB*), is capable of catalyzing both reactions, suggesting that the gene product is a

[35] J. Chappell, *Annu. Rev. Plant Physiol.* **46,** 521 (1995).
[36] H. K. Lichtenthaler, J. Schwender, A. Disch, and M. Rohmer, *FEBS Lett.* **400,** 271 (1997).
[37] V. M. Blanc, K. Mullin, and E. Pichersky, *Plant Physiol.* **111,** 652 (1966).
[38] S. M. Aitken, S. Attucci, R. K. Ibrahim, and P. J. Gulick, *Plant Physiol.* **108,** 837 (1995).
[39] G. A. Armstrong, B. S. Hundle, and J. E. Hearst, *Methods Enzymol.* **214,** 297 (1993).
[40] K. Adiwilaga and A. Kush, *Plant Mol. Biol.* **30,** 935 (1996).
[41] S. K. Math, J. E. Hearst, and C. D. Poulter, *Proc. Natl. Acad. Sci. U.S.A.* **89,** 6761 (1992).
[42] G. Sandmann, *Eur. J. Biochem.* **223,** 7 (1994).

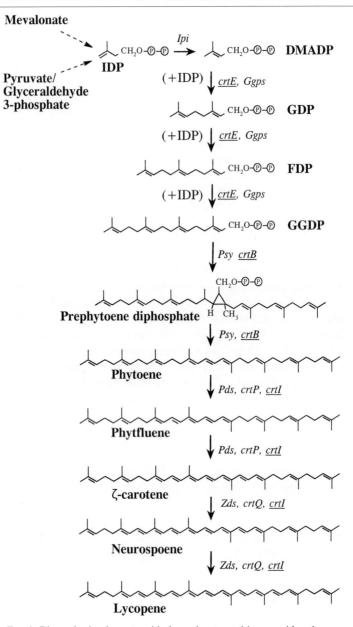

FIG. 1. Biosynthesis of carotenoids from the general isoprenoid pathway.

bifunctional enzyme. It was demonstrated in *C. annuum* that these reactions require $Mn^{2+}$.[43] In its active state *PSY* is tightly bound to the plastid membrane where it is in contact with galactolipids, the galactose residues being obligate components for the activation of the enzyme.[44] Genes for phytoene synthases were cloned from eukaryotes and prokaryotes (reviewed in Refs. 45 and 46). Their analysis indicated that these enzymes are both structurally and functionally conserved. In tomato (*L. esculentum*) two phytoene synthase genes (*Psy1* and *Psy2*), which code for polypeptides that are 95% identical in their amino acid sequence have been isolated. Both are transcribed in seedlings and leaves, whereas *Psy2* transcripts were relatively more abundant in fruits.[23] Transgenic tomato plants that expressed the anti-sense of *Psy2* cDNA, pTOM5, exhibited an altered carotenoid content in the ripe fruit and corolla but not in the leaves,[47] suggesting that *Psy1* is responsible for phytoene synthesis in green tissue.

Phytoene Desaturation

Phytoene undergoes a series of sequential desaturation reactions to give phytofluene, $\zeta$-carotene, neurosporene, and finally the maximally desaturated lycopene (Fig. 1). At each stage, two hydrogen atoms are removed by *trans* elimination from adjacent positions to introduce a new double bond, and to extend the conjugated polyene chain by two double bonds. In photosynthetic bacteria, like in other bacteria and fungi, these three or four steps are catalyzed by a single gene product, *CRTI*, to produce neurosporene or lycopene[48–51] in a reaction that requires adenosine triphosphate (ATP).[50,52] In cyanobacteria, algae, and higher plants, two enzymes— phytoene desaturase (*PDS* or *CRTP*) and $\zeta$-carotene desaturase (*ZDS* or *CrtQ*)—are responsible for the conversion of phytoene to lycopene. *PDS*

[43] O. Dogbo, A. Laferričre, A. d'Harlingue, and B. Camara, *Proc. Natl. Acad. Sci. U.S.A.* **85,** 7054 (1988).
[44] M. Schledz, S. Al-Babili, J. v.Lintig, H. Haubruck, S. Rabbani, H. Kleinig, and P. Beyer, *Plant J.* **10,** 781 (1996).
[45] S. Gradelet, P. Astorg, J. Leclerc, J. Chevalier, M. F. Vernevaut, and M. H. Siess, *Xenobiotica* **26,** 49 (1996).
[46] P. A. Scolnik and G. E. Bartley, *Plant Mol. Biol. Rep.* **14,** 305 (1996).
[47] P. Bramley, C. Teulieres, I. Blain, C. Bird, and W. Schuch, *Plant J.* **2,** 343 (1992).
[48] G. A. Armstrong, M. Alberti, F. Leach, and J. E. Hearst, *Mol. Gen. Genet.* **216,** 254 (1989).
[49] H. Linden, N. Misawa, D. Chamovitz, I. Pecker, J. Hirschberg, and G. Sandmann, *Z. Naturforsch.* **46c,** 1045 (1991).
[50] P. D. Fraser, N. Misawa, H. Linden, S. Yamano, K. Kobayashi, and G. Sandmann, *J. Biol. Chem.* **267,** 19891 (1992).
[51] M. P. Chavezmoctezuma and E. Lozoyagloria, *Plant Cell Rep.* **15,** 360 (1996).
[52] H. P. Lang, R. J. Cogdell, A. T. Gardiner, and C. N. Hunter, *J. Bacteriol.* **176,** 3859 (1994).

(or *crtP*) catalyzes the first two dehydrogenation reactions to yield ζ-carotene,[16] while *ZDS* (or *CrtQ*) catalyzes two further dehydrogenation reactions converting ζ-carotene to lycopene.[34,53] Both enzymes act symmetrically on the carotenoid substrate to introduce a new double bond into one-half of the molecule.

In *Narcissus pseudonarcissus* chromoplasts active *PDS* was shown to be tightly bound to the plastid membrane, where flavinylation of the protein was necessary for enzyme activation.[54] Phytoene desaturation is associated with a respiratory redox pathway in which quinones act as intermediate electron acceptors and oxygen as the final acceptor.[24,55] The redox mechanism and the role of quinones in phytoene desaturation have been demonstrated in experiments involving inhibitors of plastoquinone and tocopherol synthesis,[56] and in *Arabidopsis* mutants that were defective in the biosynthesis of plastoquinone and tocopherol.[54] In both cases the bleaching of the photosynthetic tissue was attributed to the indirect inhibition of phytoene desaturation.

In organisms containing 15-*cis*-phytoene, the desaturation sequence must also include a *cis* to *trans* isomerization step, since the more unsaturated carotenes are normally all-*trans*. In *Erwinia* this isomerization was shown to occur at the phytoene stage and required FAD,[48] suggesting the involvement of *CRTI* in the process.[50] In organisms possessing the *PDS/ZDS(CrtQ)* desaturation mechanism, ζ-carotene is usually in the *cis* configuration.[49] This is normally converted to all-*trans* lycopene, implying that *cis* to *trans* isomerization occurs at the level of ζ-carotene, probably in association with the ζ-carotene desaturase catalytic activity. A single-gene mutant in tomato, *tangerine,* is impaired in carotene isomerization which results in accumulation of poly-*cis* lycopene in the fruit.[57]

Cyclization Reactions

Cyclization of an acyclic carotenoid precursor, usually lycopene, is initiated by proton attack at C2 and C2′. The resulting carbonium ion is stabilized by the loss of a proton from either C1 or C4 to yield either a β-ring or a ε-ring, depending on which proton is lost[58] (Fig. 2). The lycopene β-cyclase polypeptides (*CRTL* or *LCY*) from cyanobacteria and higher

[53] H. Linden, N. Misawa, T. Saito, and G. Sandmann, *Plant Mol. Biol.* **24,** 369 (1994).
[54] S. R. Norris, T. R. Barrette, and D. Dellapenna, *Plant Cell* **7,** 2139 (1995).
[55] P. Beyer, M. Mayer, and H. Kleinig, *Eur. J. Biochem.* **184,** 141 (1989).
[56] A. Schulz, O. Ort, P. Beyer, and H. Kleinig, *FEBS Lett.* **318,** 162 (1993).
[57] L. C. Raymundo and K. L. Simpson, *Phytochemistry* **11,** 397 (1972).
[58] G. Britton, *in* "Plant Pigments" (T. W. Goodwin, ed.), p. 133. Academic Press, New York, 1988.

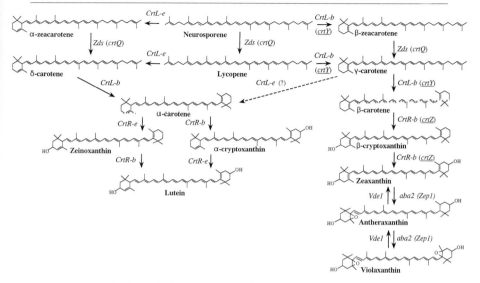

Fig. 2. Biosynthesis pathway of cyclic carotenes and xanthophylls.

plants exhibit a significant conservation of their amino acid sequences along their entire length (35–37% identities), but are distinct from the eubacterial lycopene β-cyclases (*CRTY*).[14,17,26] The two types of enzymes, *CRTY* and *CRTL*, share three short regions of sequence similarity; one of them is a characteristic dinucleotide [NAD(P)/FAD]-binding motif. These differences and those observed between *crtI* and Pds/*crtP*-type phytoene desaturases indicate that the carotenoid biosynthesis pathway in green photosynthetic organisms has evolved independently from the one in microorganisms. The higher plant lycopene cyclases have a significant sequence similarity at the amino acid sequence with the *C. annuum* capsanthin/capsorubin synthase (CCS) protein (51–54% identities).[59] CCS catalyzes the conversion of the epoxy-carotenoids antheraxanthin and violaxanthin to the ketocarotenoids capsanthin and capsorubin, respectively (see Fig. 3). CCS has also been shown to exhibit lycopene cyclase activity in *E. coli,* indicating a similar reaction mechanism for β-ring formation in β-carotene and κ-rings in capsanthin and capsorubin. The synthesis of both ring types is thought to proceed via the formation of a similar carbonium intermediate. Although lycopene is generally considered to be the usual substrate for the β-cyclases, *in vitro* experiments have shown that the *CRTL* and *CRTY* gene products from the bacteria *Erwinia uredovora* and the higher plant *C. annuum* can

[59] P. Hugueney, A. Badillo, H. C. Chen, A. Klein, J. Hirschberg, B. Camara, and M. Kuntz, *Plant J.* **8,** 417 (1995).

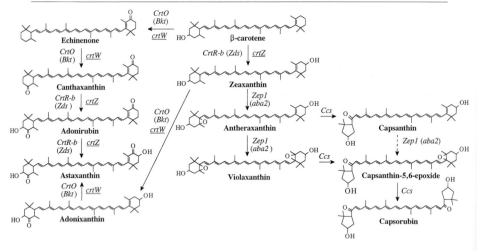

FIG. 3. Biosynthesis pathway of ketocarotenoids.

catalyze the cyclization of the 7,8-dihydro-$\psi$ end group, in addition to the usual cyclization of the $\psi$ end groups of lycopene to the $\beta$-groups.[60]

The cyclization of lycopene in photosynthetic organisms represents an important branching point in the biosynthesis pathway of carotenoids (Fig. 2). Two $\beta$-cyclization reactions at each end of the lycopene molecule result in the formation of $\beta$-carotene, which plays an essential role in the reaction centers of photosystem II (PSII) and Photosystem I (PSI). $\beta$-Carotene is also the precursor for a number of oxygenation reactions, which give rise to several indispensable xanthophylls such as zeaxanthin, violaxanthin, and neoxanthin. Alternatively, the $\psi$ end group of lycopene can undergo a second type of cyclization resulting in the formation of an $\varepsilon$-ring. The majority of cyclic carotenoids encountered in photosynthetic organisms that contain an $\varepsilon$-ring usually contain a $\beta$-ring at the other end of the molecule. The $\beta,\varepsilon$-carotene, called $\alpha$-carotene, is the precursor for the xanthophyll lutein, which is ubiquitous in the light-harvesting complexes of most higher plants. $\varepsilon$-Ring formation is catalyzed by the lycopene $\varepsilon$-cyclase (*CRTL-E*), which was cloned from *Arabidopsis*[26] and tomato (our unpub-

[60] S. Takaichi, G. Sandmann, G. Schnurr, Y. Satomi, A. Suzuki, and N. Misawa, *Eur. J. Biochem.* **241**, 291 (1996).

lished data). In tomato, the amino acid sequence of *CRTL-E* is 38% identical to that of *CRTLB* and 35% identical to the capsanthin–capsorubin synthase (CCS) of pepper.

Unlike the $\beta$-cyclases, $\varepsilon$-cyclase only catalyzes a single ring formation of the acyclic carotenoid substrate, leaving the other half of the molecule available for $\beta$-ring cyclization. It appears that the presence of a $\beta$- or $\varepsilon$-ring on one-half of the carotenoid substrate prevents the formation of a second $\varepsilon$-ring. It has been demonstrated the $\alpha$-carotene is synthesized in transformed *E. coli* cells by the combined activity of *CRTL-B* and *CRTL-E*[26] (and our unpublished results). Thus, the activity of *CRTL-E* relative to *CRTL-B* could indicate a major regulatory mechanism in the biosynthesis of cyclic carotenoids, having implications in the adaptation of plants to differing light regimes. The $\alpha$-carotene-derived xanthophyll lutein plays a major role in light harvesting, whereas the $\beta$-carotene-derived xanthophyll zeaxanthin is essential for protection against photooxidative damages. It has been shown in plants that the composition of carotenoids in the leaves changes in response to light intensity. Under low-light conditions plants accumulate higher levels of lutein relative to the xanthophyll-cycle components, violaxanthin, antheraxanthin, and zeaxanthin, probably in order to increase the efficiency of photon capture under limiting light intensities.[61,62] However, exposure to high light increases the ratio of $\beta,\beta$-carotenoids relative to $\beta,\varepsilon$-carotenoids. A similar influence of light intensity on composition of $\alpha$-carotene and $\beta$-carotene and their xanthophylls has been observed in green algae.[63] It is possible that the preferential route of xanthophyll biosynthesis is determined by the relative expression and/or activity of *CRTL-E,* which dictates to which pathway the lycopene precursor will be channeled in the biosynthesis of cyclic carotenoids. In that case, the regulatory mechanism of *CRTL-E* expression and/or activation is very likely to be influenced by external environmental parameters resulting in an increase or decrease of $\varepsilon$-cyclase expression/activity.

Xanthophyll Biosynthesis

The oxidation of $\beta$- and $\alpha$-carotenes results in the formation of the structurally diverse array xanthophylls utilized by photosynthetic organisms. The xanthophylls play an essential role in the nonphotochemical quenching of chlorophyll excited state in conditions of excessive photon

[61] S. S. Thayer and O. Björkman, *Photosynth. Res.* **23,** 331 (1992).
[62] B. Demmig-Adams and W. W. Adams, *Photosynth. Res.* **23,** 331 (1992).
[63] H. Senger, C. Wagner, D. Hermsmeier, N. Hohl, T. Urbig, and N. I. Bishop, *J. Photochem. Photobiol. B-Biol.* **18,** 273 (1993).

capture by the photosynthetic apparatus and in light-harvesting complexes at both the structural and functional level. Zeaxanthin is synthesized from $\beta$-carotene by the hydroxylation of C3 and C3' of the $\beta$-rings via the monohydroxylated intermediate $\beta$-cryptoxanthin, a process requiring molecular oxygen in a mixed-function oxidase reaction.[58] The gene encoding $\beta$-carotene hydroxylase (crtZ) has been cloned from a number of nonphotosynthetic prokaryotes (reviewed in Ref. 8). The $\beta$-carotene hydroxylase gene (CrtR-b) of a photosynthetic organism was cloned from Arabidopsis[27] and tomato (our unpublished data), where two copies of CrtR-b have been identified. The product of this cDNA clone was able to convert $\beta$-carotene to zeaxanthin in E. coli that was genetically engineered to produce $\beta$-carotene.

Lutein is synthesized by the hydroxylation of C3 and C3' of the $\beta$ and $\varepsilon$-rings of $\alpha$-carotene, respectively. The $\beta$-carotene hydroxylase gene product from Arabidopsis was unable to catalyze the hydroxylation of the $\varepsilon$-ring of $\delta$-carotene in E. coli,[27] indicating the existence of a second hydroxylase that catalyzes the hydroxylation of the $\varepsilon$-ring in the biosynthesis of lutein. A lutein-deficient Arabidopsis mutant, lut1, lacked hydroxylation activity of the $\varepsilon$-ring of $\alpha$-carotene, resulting in the accumulation of zeinoxanthin. This provides further evidence for the existence of a specific $\varepsilon$-ring hydroxylase.[64] A second lutein-deficient mutant, lut2, lacked $\varepsilon$-cyclase activity and accumulated relatively high levels of $\beta$-carotene. Despite the absence of lutein in the leaves of both mutants, they were photosynthetically active, and showed no visible altered phenotype or modified chlorophyll content. However, both mutants had significantly higher levels of $\beta,\beta$-carotenoids, particularly the levels of the xanthophyll-cycle components.[64] Similar mutations were also characterized in green algae.[65]

Zeaxanthin acts as the carotenoid substrate for the biosynthesis of violaxanthin (Fig. 2). This reaction is catalyzed by zeaxanthin epoxidase, which epoxidizes both $\beta$-rings of zeaxanthin at the 5,6 positions. The reaction proceeds via the formation of the monoepoxide intermediate antheraxanthin, which is further epoxidized to violaxanthin, a reaction requiring $O_2$ and NAD(P)H and FAD.[66,67] This reaction is also the first step in the biosynthesis of the important plant hormone abscisic acid (ABA).[68] The zeaxanthin epoxidase gene of Nicotiana plumbaginifolia was cloned by

[64] B. Pogson, K. A. Mcdonald, M. Truong, G. Britton, and D. Dellapenna, Plant Cell 8, 1627 (1996).

[65] N. I. Bishop, T. Urbig, and H. Senger, FEBS Lett. 367, 158 (1995).

[66] K. Büch, H. Stransky, H. J. Bigus, and A. Hager, J. Plant Physiol. 144, 641 (1994).

[67] K. Büch, H. Stransky, and A. Hager, FEBS Lett. 376, 45 (1995).

[68] J. A. D. Zeevaart and R. A. Creelman, Ann. Rev. Plant Physiol. Plant Mol. Biol. 39, 439 (1988).

transposon tagging that generated an ABA-deficient mutant.[29] The homologous gene of pepper has also been cloned.[69] The cloned cDNA encodes a protein of 72.5 kDa, which shares significant sequence similarities to other mono-oxygenases and oxidases of bacterial origin and contains an ADP-binding fold and an FAD-binding domain. Violaxanthin is considered to be the carotenoid substrate for the synthesis of neoxanthin. However, no protein or gene has yet been isolated which catalyzes the conversion of violaxanthin to neoxanthin. The products of violaxanthin/neoxanthin cleavage are the substrates for ABA synthesis. The gene for the cleaving enzyme has been recently cloned.[70]

The exposure of photosynthetic tissues to light in excess of that which can be utilized in photosynthesis can have deleterious effects on the photosynthetic apparatus. In higher plants and some algae the zeaxanthin and antheraxanthin-dependent dissipation of energy in the light-harvesting complexes of PSII is correlated with the nonphotochemical quenching of chlorophyll fluorescence.[5,6] Following the exposure to excessive photon fluxes, zeaxanthin and antheraxanthin are re-formed from violaxanthin by violaxanthin deepoxidase (VDE). The conversion of violaxanthin to zeaxanthin by VDE is activated by a decrease in the luminal pH.[71] Under these conditions VDE, which is normally mobile within the thylakoid lumen, becomes tightly bound to the membrane.[72] This allows VDE access to its substrate, violaxanthin. Purification of the enzyme has revealed it to be a 43-kDa protein.[73] The VDE cDNA was cloned from lettuce and the *E. coli*-expressed enzyme exhibited VDE activity that requires ascorbate and monogalactosyldiacylglycerol for activity.[74] The sequence data indicated three important domains in VDE: (1) a cysteine-rich region, (2) a lipocalin signature, and (3) a highly charged region.

In the ripening chromoplasts of *Capsicum* fruit, capsanthin and capsorubin accumulate as the major pigments. These oxocarotenoids are formed from antheraxanthin in the case of capsanthin and violaxanthin via capsanthin-5,6-epoxide in the case of capsorubin. The oxo groups are formed by the rearrangement of the epoxy groups of the β-rings into cyclopentane

[69] F. Bouvier, A. Dharlingue, P. Hugueney, E. Marin, A. Marionpoll, and B. Camara, *J. Biol. Chem.* **271**, 28861 (1996).

[70] S. H. Schwartz, B. C. Tan, D. A. Gage, J. A. D. Zeevaart, and D. R. McCarty, *Science* **276**, 1872 (1997).

[71] C. E. Bratt, P. O. Arvidsson, M. Carlsson, and H. E. Akerlund, *Photosynth. Res.* **45**, 169 (1995).

[72] A. Hager and K. Holocher, *Planta* **192**, 581 (1994).

[73] P. O. Arvidsson, C. E. Bratt, M. Carlsson, and H. E. Akerlund, *Photosynth. Res.* **49**, 119 (1996).

[74] R. C. Bugos and H. Y. Yamamoto, *Proc. Natl. Acad. Sci. U.S.A.* **93**, 6320 (1996).

rings (Fig. 3). The gene product CCS catalyzes this reaction in which proton attack of the epoxy groups and subsequent pinacolic arrangement convert the epoxy group into an oxo group.[32]

In some species of unicellular green algae, specific carotenoids are accumulated outside the chloroplast when the cells are exposed to adverse nutritional and environmental conditions. These so-called secondary carotenoids are usually $\beta$-carotene and its ketonic derivatives echinenone, canthaxanthin, adonirubin, and astaxanthin (Fig. 3). Echinenone and canthaxanthin also accumulate in some species of cyanobacteria, a process also influenced by the nutritional and environmental conditions encountered by the cells. A cDNA encoding $\beta$-C-4-oxygenase (*CrtO, Bkt*), has been cloned from the green alga *Haematococcus pluvialis*,[25,75] which accumulates astaxanthin under stress conditions. The enzyme catalyes the introduction of keto groups to C4 and C4' to $\beta$-carotene to form canthaxanthin via the mono keto intermediate echinenone. The $\beta$-C-4-oxygenase from *H. pluvialis* is also able to function in conjunction with bacterial and eukaryotic derived $\beta$-carotene hydroxylases, which introduce hydroxyl groups at positions C3 and C3', resulting in the synthesis of a number of ketocarotenoids including astaxanthin.[75,76]

Acyclic Xanthophyll Biosynthesis in *Rhodobacter*

The formation of acyclic xanthophylls in the nonsulfur photosynthetic bacteria *Rhodobacter capsulatus* and *Rhodobacter sphaeroides* has been genetically analyzed. The early steps of the pathway, including the formation and desaturation of phytoene to produce neurosporene, are similar to those described earlier for cyclic carotenoids. The desaturation of phytoene stops at the stage of neurosporene, which is converted to 1-hydroxyneurosporene (Fig. 4). One of two reactions can then occur, either 1-hydroxyneurosporene undergoes methylation of the hydroxy group to form methoxyneurosporene, or, alternatively, a double bond is introduced at position 3,4 to produce demethylspheroidene. Both intermediates can be converted to spheroidene. In the presence of oxygen, speroidene is converted to spheroidenone by the addition of a keto group at position C2. Demethylspheroidene can also be oxygenated at C2, producing demethylspheroidenone. Spheroidene can undergo a further hydroxylation reaction to form hydroxyspheroidene.

[75] S. Kajiwara, T. Kakizono, T. Saito, K. Kondo, T. Ohtani, N. Nishio, S. Nagai, and N. Misawa, *Plant Mol. Biol.* **29**, 343 (1995).
[76] M. Harker and J. Hirschberg, *FEBS Lett.* **404**, 129 (1997).

## Common crotenoid biosynthesis pathway

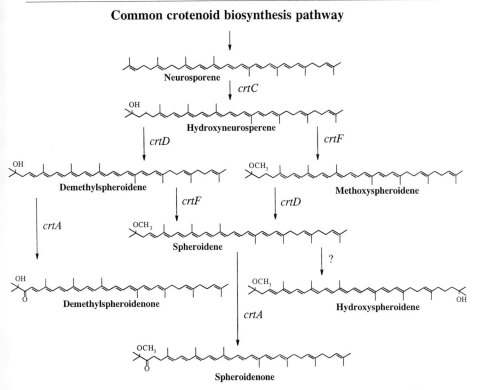

FIG. 4. Biosynthesis pathway of acyclic xanthophylls in *Rhodobacter*.

The nucleotide sequences for all the *R. capsulatus crt* genes (*crtA, B, C, D, E, F,* and *I*) have been determined,[48] and *crtD, I, B,* and part of *C* in *R. sphaeroides*.[77,78] In *R. capsulatus* the carotenoid biosynthesis genes are clustered in a ~11.0-kb DNA region, with *crtEF* and *crtIB* genes forming multigene operons. *Escherichia coli*-like $\sigma^{70}$ promoter sequences, rho-independent transcription terminators, and recognition sites for DNA-binding transcriptional factors were observed within the *crt* gene cluster. The predicted amino acid sequences of *crtB, crtE,* and *crtI* are conserved to some extent with those found in other carotenoid producing organisms. *CRTD*, a specific dehydrogenase of hydroxyneurosporene and methoxyneurosporene, is highly conserved with the phytoene desaturase *CRTI*.[48] Carotenoid genes implicated in the spirilloxanthin biosynthesis in *Rubrivivax gelatino-*

[77] E. Gari, J. C. Toledo, I. Gilbert, and J. Barbé, *FEMS Microbiol. Lett.* **93**, 103 (1992).
[78] H. P. Lang, R. J. Cogdell, S. Takaichi, and C. N. Hunter, *J. Bacteriol.* **177**, 2064 (1995).

*sus* have been identified.[79] It was determined that *crtD* and *crtC* genes are involved not only in the hydroxyspheroidene biosynthesis pathway as in *Rhodobacter* species, but also in the spirilloxanthin biosynthesis pathway.

Regulation of Carotenoid Biosynthesis

Carotenoid biosynthesis is a highly regulated process, both in prokaryotes and eukaryotes. The regulation of carotenoid biosynthesis in *Rhodobacter* is dependent on external stimuli, including light intensity and oxygen availability. The two major carotenoids in *R. sphaeroides,* spheroidene and spheroidenone, are differentially synthesized depending on the growth conditions. Spheroidene prevails during growth under anaerobic conditions and low light intensities, whereas spheroidenone is predominant in semiaerobically grown cells or during anaerobic growth at high light.[80] Transcription of genes involved in carotenoid biosynthesis was measured in *R. capsulatus* using gene-specific probes and promoter–reporter gene fusions.[81] The onset of illumination together with a decrease of oxygen caused the upregulation of *crtA, D, E,* and *F* by approximately 2 to 12-fold; however, the expression of *crtB* and *I* remained unaffected. The results indicate that carotenoid biosynthesis in *R. capsulatus* is under transcriptional control, but not the conversion of GGPP to neurosporene. It is interesting to note that also in nonphotosynthetic organisms, such as fungi, carotenogenesis is regulated by gene expression at the transcriptional level that is influenced both by external stimuli, for example, light and developmental processes.[82]

In plants carotenoid biosynthesis has been shown to be necessary for the development of chloroplasts[83] where the light-induced formation of the photosynthetic complexes is mediated by transcriptional and posttranscriptional regulation. However, expression of the nuclear encoded genes for carotenoid biosynthesis enzymes is not light dependent. During deetiolation in tomato seedlings carotenoid levels increase several-fold, while the transcript levels of *Pds* and *Psy1* remain relatively unchanged.[84] This indicates that transcriptional control of *Pds* and *Psy1* does not play a major role in the regulation of carotenoid biosynthesis during deetiolation in tomato.

[79] S. Ouchane, M. Picaud, C. Varnotte, F. Reissshusson, and C. Astier, *J. Biol. Chem.* **272,** 1670 (1997).
[80] A. A. Yeliseev, J. M. Eraso, and S. Kaplan, *J. Bacteriol.* **178,** 5877 (1996).
[81] G. A. Armstrong, D. N. Cook, D. Ma, M. Alberti, D. H. Burke, and J. E. Hearst, *J. Gen. Microbiol.* **139,** 897 (1993).
[82] G. Arpaia, A. Carattoli, and G. Macino, *Dev. Biol.* **170,** 626 (1995).
[83] B. Dreyfuss and J. Thornber, *Plant Physiol.* **106,** 829 (1994).
[84] G. Giuliano, G. E. Bartley, and P. Scolnik, *Plant Cell* **5,** 379 (1993).

In chromoplasts of *Narcissus pseudonarcissus,* the enzymes *PSY* and *PDS* were detected in an inactive form in the soluble fraction and in an active form when membrane bound.[44,85] This excess of inactive protein indicates a posttranslational regulation mechanism in which the role of events such as the redox-state of the membrane-bound electron acceptor appear to be major components influencing the biosynthesis of carotenoids.

Biochemical or genetic blocks in carotenoid biosynthesis result in photo-oxidation and bleaching of photosynthetic tissues. The inhibition of lyco-pene cyclization in the photosynthetic tissue of tomato caused the expression of *Pds* and *Psy1* to be up-regulated.[84,86] However, this up-regulation was not a direct response to photooxidation, since up-regulation was also observed in the absence of light and of the major photosensitizer, chlorophyll. These data suggest that the carotenoid pathway may be regulatd by responding to the concentration of its end products. This hypothesis is supported by further inhibitor studies where treated photosynthetic tissues accumulated higher molar concentrations of carotenoids than untreated tissues.[87] The mechanism of this regulation is unknown. Although ABA, which is synthesized from cyclic carotenoids, has also been suggested as a possible activator, *Arabidopsis* mutants that contained cyclic carotenoids but inhibited in ABA synthesis did not exhibit a significant increase carotenoid biosynthesis.[88]

The accumulation of carotenoids in chromoplasts of fruits and flowers is developmentally regulated. During tomato fruit ripening the concentration of lycopene increases by 500-fold, causing the carotenoid concentration in the fruit to increase 10- to 15-fold. During this process it has been shown that the mRNA levels of *Psy1* and *Pds* increase significantly.[16,23,84] Transcriptional control was demonstrated in this case.[86] In contrast, the mRNA levels of the lycopene $\beta$- and $\varepsilon$-cyclases decrease during ripening to almost an undetectable concentration[17,89] (and unpublished results), thus causing the accumulation of lycopene. During the fruit ripening of pepper an increase in the carotenoid concentration is observed that is concomitant with an increase in the transcript levels of *Ggpps, Pds,* and *Ccs.*[90] However,

[85] S. Albabili, J. Vonlintig, H. Haubruck, and P. Beyer, *Plant J.* **9,** 601 (1996).
[86] V. Corona, B. Aracri, G. Kosturkova, G. E. Bartley, L. Pitto, L. Giorgetti, P. A. Scolnik, and G. Giuliano, *Plant J.* **9,** 505 (1996).
[87] D. J. Simpson, F. M. M. Rahman, K. A. Buckle, and T. H. Lee, *Aust. J. Plant Physiol.* **1,** 135 (1974).
[88] C. D. Rock and J. A. D. Zeevaart, *Proc. Natl. Acad. Sci. U.S.A.* **88,** 7496 (1991).
[89] J. Hirschberg, M. Cohen, M. Harker, T. Lotan, V. Mann, and I. Pecker, *Pure Appl. Chem.* **69,** 2152 (1997).
[90] P. Hugueney, F. Bouvier, A. Badillo, J. Quennemet, A. Dharlingue, and B. Camara, *Plant Physiol.* **111,** 619 (1996).

*Psy* mRNA levels only increase slightly during ripening. This may indicate that phytoene synthesis represents a rate-limiting step in the synthesis of carotenoids during fruit ripening in pepper. Alternatively, fruit ripening may trigger posttranscriptional events resulting in increased PSY activity.

### Inhibitors of Carotenoid Biosynthesis

Many compounds have been identified that interfere with carotenoid biosynthesis, the effects of which were discussed in the previous section. Several compounds are known that inhibit the plant-type phytoene desaturase (PDS), resulting in the accumulation of phytoene, which normally is not detected in green tissues.[91-93] Two of the best known examples are the phenylpyridazones, norflurazon [SAN 9879; 4-chloro-5-methylamino-2-(3-trifluoromethylphenyl)-pyridazin-3(2H)one], and metflurazon (SAN 6707). They act by directly interfering with PDS, resulting in decreased enzyme activity. Mutations in *Pds* that lead to norflurazon resistance have been described.[94-96] Other phytoene desaturase inhibitors include fluridone [1-methyl-3-phenyl-5-(3-trifluoromethylphenyl)-4(1 H)-pyridone] and diflufenican [MB 38544; *N*-(2,4-difluorophenyl)-2(3-trifluoromethylphenoxy)nicotinamide]. Few compounds are known that inhibit the desaturation of ζ-carotene, although amitrole (3-amino-5-triazole) and dichlormate have been reported to cause the accumulation of ζ-carotene in whole plants. There are relatively few inhibitors of lycopene cyclase, the only known compounds are the triethylamines, for example, CPTA [2-(4-chlorophenylthio)triethylamine HCl) and nicotine. There are also only a few inhibitors specific for the mixed-function oxygenase reactions of xanthophyll formation. These include nicotine, diphenylamine, and 1-aminobenzotriazole. One of the first compounds known to inhibit the desaturation of phytoene in prokaryotes were bis-phenyl derivatives such as diphenylamine (DPA).

### Genetic Manipulation of Carotenoid Biosynthesis

Carotenoid biosynthesis was introduced in *E. coli,* indicating the possibilities for genetic manipulations of the pathway *in vivo.* Constitutive ex-

[91] P. Böger, *in* "Target Sites of Herbicide Action" (P. Böger and G. Sandmann, eds.), p. 247. CRC Press, Inc., Boca Raton, 1989.
[92] G. Britton, P. Barry, and A. J. Young, *in* "Herbicides and Plants Metabolism" (A. D. Dodge, ed.), p. 51. Cambridge University Press, 1989.
[93] P. M. Bramley, *in* "Carotenoids in Photosynthesis" (A. Young and G. Britton, eds.), p. 127. Chapman and Hall, London, 1993.
[94] D. Chamovitz, I. Pecker, and J. Hirschberg, *Plant Mol. Biol.* **16,** 967 (1991).
[95] D. Chamovitz, G. Sandmann, and J. Hirschberg, *J. Biol. Chem.* **268,** 17348 (1993).
[96] I. M. Martinez-Ferez and A. Vioque, *Plant Mol. Biol.* **18,** 981 (1992).

pression of pTOM5, the tomato cDNA for phytoene synthase, in transgenic tobacco has led to dwarfism due to redirecting metabolites from the gibberellin pathway.[97] Transformation of tobacco with the *Erwinia crtI* gene for phytoene desaturase conferred tolerance to the herbicide norflurazon and caused a small increase of $\beta$-carotene and minor changes in the composition of xanthophylls in the leaves.[98] Expression of the daffodil cDNA for *Psy* under the regulation of an endosperm-specific promoter was achieved in rice and resulted in phytoene accumulation in the seeds of the transgenic plants.[99] The gene *crtO* from the green alga *Haematococcus pluvialis,* encoding $\beta$-C-4-oxygenase which converts $\beta$-carotene to canthaxanthin, was expressed in the cyanobacterium *Synechococcus* PCC7942.[100] The genetically engineered cyanobacteria produced astaxanthin as well as other ketocarotenoids. Transformation of the same cDNA to tobacco plants induced the accumulation of a high concentration of ketocarotenoids, including astaxanthin, in the chromoplasts of the nectary tissue (Mann *et al.*, submitted for publication). However, only low concentrations of ketocarotenoids were detected in the leaves of these plants. These results provide evidence for the future possibilities of genetic engineering of the carotenoid biosynthesis pathway for biotechnology, as well as studying of the functions of carotenoids in photosynthesis.

[97] R. G. Fray, A. Wallace, P. D. Fraser, D. Valero, P. Hedden, P. M. Bramley, and D. Grierson, *Plant J.* **8,** 693 (1995).
[98] N. Misawa, K. Masamoto, T. Hori, T. Ohtani, P. Boger, and G. Sandmann, *Plant J.* **6,** 481 (1994).
[99] P. K. Burkhardt, P. Beyer, J. Wunn, A. Kloti, G. A. Armstrong, M. Schledz, J. Vonlintig, and I. Potrykus, *Plant J.* **11,** 1071 (1997).
[100] M. Harker and J. Hirschberg, *FEBS Lett.* **404,** 129 (1997).

# [18] Copper-Responsive Gene Expression during Adaptation to Copper Deficiency

*By* Jeanette M. Quinn and Sabeeha Merchant

## Introduction

Copper is an essential micronutrient for all organisms because of its role as a redox active cofactor in many essential enzymes and electron transfer catalysts.[1] However, because excess copper is toxic, organisms

[1] M. C. Linder, *in* "Biochemistry of Copper." Plenum Press, New York, 1991.

maintain copper homeostasis by regulating copper uptake and utilization. The unicellular green alga *Chlamydomonas reinhardtii* has proved to be a useful model system for studying one aspect of copper metabolism, specifically, adaptations to copper deficiency. Like other green algae, *C. reinhardtii* does not appear to contain copper enzymes such as copper/zinc superoxide dismutase or polyphenol oxidase that are abundant in vascular plants. The major copper proteins of Chlamydomonas are plastocyanin, utilized in the photosynthetic apparatus where it functions to transfer electrons from the cytochrome $b_6f$ complex to Photosystem I and cytochrome (cyt) oxidase, which is required for respiration in the mitochondrion. In *C. reinhardtii*, plastocyanin is estimated to be at least 10-fold more abundant than cytochrome oxidase, which makes it the major metabolic sink for copper.

When copper is limiting, certain green algae and cyanobacteria synthesize cytochrome $c_6$ (a heme-containing protein), which can substitute functionally for plastocyanin. The accumulation of one or the other protein is reciprocally dependent on the presence of copper in the growth medium. In *C. reinhardtii*, this is effected by enhanced degradation of plastocyanin in copper-deficient cells[2] and activation of transcription of the *Cyc6* gene (encoding cyt $c_6$).[3] The extent of transcription of the *Cyc6* gene is directly proportional to the perceived copper deficiency.[4] Thus, measurement of *Cyc6* expression is generally a good assay for copper deficiency. Other changes in gene expression include induction of coprogen oxidase synthesis (encoded by the *Cpx1* gene[5]), which is required to meet the increased demand for heme in copper-deficient cells. Copper deficiency also induces a cell surface cupric reductase and a high-affinity copper transporter; these are likely to be components of an assimilatory uptake system.[6] The regulatory system described is not only a valuable system for the study of copper-responsive gene expression but it may also be useful for the study of mRNA decay mechanisms: the transcription of genes controlled by the copper-responsive elements of the *Cyc6* gene can be selectively and completely turned off simply by the addition of copper to the growth medium. The selectivity of the *Cyc6* promoter can be exploited also to control gene expression in transgenic organisms.

This chapter describes the preparation and use of copper-deficient media to study copper-responsive gene expression and copper metabolism with emphasis on *C. reinhardtii* as an experimental system. The principles

[2] H. H. Li and S. Merchant, *J. Biol. Chem.* **270**, 23504 (1995).
[3] J. M. Quinn and S. Merchant, *Plant Cell* **7**, 623 (1995).
[4] K. L. Hill and S. Merchant, *Plant Physiol.* **100**, 319 (1992).
[5] K. L. Hill and S. Merchant, *EMBO J.* **14**, 857 (1995).
[6] K. L. Hill, R. Hassett, D. Kosman, and S. Merchant, *Plant Physiol.* **112**, 697 (1996).

discussed in this chapter can also be applied to the study of other trace metal responsive processes.[7] General methods for growth and maintenance of Chlamydomonas strains are described in *The Chlamydomonas Sourcebook*.[8] Methods for the measurement of copper metabolism (e.g., uptake of $^{64}Cu$, cupric reduction[6]) are not described in this article, because they may not be generally applicable. Although the literature contains a substantial body of work on copper-responsive gene expression in cyanobacteria[9–12] and other algae,[13] we have limited our descriptions to methods that we have experience with.

Growth Media

For growing cells under copper-deficient conditions, it is necessary to use several precautions.

1. All glassware, plasticware (with the exception of colorless disposable plastic pipettes), and stirbars must be washed with 6 $N$ hydrochloric acid to displace any residual copper ions. All surfaces that will contact chemicals and solutions are washed with 6 $N$ hydrochloric acid, followed by seven washes with distilled water and three washes with MilliQ-purified (Millipore Corp., Bedford, MA) water. Used hydrochloric acid should be treated as hazardous chemical waste and disposed of appropriately.

2. High-purity chemicals should be used for making "copper-free" stock solutions for media preparation. It is recommended that these chemicals be kept separate from other laboratory chemicals to avoid metal contamination from use of metal spatulas and to avoid use of these higher priced chemicals in less stringent applications. Aldrich gold label chemicals (Aldrich Chemical Company, Milwaukee, WI) or other special trace metal grade chemicals (Fisher Scientific, Tustin, CA; Sigma Chemical Company, St. Louis, MO) are recommended. A trace element composition analysis is generally provided with the chemical. The source of chemicals we use are indicated below.

3. MilliQ-purified water (or the equivalent analytical grade water, <0.02 ppb $Cu$[14]) should be used for preparation of media and all stock solutions

[7] C. D. Cox, *Methods Enzymol.* **235,** 315 (1994).

[8] E. H. Harris, *in* "The *Chlamydomonas* Sourcebook: A Comprehensive Guide to Biology and Laboratory Use," p. 25. Academic Press, San Diego, California, 1988.

[9] L. M. Briggs, V. L. Pecoraro, and L. McIntosh, *Plant Mol. Biol.* **15,** 633 (1990).

[10] A. Bovy, G. deVrieze, M. Borrias, and P. Weisbeek, *Mol. Microbiol.* **6,** 1507 (1992).

[11] L. Zhang, B. McSpadden, H. B. Pakrasi, and J. Whitmarsh, *J. Biol. Chem.* **267,** 19054 (1992).

[12] M. Ghassemian, B. Wong, F. Ferriera, J. L. Markley, and N. A. Straus, *Microbiol.* **140,** 1151 (1994).

[13] M. Nakamura, M. Yamagachi, F. Yoshizaki, and Y. Sugimura, *J. Biochem.* (*Tokyo*) **111,** 219 (1992).

[14] J. Mather, F. Kaczarowski, R. Gabler, and F. Wilkins, *Biotechniques* **4,** 56 (1986).

to minimize copper contamination from the water source (especially if the house water is supplied via copper pipes).

### Source and Preparation of Reagents

The standard recipes for *C. reinhardtii* growth medium are described by Harris.[8] Below we list special sources or special preparation methods.

#### Beijerinck's Solution (1 liter)

8 g Ammonium chloride, 99.998% (Aldrich)
1 g Magnesium sulfate, 99.99+% (Aldrich)
1 ml 1 g/ml Calcium chloride, 99.99+% (Aldrich)

(The gold label CaCl$_2$ is deliquescent and is difficult to weigh. A 1 g/ml solution is prepared and stored at 4°.)

#### Tris-Acetate (1 liter)

242.2 g Tris base (Fisher Scientific, or other biotechnology grade)
100 ml Glacial acetic acid (Fisher Scientific, trace metal grade)

#### Phosphate Solution (1 liter)

18.5 g Potassium phosphate, monobasic, 99.99% (Aldrich)

Dissolve 18.5 g potassium phosphate in ~800 ml MilliQ water. Adjust the pH of the solution to 7.1 using a 3 $N$ copper-free solution of potassium hydroxide, prepared by dissolving potassium hydroxide pellets (Aldrich, 99.99%) in water. Bring to a final volume of 1 liter. Store at 4°.

#### Trace Element Solution (1 liter)

50 g Ethylenediaminetetraacetic acid (EDTA) (Fisher Scientific or other ACS grade)
22 g Zinc sulfate, heptahydrate, 99%, ACS reagent (Aldrich)
11.4 g Boric acid, 99.99% (Aldrich)
5.1 g Manganese chloride, tetrahydrate, 99.99% (Aldrich)
5 g Ferrous sulfate, heptahydrate, 99+%, ACS reagent (Aldrich)
1.6 g Cobalt chloride, hexahydrate, 98%, ACS reagent (Aldrich)
1.1 g Ammonium molybdate, tetrahydrate, 99.98% (Aldrich)

The solution should be prepared as detailed by Harris[8] except that the color of the solution should turn from orangy red to burgundy red. The orange precipitate is removed by filtration through a scintered glass filter (washed with 6$N$ HCl as described earlier). The solution is stable for at least 1 year at 4°.

To make stock solutions for copper-supplemented media, the same recipes are followed for Beijerinck's, Tris-acetate, and phosphate solutions with the usual ACS grade laboratory chemicals, and 1.6 g of copper chloride pentahydrate is added to the trace element recipe. Copper-supplemented trace elements will turn from olive green to purple on standing.

For radiolabeling experiments, the sulfate content of the medium in which the cells are grown can be reduced (if 35S-sulfate is used as the source of label) by substituting 1.63 g of $MgCl_2 \cdot 6H_2O$ (Aldrich, 99.995%) for 1 g of $MgSO_4$ in Beijerinck's solution. For the actual labeling procedure, the cells are resuspended in sulfate-free medium (prepared by using sulfate-free Beijerinck's solution and omitting the trace elements).

The estimated copper ion concentration of copper-deficient media prepared with the above specified chemicals is less than 3 n$M$.

*Solid Medium*

For preparation of copper-deficient agar plates, the agar must be washed. Twenty grams of TC agar (JRH Biosciences, Lenexa, KS) is added to an acid-washed 2-liter flask or fleaker (on which the 1-liter level has been marked) containing a stir bar. MilliQ-purified water and EDTA (from a 0.5 $M$, pH 8.0, stock) are added to a final volume of 2 liters and a final EDTA concentration of 20 m$M$. The agar is stirred gently at 4°. Stirring should be just fast enough to maintain suspension of agar. If the agar is stirred too vigorously, the particles will break up and they will not settle well between washes, which leads to unacceptably high losses during the decanting. After stirring for 8–16 hr, the agar is allowed to settle for 1–2 hr and the liquid is decanted carefully, retaining as much agar as possible. The EDTA wash is repeated three more times, followed by four washes performed in the same manner with MilliQ-purified water. This serves to remove any residual EDTA which inhibits growth. The final settled agar is used to make Tris-acetate phosphate (TAP) or minimal media in the same flask using the 1-liter mark to bring the solution to volume. For any comparative studies of cell growth, enzyme expression, etc., between copper-supplemented and copper-deficient cells it is recommended that the agar used for copper-supplemented plates also be prepared by washing with EDTA because this treatment undoubtedly removes many other (potentially inhibitory) compounds from the agar in addition to divalent metal ions.

*Use of Chelators*

An alternative or additional method for preparing metal-deficient solutions (such as the Tris-acetate or phosphate stock, that do not contain other metal ions) when high-purity chemicals or solvents are not available is to treat the solution with dithizone (1,5-diphenylthiocarbazone).[15] In a chemi-

---

[15] T. Matsubara, K. Frunzke, and W. G. Zumft, *J. Bacteriol.* **149**, 816 (1982).

cal fume hood, 5 ml of a 0.05% solution of dithizone in chloroform is added per liter of solution in an acid-washed 1-liter separatory funnel with a glass or Teflon stopcock. The phases are mixed thoroughly. The chloroform phase settles to the bottom from where it is removed. The aqueous phase will be pink from residual dithizone, which is removed by repeated extractions with 5-ml aliquots of chloroform until the aqueous and organic phases are colorless. After two additional extractions, the final aqueous phase is left in an open container in the hood to allow the residual chloroform to evaporate. If required, dithizone-extracted solutions can be supplemented with other metal ions from high-purity sources.

Other methods for removing ions can also be used. Chelex resin (Bio-Rad, Richmond, CA) exhibits high selectivity for divalent versus monovalent ions (5000:1). The solution to be deionized should be mixed gently for >1 hr with the resin. The resin can be removed by filtration, and salts (of high purity) can be added back to the solution as desired. In our system, Chelex treatment of the medium does not enhance the expression of the $Cyc6$ gene compared to medium prepared with "gold label" chemicals (see above); nevertheless, the treatment may be useful for demetalation of solutions when suitably pure ingredients are not available. In theory, it should also be possible to use soluble copper-selective chelators, e.g., $o$-phenanthroline or CDTA, to deplete the medium of available copper. In this case, the concentration of chelator should be calculated from the stability constants of the chelator–metal complexes (Data for Biochemical Research) such that copper is chelated but other essential cations (e.g., $Zn^{2+}$) are not. We have tested bathocuproinedisulfonate in copper-supplemented TAP medium and found it to be ineffective at preventing copper utilization in that medium. However, we have not tested $o$-phenanthroline or CDTA. Because C. reinhardtii can utilize copper from a copper–EDTA complex, the utility of the chelators needs to be experimentally determined.

*Culturing*

To generate copper-deficient cultures, cells in the pretransfer culture are grown to mid to late log phase ($10^6$–$10^7$ cells per ml) and diluted 1/100 to 1/200 into fresh copper-deficient medium. One transfer can be sufficient, but copper carryover (e.g., from internal stores or from cell-wall bound forms) can vary from strain to strain. Therefore, three sequential transfers (from $10^6$–$10^7$ cells/ml into fresh medium) are recommended to ensure that the residual copper in the copper-deficient culture is reduced to <3 n$M$. After the third transfer, the culture is generally copper deficient. Nevertheless, it is advisable to confirm this (see below). On solid medium, cells appear (on the basis of plastocyanin and cyt $c_6$ accumulation) to become copper deficient after a single transfer.[16] Note that many photosynthetic mutants grow poorly in copper-deficient medium: copper-deficient cultures of such mutants may never exceed $10^6$ cells/ml. To obtain sufficient quantities of cells for analyses, it may be necessary to grow larger cultures that can be harvested at lower densities. For certain cell-wall deficient strains we have noted during collection of cells that the cell pellet from copper-supplemented cultures is more tightly packed in the bottom of the microcentrifuge tube than the pellet from copper-deficient cells (which tends to smear against the wall of the tube). The reason for this difference is not known to us, but the phenomenon does not affect any of the assays described later.

Even when media are prepared with scrupulous attention to sources of potential metal contamination, fresh copper-deficient medium generally contains some amount of residual copper ions (see below for measurement by spectroscopy). At low cell densities, this amount is sufficient to satisfy the plastocyanin biosynthetic pathway.[17] The low level of residual copper is also sufficient to cause transient changes in gene expression when cells are transferred to fresh medium (e.g., prior to radiolabeling experiments), and if this is a consideration, the experiment should be designed to avoid transfer to fresh medium.

*Scenedesmus obliquus* is another green alga that is closely related to *C. reinhardtii*. This organism also responds to copper-deficient growth conditions with alteration in the accumulation of cyt $c_6$ and plastocyanin. However, in *S. obliquus*, copper-responsive regulation of plastocyanin expression occurs via mechanisms that affect protein and also messenger RNA levels.[18] In the copper-supplemented TAP medium described earlier, copper ions are chelated by EDTA but this chelated form is available to *C. reinhardtii* for plastocyanin synthesis. However, this chelated form of copper appears to be invisible to *S. obliquus* cells. Specifically, *S. obliquus* cells grown in the usual TAP medium behave as if they were copper deficient (no plastocyanin, high levels of cyt $c_6$). To grow *S. obliquus* in fully copper-supplemented medium, copper salts (20 $\mu M$ copper sulfate or chloride) must be added. This difference between *C. reinhardtii* and *S. obliquus* with respect to their ability to utilize chelated copper suggests that it is important to test for appropriate copper-supplemented versus copper-deficient growth conditions when initiating studies on other organisms.

[16] Z. Xie, D. Culler, B. W. Dreyfuss, R. Kuras, F.-A. Wollman, J. Girard-Bascou, and S. Merchant, *Genetics* **148,** 681 (1998).
[17] S. Merchant, K. Hill, and G. Howe, *EMBO J.* **10,** 1383 (1991).
[18] H. H. Li and S. Merchant, *J. Biol. Chem.* **257,** 9368 (1992).

Sample Preparation

## Chlamydomonas reinhardtii

A soluble protein fraction containing plastocyanin and cyt $c_6$ can be prepared as described by Howe and Merchant.[19] Special precautions to maintain the cells in a copper-free state are not necessary for preparation of protein fractions because the proteins are stable for hours. The supernatants may be analyzed as described below.

For the preparation of RNA from copper-deficient *C. reinhardtii* cultures, it is important to remember that the $t_{1/2}$ of the mRNA encoding cytochrome $c_6$ can be equivalent to the RNA preparation time (~45 min), and that of the copper-deficient form of the *Cpx1* RNA appears to be even shorter. Therefore, in order to avoid loss of the RNA of interest, it is essential to maintain "copper-free" conditions while collecting the cells and also during the lysis step. Concentrations as low as 2–5 n$M$ Cu ions result in changes in the abundance of *Cyc6* and *Cpx1* transcripts.[5,17] Therefore, solutions are prepared with gold label chemicals, where possible, and all glassware and other material used for preparation of the solutions and collection of cells are acid washed.

For RNA blot analysis of the *Cyc6* and *Cpx1* mRNAs, we have used the hybridization conditions described by Church and Gilbert[20] and hybridization and wash temperatures of 65°. With oligo-labeled probes[21] of specific activity in the range of 1.5 to 3 × $10^8$ cpm/mg and 5–10 μg of total RNA per lane, an excellent signal is obtained for the *Cyc6* and *Cpx1* mRNAs with an overnight exposure at −80° to Kodak XAR-5 film (two intensifying screens) or a PhosphorImager (Molecular Dynamics, Sunnyvale, CA) screen.

Nuclei can be isolated from copper-supplemented and copper-deficient cultures of cell-wall deficient strains of *C. reinhardtii* for analysis of transcription of genes of interest (e.g., *Cyc6* or *Pcy1*) by nuclear run-on assays.[17,22] All solutions and materials used during the preparation of nuclei should be "copper free" to minimize copper contamination (see RNA isolation just discussed). Transcription of *Cyc6* is not affected by the addition of copper to isolated nuclei; thus, it is not necessary to maintain copper-free conditions during the assay of the nuclei.

## Scenedesmus Obliquus

*Scenedesmus obliquus* cells are somewhat hardier than *C. reinhardtii* cells; thus, the procedures described above must be modified slightly. For

[19] G. Howe and S. Merchant, *EMBO J.* **11**, 2789 (1992).
[20] G. M. Church and W. Gilbert, *Proc. Natl. Acad. Sci. U.S.A.* **81**, 1991 (1984).
[21] A. P. Feinberg and B. Vogelstein, *Anal. Biochem.* **137**, 266 (1984).

preparation of soluble proteins, the cells are broken by sonication in a microcentrifuge tube (microtip probe, 60% intensity, 5 min, Sonic Dismembrator, Fisher Scientific).[18] Other methods for breaking *S. obliquus* cells involve the use of glass beads in a Vibrogen (Buhler, Germany) cell mill[23–25] or a Bead Beater[26] (Biospec Products, Bartlesville, OK). We have used the Bead Beater to lyse *S. obliquus* cells for large-scale purification of plastocyanin.

For preparation of RNA, the cell pellet is resuspended in water, transferred to a mortar containing liquid nitrogen and ground to a fine powder under liquid nitrogen.[18,27] The paste, on thawing, is transferred immediately to a disposable centrifuge tube, diluted with one volume of 2× lysis buffer, and the preparation continued as for *C. reinhardtii* cells.

Testing for Copper Deficiency in the Culture

Copper deficiency in the context of the photosynthetic apparatus should be assessed by measuring plastocyanin and/or cyt $c_6$ abundance in the culture of interest. Several different methods have been employed; the sensitivity of the method must be balanced with the simplicity and speed of the procedure. The extent of expression of the cyt $c_6$-encoding gene is directly proportional to the copper content in the medium on a per cell basis (up to $10^7$ Cu ions per cell; see Fig. 1).[11,17] Thus, measurement of cyt $c^6$ content of cells is a good measure of the copper deficiency. Likewise, holoplastocyanin formation is directly proportional to the amount of copper available to the cell; measurement of plastocyanin content relative to a fully copper-supplemented culture can give a good indication of the copper status of the cell.[28] Of the methods described later, heme and Coomassie staining are the easiest and use common laboratory reagents. Although immunoblots have the potential to provide greater sensitivity, appropriate antisera are required. The spectrophotometric method is the cheapest (assuming a suitable instrument is available) but much larger amounts of biological material are required.

*Spectroscopic Analysis of Soluble Extracts*

Plastocyanin and cytochrome $c_6$ display unique visible spectra; thus, difference spectroscopy can be applied to estimate the amount of the two

[22] L. R. Keller, J. L. Schloss, C. D. Silflow, and J. L. Rosenbaum, *J. Cell Biol.* **98,** 1138 (1984).
[23] H. Bohner and P. Boger, *FEBS Lett.* **85,** 337 (1978).
[24] V. Breu and D. Dornemann, *Biochim. Biophys. Acta* **967,** 135 (1988).
[25] J. Schnackenberg, R. Schulz, and H. Senger, *FEBS Lett.* **327,** 21 (1993).
[26] B. A. Diner, D. F. Ries, B. N. Cohen, and J. G. Metz, *J. Biol. Chem.* **263,** 8972 (1988).
[27] D. Hermsmeier, R. Schulz, and H. Senger, *Planta* **193,** 406 (1994).
[28] P. M. Wood, *Eur. J. Biochem.* **87,** 9 (1978).

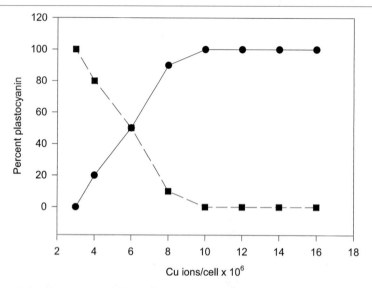

Fɪɢ. 1. Reciprocal expression of cyt $c_6$ and plastocyanin in response to copper ions in the medium.

proteins in cell extracts. Soluble protein extracts (from cells equivalent to 1 mg/ml chlorophyll) are reduced with sodium ascorbate (to 1 m$M$) or oxidized with potassium ferricyanide (1 m$M$). The samples are scanned from 500 to 650 nm and the signal difference is recorded. For *C. reinhardtii* cytochrome $c_6$, a $\Delta\varepsilon$ (reduced–oxidized) at 552.5 nm of 20 m$M^{-1}$ is used and for plastocyanin, a $\Delta\varepsilon$ (oxidized–reduced) at 597 nm of 4.5 m$M^{-1}$ is used. Although spectroscopic analysis is simple and rapid and does not require specialized reagents, it is several orders of magnitude less sensitive than the chemical or immunochemical methods described later. For instance, for cyt $c_6$, an extract from fully copper-deficient wild-type cells equivalent to 1 mg/ml chlorophyll yields a maximum absorbance of only $4 \times 10^{-2}$ at 552.5 nm. The lower detection limit for cyt $c_6$ corresponds to an amount equivalent to 10% of the amount in a fully copper-deficient wild-type cell owing to the presence of mitochondrial cyt $c$ ($\lambda_{max}$ = 550 nm).[19] The lower detection limit for plastocyanin is reported to be 2% of wild-type levels[29] but it should be noted that the samples need to be at least 5- to 10-fold more concentrated for a useful measurement owing to the lower extinction coefficient for plastocyanin. Spectroscopic analyses have been used to estimate cyt $c_6$ and plastocyanin content in several other

[29] D. S. Gorman and R. P. Levine, *Plant Physiol.* **41**, 1648 (1966).

organisms. Bohner and Boger[23] used an extinction coefficient of 17.3 m$M^{-1}$ cm$^{-1}$ at 553 nm for cyt $c_6$ (determined for the *S. obliquus* protein) and a $\Delta\varepsilon$ of 4.9 m$M^{-1}$ cm$^{-1}$ for plastocyanin (determined for the spinach protein). Sandmann *et al.*[30] used the same values to evaluate cultures of numerous different algae, and Zhang *et al.*[11] used the same value for cyt $c_6$ for studies with *Synechocystis*. On the basis of direct measurement, Kong and Whitmarsh[31] suggest a value of 19.5 ± 0.5 m$M^{-1}$ cm$^{-1}$.

### Immunoblot Analysis for Estimation of Plastocyanin and Cytochrome $c_6$ Content[19]

Plastocyanin and cyt $c_6$ are small proteins (10.3 and 9.7 kDa, respectively) and are therefore most conveniently separated by electrophoresis through gels polymerized from higher concentrations of acrylamide (12–15%). For immunoblot analysis, denaturing gel electrophoresis is recommended.

Traditional denaturing electrophoresis[32] (running gel = 12% acrylamide, 0.33% bis) in the MiniProtein II gel system (BioRad, Richmond, CA) is the routine method in our laboratory. The samples are separated by electrophoresis until the tracking dye (bromophenol blue) reaches the bottom. The separated proteins may be transferred to 0.1-$\mu$m nitrocellulose or polyvinylidene fluoride (PVDF) membranes (50 V, 2 hr, 4°) and analyzed by heme staining or immunedecoration. For effective transfer of plastocyanin and cyt $c_6$, it is important to omit sodium dodecyl sulfate (SDS) from the standard Tris-glycine transfer buffer and to include methanol (to 20%). Several different anti-plastocyanin antisera preparations from our laboratory have shown weak cross-reactivity with cytochrome $c_6$ (despite high purity of the immunizing antigen), but not vice versa. To avoid visualization of the cross-reactive signal, a lower sensitivity assay (nitrocellulose membrane, HRP-conjugated secondary antibody) may be used. When greater sensitivity is required, PVDF membranes and AP-conjugated secondary are preferred.

Typically, we load soluble extract equivalent to 20 $\mu$g of chlorophyll per lane, and obtain a suitable signal on PVDF membranes with an alkaline-phosphatase conjugated secondary antibody within 1–2 min of exposure to the developing solution, or with a horseradish peroxidase conjugated secondary antibody within 5–10 min. The use of chemiluminescent reagents in place of chromogenic substrates will increase the sensitivity. For instance, after treatment of the membrane with the luminol reagents for the peroxi-

[30] G. Sandmann, H. Reck, E. Kessler, and P. Boger, *Arch. Microbiol.* **134,** 23 (1983).
[31] Y. Kong and J. Whitmarsh, personal communication (1997).
[32] U. K. Laemmli, *Nature (Lond.) New Biol.* **227,** 680 (1970).

dase assay, a 60-sec exposure to film is sufficient to detect signals from extracts equivalent to 0.2 $\mu$g of chlorophyll.

*Visualization of Cytochrome $c_6$ on the Basis of Its Heme-Dependent Peroxidase Activity and Coomassie Strain Analysis of Plastocyanin and Cytochrome $c_6$*

A sensitive and quantitative method for estimating holocytochrome $c_6$ exploits the peroxidase activity of the heme cofactor. In this case, the samples (from cells equivalent to 20 $\mu$g of chlorophyll) are separated by electrophoresis in a nondenaturing gel (see below). Alternatively, if dena- turing gels are used, 2-mercaptoethanol should not be added and the sam- ples should not be heated. Heme staining can either be performed directly on the gel after electrophoresis with TMBZ (3,3′,5,5′-tetramethylbenzidine) as the substrate,[33] and this traditional method is the fastest; or for a more sensitive assay that is also amenable to quantitation, the heme stain can be performed after transfer to a PVDF membrane with a chemiluminescent substrate.[34,35] In either case, plastocyanin content of the sample can also be assessed after heme staining by either staining the gel with Coomassie Blue or by immunedecoration of the PVDF membrane (Fig. 2).

For the TMBZ heme stain, the gel is soaked (in the dark at room temperature) for 1–3 hr immediately following electrophoresis in a solution containing 0.045% w/v TMBZ in 0.175 $M$ sodium acetate, pH 5.0, 30% methanol. Hydrogen peroxide is added to 30 m$M$ to initiate staining; blue- green bands corresponding to heme proteins appear in 5–10 min. The gel can be photographed (Polaroid Type 667, shutter speed 1/125, $f$-stop 11–16) under tungsten lamp illumination for a permanent record. In samples from copper-deficient cells, the signal corresponding to cyt $c_6$ is the strongest; other cytochromes give weaker reactions. Following photography, the gel can be stained with Coomassie Blue as usual. The detection limit corre- sponds to 1% of wild-type levels (extract equivalent to 0.2 $\mu$g of chloro- phyll).

For the chemiluminescent heme stain, the proteins are transferred to PVDF as described earlier, the membrane is rinsed in TBS and wetted rapidly in chemiluminescent reagent (described below) ($\sim$0.1 ml/cm$^2$). The same preparation of reagent can be used immediately for several separate membranes. The reagent is distributed by vigorous agitation for $\sim$30 sec. The excess liquid is blotted off the membrane with laboratory wipes, and the membrane is wrapped in plastic wrap and exposed to film (5 min to 1

[33] P. E. Thomas, D. Ryan, and W. Levin, *Anal. Biochem.* **75,** 168 (1976).
[34] D. W. Dorward, *Anal. Biochem.* **209,** 219 (1993).
[35] C. Vargas, A. G. McEwan, and J. A. Downie, *Anal. Biochem.* **209,** 323 (1993).

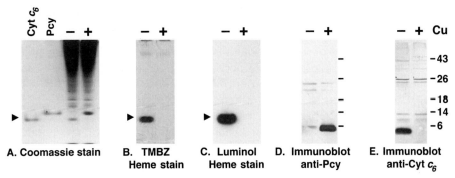

FIG. 2. Visualization of plastocyanin and cyt $c_6$ in extracts of soluble proteins from *C. reinhardtii*. Extracts equivalent to 20 (A, B, D, and E) or 5 (C) μg of chlorophyll were prepared from copper-supplemented (+) or copper-deficient (−) cultures and separated by electrophoresis under nondenaturing conditions (A–C) or denaturing conditions (D and E). (A) Purified plastocyanin (Pcy) and cytochrome (Cyt) $c_6$ were used as markers in the left-hand lanes. (B) The gel was stained with TMBZ as described in the text. Note the presence of other heme proteins in both samples. These can serve as a loading control. (C) The proteins were transferred to PVDF membranes and the membrane was treated with chemiluminescence reagents for 30 sec as described in the text, and immediately exposed to NEN/Dupont-Reflection film for 5 min. (D and E) The proteins were transferred to PVDF membranes. The membranes were decorated with antibodies raised against plastocyanin (1:500 dilution) or cyt $c_6$ (1:500 dilution) as indicated. An alkaline phosphatase-conjugated secondary antibody and chromogenic substrates were used for visualization. Plastocyanin (molecular weight 10,300) migrates with the 6000 molecular weight marker and cyt $c_6$ (9800) migrates a little faster.

hr). The peak light output occurs within the first 5 min and decays thereafter. To quantitate cyt $c_6$, a dilution series of purified cyt $c_6$ can be analyzed in parallel under exactly the same conditions. With a 5-min exposure, cyt $c_6$ can be detected in copper-deficient, wild-type extracts corresponding to 0.05 μg of chlorophyll. In principle, it should be possible to increase the sensitivity by one or two orders of magnitude by increasing exposure time and by increasing the amount of sample analyzed. Several different exposures of the membrane may be necessary to give a signal in the linear range of the film. Several companies market chemiluminescent reagents, including Pierce Chemical Company, Rockford, IL (Supersignal CL-HRP substrate) and Dupont–NEN, Boston, MA (Renaissance Western blot chemiluminescent reagent). The reagent can also be made in the laboratory with luminol (3-aminophthalhydrazide) and *p*-coumaric acid.[36] The staining procedure cannot be applied twice to the same membrane.

[36] I. Durrant, *Nature* (*Lond.*) New Biol. **346,** 297 (1990).

## Coomassie Stain Analysis of Plastocyanin and Cytochrome $c_6$ in Native Polyacrylamide Gels

When extracts of soluble protein are separated in nondenaturing poly-acrylamide gels[37] made with 15% acrylamide monomer, cytochrome $c_6$ and plastocyanin can be resolved from each other and from most of the other proteins in the extract (owing to their acidic p$I$'s and small size). (Fig. 2A) The samples are separated by electrophoresis (90 V running) on a Bio-Rad MiniProtein II system until 30 min *after* the tracking dye has migrated off the bottom. The gel can be stained with Coomassie Blue or used for transfer.

## Reporter Gene Assay

A reporter gene that is controlled by the promoter of the *Cyc6* gene can be introduced into *C. reinhardtii* cells to provide an alternate method for assay of copper content in the medium, or can be used for studies of copper-responsive gene expression. The gene encoding arylsulfatase is the reporter gene used in our laboratory. It was described by Davies *et al.*[38] The assays performed in our laboratory reproduce exactly the methods described in that work.[38,39] Arylsulfatase is not normally expressed in cells grown in standard laboratory media (TAP or minimal) and is therefore very useful for the study of gene expression in *C. reinhardtii*. It has been exploited also for the study of development in *Volvox carterii.*[40]

Two assays are described. The plate assay is useful for screening large numbers of *C. reinhardtii* colonies for transgenic ones that express the reporter gene construct of interest[3,38] or for isolating regulatory mutants.[41] The liquid assay is employed when quantitation is necessary. Both assays are simplified by the fact that the gene product (arylsulfatase) is located in the periplasm and is secreted into the medium in cell-wall deficient strains.[39] Both assays can be performed with both walled and wall-less strains.[39,42–44]

Since the reporter gene integrates into the *C. reinhardtii* genome by nonhomologous recombination, the expression of the reporter gene can vary between independent transgenic strains. Therefore, for studies of cop-

[37] B. J. Davis, *Ann. N.Y. Acad. Sci.* **121,** 404 (1964).
[38] J. P. Davies, D. P. Weeks, and A. R. Grossman, *Nucl. Acids Res.* **20,** 2959 (1992).
[39] E. L. deHostos, R. K. Togasaki, and A. Grossman, *J. Cell Biol.* **106,** 29 (1988).
[40] A. Hallmann and M. Sumper, *Proc. Natl. Acad. Sci. U.S.A.* **91,** 11562 (1994).
[41] J. P. Davies, F. Yildez, and A. Grossman, *Plant Cell* **6,** 53 (1994).
[42] T. Lien and O. Schreiner, *Biochim. Biophys. Acta* **384,** 168 (1975).
[43] F. H. Yildez, J. P. Davies, and A. R. Grossman, *Plant Physiol.* **104,** 981 (1994).
[44] J. P. Davies, personal communication (1997).

per-responsive gene expression, it is more important to measure copper-responsive arylsulfatase activity as the difference between the activity in copper-supplemented versus copper-deficient cells of a single transgenic strain than to make comparisons between strains. It is essential that several independently generated transgenic strains be analyzed to ensure that the result is independent of the site of integration.

*Plate Assay.* Colonies of interest are transferred to a gridded plate. A grid of about 64 squares on a 100-mm plate is optimal. When the colonies are about 1–2 mm in diameter (4–6 days if grown at 22° under a light intensity of 10–40 $\mu$mol/m$^2$/s), they are transferred, in duplicate, to +Cu and −Cu plates. The colonies are transferred two more times following growth for a total of three serial transfers. Before testing the colonies resulting from the third transfer, a small amount of cells from each colony should be transferred to a fresh plate to create a master because it may not be possible to recover viable cells from the test plate. The test plates are sprayed with a solution of 10 m$M$ 5-bromo-4-chloro-3-indolyl sulfate (X-SO$_4$) in 0.1 $M$ Tris-Cl, pH 7.5, and left at room temperature for 16–24 hr. Colonies that express arylsulfatase activity will display blue halos around and beneath the colony. It is advisable to score for expression on the basis of comparison between the −Cu versus +Cu staining to ensure that the assay measures *copper-responsive* expression. Colonies of interest can be recovered from the master and tested by the liquid assay (see below) for a more quantitative assessment of reporter gene activity. The expression of the same construct in independent isolates can be quite varied: some colonies display differential expression after a single transfer to +Cu versus −Cu plates, while other isolates may require up to three transfers through copper-deficient versus copper-supplemented medium to maximize the difference in expression.

*Quantitative Assay on Liquid Cultures.* Colonies to be tested are adapted to copper-deficient and copper-supplemented growth conditions by serial transfer (three times) in liquid medium (as described previously). Arylsulfatase can be assayed as described by deHostos *et al.*[39]

## Direct Measurement of Copper in Cell, Media, and Solutions

### Atomic Absorption Spectroscopy

The concentration or amount of copper in samples of cells or solutions can be measured directly by atomic absorption (AA) spectroscopy using either a flame or graphite furnace atomizer. The choice of atomizer depends on the sensitivity required and the amount of sample available. Both procedures can be applied to samples of media (acidified by the addition of nitric acid to 0.1 $N$) or cells/tissues. We have not tested methods for disrupting *C. reinhardtii* cells for AA spectroscopy but methods have been developed for dissolution of other walled microorganisms including bacteria[45] (where dried cells are resuspended in 10% nitric acid + 10% perchloric acid and heated to 100° for 30 min) and yeast[46] (where the paste is refluxed in 0.1% (w/v) nitric acid at 100°C overnight and the final supernatant analyzed), and these methods should be applicable to photosynthetic microorganisms as well. It is important that digestion be complete and metal contamination avoided for reproducible results. A spectrometer equipped with a flame atomizer has a detection limit for copper of 1 ppb, while for one with a graphite furnace the limit is 0.02 ppb. The graphite furnace offers the additional advantage of requiring as little as 20 $\mu$l of liquid sample. To avoid unacceptable background noise in applications that require the higher sensitivity of the graphite furnace, all glassware and plasticware used in the preparation or handling of samples must be pretreated by soaking for 2 days in 2.4 $N$ hydrochloric acid followed by another 2 days of soaking in 3.2 $N$ nitric acid. The soaked labware must be rinsed thoroughly in MilliQ-purified water and covered to prevent airborne particles from falling within. The standard curve is prepared by diluting a commercially available copper standard in 0.1 $N$ nitric acid. For the graphite furnace, the standard should be prepared just before use. With either method, the working concentration range is only two orders of magnitude.

*Inductively Coupled Plasma Mass Spectroscopy*

Inductively coupled plasma mass spectroscopy (ICP-MS) is the detection method of choice when high precision and accuracy ($\pm 5\%$) are required. ICP-MS analysis can detect as little as 0.07 ppb in a 0.1% nitric acid solution. Unlike AA in which the concentration of only a single element can be determined in a sample, multiple elements can be measured by ICP-MS. This may be advantageous for studies involving multiple metal measurements or when the abundance of another element is used as an internal control for the comparison of copper concentrations in multiple samples. The concentration range for ICP-MS is also wider (three to four orders of magnitude, compared to two for AA). While many institutions have AA instruments on site, the availability of ICP-MS instruments is rare. The University of Wisconsin Soil and Plant Analysis Laboratory, Soil Science Department 5711 Mineral Point Road, Madison, WI 53705, (608)262-4364, has performed our analyses (for a reasonable fee).

---

[45] A. Odermatt, R. Krapf, and M. Solioz, *Biochem. Biophys. Res. Commun.* **202,** 44 (1994).
[46] J. Goto and E. Gralla, personal communication (1997).

Using the *Cyc*6 Promoter to Control Gene Expression

Copper-responsive sequences are located upstream of the *Cyc*6 transcription start site and can be used to direct copper-responsive gene expression in transgenic *C. reinhardtii*.[3] A fragment corresponding to *Cyc*6 sequences from −852 to −7 can be cloned upstream of the gene of interest. If a promoterless fragment of the gene of interest is used, the resulting construct will show minimal expression in +Cu medium. For convenience, we have also mutated the start site of translation in order to generate a restriction site at that position (*Nde*I). This makes it convenient to construct ATG fusions. In this case, the chimeric construct would contain *Cyc*6 sequences from −852 to +80. Similar levels of expression result from either construct. Smaller promoter fragments (from −127 to −110 or from −110 to −55) can also be used but the level of expression appears to be more sensitive to context. Further, it may be necessary to use the small fragments in multiple copies (at least two) in order to obtain reliably high levels of expression.

Acknowledgments

The work in our laboratory is supported by the NIH and USDA (SM) and NSF MCB-9306694 (JQ). We thank the members of the group for the development of the methods described here, especially Duane Culler, Beth Dreyfuss, Gregg Howe, Kent Hill, and Jeff Moseley. We are grateful to them and Ginger Armbrust, John Davies, and Edie Gralla for reviewing the manuscript, and to John Whitmarsh for communicating unpublished data.

# [19] Use of Molecular Genetics to Investigate Complementary Chromatic Adaptation: Advances in Transformation and Complementation

*By* DAVID M. KEHOE and ARTHUR R. GROSSMAN

Study of Complementary Chromatic Adaptation in Cyanobacteria

It is essential that photosynthetic organisms sense and respond to changes in their light environment. Many cyanobacteria adapt to changes in light quality by altering the structure of their Photosystem II (PSII) light-harvesting antenna, called *phycobilisomes* (PBS), through a process known as complementary chromatic adaptation (CCA).[1,2] During CCA the absorp-

---

[1] A. Bennett and L. Bogorad, *Biochemistry* **10,** 3625 (1971).
[2] A. Bennett and L. Bogorad, *J. Cell Biol.* **58,** 419 (1973).

tion properties of the PBS are altered. In green light, high levels of the green light-absorbing chromoprotein phycoerythrin and its associated structural proteins (linkers) are synthesized and assembled into the PBS rods. In contrast, in red light, high levels of the red light-absorbing chromoprotein phycocyanin and its linker proteins are synthesized and assembled into the PBS rods.[3-5] Thus CCA, which is a red light–green light photoreversible process, helps certain cyanobacteria maximize the absorption of the predominant wavelength of light. The control of phycocyanin and phycoerythrin synthesis during CCA is regulated primarily at the level of transcription; *cpeBA* (encoding apophycoerythrin) is transcriptionally active in green light and inactive in red light and *cpcB2A2* (encoding inducible apophycocyanin) is transcriptionally active in red light and inactive in green light.[6]

The study of CCA serves as a model to help elucidate mechanisms of light regulation in photosynthetic organisms. Recently, we have found that at least one component of the CCA signal transduction pathway resembles a family of plant photoreceptors called the *phytochromes*.[7] Hence, elucidation of the mechanisms that govern CCA are contributing to our understanding of both the origin and the mechanisms of action of the phytochrome family of photoreceptors.

In-depth studies of CCA have used two closely related filamentous cyanobacteria, *Calothrix* spp. PCC 7601 and *Fremyella diplosiphon*. Several advantages accrue from working with these species. Most importantly, in these cyanobacteria both the *cpeBA* and *cpcB2A2* operons (and the genes encoding their associated linkers) are regulated during CCA. In many other cyanobacteria, only one of these operons is regulated during chromatic adaptation.[8] This "two-operon" regulation makes the process of CCA dramatically colorful: in red light, the cyanobacterial filaments are blue-green in color and in green light, the filaments are red. The distinct pigment phenotypes of red light- and green light-grown cells make it easy to distinguish mutants with aberrant CCA. In *F. diplosiphon,* a strain with shortened filaments[9] has been extremely useful in separating individual colonies for working with clonal populations. Finally, many laboratories have contrib-

[3] S. Diakoff and S. Scheibe, *Plant Physiol.* **51,** 382 (1973).

[4] J. F. Haury and L. Bogorad, *Plant Physiol.* **60,** 835 (1977).

[5] T. C. Vogelmann and J. Scheibe, *Planta* **143,** 233 (1978).

[6] R. Oelmüller, P. B. Conley, N. Federspiel, W. R. Briggs, and A. R. Grossman, *Plant Physiol.* **88,** 1077 (1988).

[7] D. M. Kehoe and A. R. Grossman, *Science* **273,** 1409 (1996).

[8] N. Tandeau de Marsac, *Bull. Inst. Pasteur* **81,** 201 (1983).

[9] J. G. Cobley, E. Zerweck, R. Reyes, A. Mody, J. R. Seludo-Unson, H. Jaeger, S. Weerasuriya, and S. Navankasattusas, *Plasmid* **30,** 90 (1993).

uted to our understanding of the physiology, photobiology, and molecular biology of CCA (reviewed in Refs. 10–12).

There are also some difficulties in working with *F. diplosiphon* and *Calothrix* spp. PCC 7601. First, unlike many cyanobacteria, homologous recombination does not occur at a high frequency; thus targeted gene inactivation is not currently feasible. This has made the use of reverse genetics difficult, although this drawback has been overcome to some extent by the generation and molecular characterization of large numbers of mutants. Another drawback is the lack of a reliable, high-frequency transformation method for these cyanobacteria. Our recent efforts to resolve this problem have resulted in an improvement in transformation efficiencies as much as 100-fold, and has allowed us to complement a number of new CCA mutants. This review summarizes the methods that we have used to generate and complement mutants that are defective in CCA.

Generation of Mutants in CCA

Mutants with altered CCA responses have been generated as (1) spontaneous mutants, (2) by chemical mutagens, and (3) as a result of electroporation.[8,11,13,14] The effect of both electroporation and chemical mutagens appears to be indirect in the mutants isolated thus far; this is discussed further below. The CCA mutants that have been obtained can be grouped into two distinct classes, those in which both *cpeBA* and *cpcB2A2* expression is abnormal and those in which CCA regulation of only one of these operons is aberrant. Although members of both of these classes have been analyzed and complemented[7,15,16] we have primarily focused on those containing single lesions that affect both operons, with the expectation that these may be in genes encoding proteins that act in the early steps of the CCA signal transduction pathway.

Regardless of the method used for mutant generation, all of the mutations affecting CCA that have been complemented and characterized at the molecular level appear to be the result of the insertion of mobile elements in the cyanobacterial genome.[7,15,16] It has been suggested that the stresses imposed by electroporation or chemical mutagens result in

[10] N. Tandeau de Marsac and J. Houmard, *FEMS Microbiol. Rev.* **104,** 119 (1993).
[11] D. M. Kehoe and A. R. Grossman, *Sem. Cell Biol.* **5,** 303 (1994).
[12] A. R. Grossman, D. Bhaya, K. E. Apt, and D. M. Kehoe, *Annu. Rev. Genet.* **29,** 231 (1995).
[13] J. G. Cobley and R. D. Miranda, *J. Bacteriol.* **153,** 1486 (1983).
[14] B. Bruns, W. R. Briggs, and A. R. Grossman, *J. Bacteriol.* **171,** 901 (1989).
[15] G. G. Chiang, M. R. Schaefer, and A. R. Grossman, *Proc. Natl. Acad. Sci. U.S.A.* **89,** 9415 (1992).
[16] D. M. Kehoe and A. R. Grossman, *J. Bacteriol* **179,** 3914 (1997).

large-scale mobilization of these genomic sequences. At least one transposon from *F. diplosiphon* is currently being used as a "tag" in an attempt to identify genes that may be involved in the regulation of CCA (M. Schaefer, personal communication).

## Construction of a Plasmid Library

To complement CCA mutants, it was necessary to construct a complete genomic library from wild-type *F. diplosiphon*. This plasmid library contains approximately $3 \times 10^5$ individual recombinant clones with an average insert size of 6–7 kilobase pairs (kb). This library was constructed by partially cutting *F. diplosiphon* genomic DNA with *Sau*3AI, size fractionating the digested DNA, and inserting the fragments into the shuttle plasmid pPL2.7 digested with *Bam*HI. pPL2.7 contains the *Escherichia coli* origin of replication (*oriV*) and origin of mobility (*bom*) region, the kanamycin resistance gene *nptI* from the transposon Tn903,[17] and a region of DNA containing an origin of replication from the plasmid pFdA,[9] which is an endogenous *F. diplosiphon* plasmid[18] (see Ref. 19 for details). The plasmid is approximately 5.6 kb and contains unique *Bam*HI and *Pst*I sites. A map of pPL2.7 with extensive restriction enzyme cleavage information is shown in Fig. 1. A detailed plasmid library construction protocol is provided below.

### *Insert DNA Preparation*

1. Prepare genomic DNA that can be easily cut; the size should be in the range of 25–30 kb when checked on a 0.6% gel.

2. Do test digests of the genomic DNA with *Sau*3AI to determine the concentration and incubation time needed to generate fragments of the appropriate size (6–20 kb). Typical initial test conditions:

| | |
|---|---|
| High molecular weight genomic DNA (0.2 $\mu$g/$\mu$l) | 60 $\mu$l (12 $\mu$g DNA total) |
| 10× *Sau*3AI restriction enzyme buffer | 18 $\mu$l |
| Distilled water | 102 $\mu$l |

3. Set up a dilution series using 10 microfuge tubes. Put 30 $\mu$l of the above mixture into the first tube and 15 $\mu$l into each of the rest, then add 4 U of *Sau*3AI to the first tube, mix, and transfer 15 $\mu$l to the second tube. Repeat through the tenth tube. Incubate the tubes for 1 hr at 37°.

4. Stop the digestion reactions by adding 1 $\mu$l of 0.5 $M$ EDTA (pH 8.0) and 10 $\mu$l of agarose gel dye to each tube and place the tubes on ice. Run each sample on a 0.6% agarose TAE {40 m$M$ Tris-acetate, 1 m$M$ EDTA,

[17] A. Oka, H. Sugisaki, and M. Takanami, *J. Mol. Biol.* **147,** 217 (1981).
[18] S. M. Gendel, *Curr. Microbiol.* **17,** 23 (1988).
[19] G. G. Chiang, M. R. Schaefer, and A. R. Grossman, *Plant Physiol. Biochem.* **30,** 315 (1992).

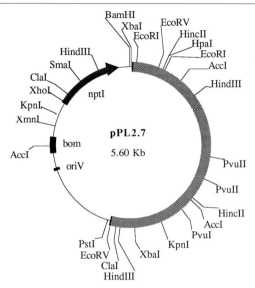

Fig. 1. Restriction digest map of pPL2.7 showing the region of *F. diplosiphon* DNA containing a cyanobacterial origin of replication (gray area), the gene conferring kanamycin resistance (*nptI,* thick dark arrow), and the *E. coli* origin of replication (*oriV*) and basis of mobility (*bom*) regions. Known restriction enzyme sites are indicated. The following enzymes do not cut pPL2.7: *Aat*II, *Apa*I, *Bcl*I, *Bgl*II, *Bst*BI, *Bst*XI, *Not*I, *Sal*I, *Sca*I, *Sph*I.

(pH 8.0)}[20] gel at 100 V for 3–5 hr to size the digestion products. Select the conditions that produce a range of fragment sizes from 6 to 20 kb and use these initially to scale-up the digestion. These conditions may need to be modified when digesting in larger volumes. Four times more *Sau*3AI was required in our large-scale digestions to achieve the same degree of cutting.

5. For the large-scale digestions, eight individual tubes were each set up as follows:

| | |
|---|---|
| High molecular weight genomic DNA (0.2 $\mu$g/$\mu$l) | 125 $\mu$l |
| 10× *Sau*3AI restriction enzyme buffer | 30 $\mu$l |
| 0.0625 U/$\mu$l *Sau*3AI (dilute in 1× enzyme buffer) | 10 $\mu$l |
| Distilled water | 135 $\mu$l |

6. Incubate for 1 hr at 37°, pool the reactions, extract the reaction mixture with phenol/chloroform, then chloroform, ethanol precipitate the DNA, and wash the pellet once with 70% ethanol. Resuspend the cut DNA in 200 $\mu$l of TE (pH 8.0) if it is to be size fractionated on an agarose gel, or in 400 $\mu$l of TE (pH 8.0) if it is to be fractionated on a sucrose gradient (see step 7).

[20] J. Sambrook, E. F. Fritsch, and T. Maniatis, *in* "Molecular Cloning: A Laboratory Manual," 2nd ed., Vol. I. p 6.3. Cold Spring Harbor Laboratory Press, Cold Spring Harbor, New York, 1989.

7. Check the degree of digestion by electrophoresing 2–4 $\mu$l of digested DNA on a 0.6% agarose TAE gel. If it is appropriately digested, size fractionate the remainder of the DNA on a 0.6% TAE agarose gel and excise the region of the gel containing DNA of the correct size (see step 8). The DNA in the gel slices can be purified by standard methods such as with GeneClean (Bio101, LaJolla, CA) or Qiaex II (Qiagen, Chatsworth, CA). Resuspend the purified DNA in 50–100 $\mu$l of TE (pH 8.0) and use directly in ligation reactions. Alternatively, DNA of the correct size can be isolated on a sucrose 10–40% continuous gradient as follows:

10% Sucrose:    1.5 g/15 ml [10 m$M$ Tris (pH 8)/10 m$M$
                NaCl/1 m$M$ EDTA (pH 8)]
40% Sucrose:    6.0 g/15 ml [10 m$M$ Tris (pH 8)/10 m$M$
                NaCl/1 m$M$ EDTA
                (pH 8)]

Prepare the gradient in Beckman SW41 tubes or the equivalent (SW41 tubes hold 13.2 ml). Warm the DNA to 65° for 10 min, slowly cool to 20°, and then load 200 $\mu$l of the DNA (50–100 $\mu$g) onto each gradient. Centrifuge the tubes at 20° for 20 hr at 60,000$g$. Fractionate the gradient into 1-ml aliquots and check a sample of each on a 0.6% agarose TAE gel (combine 40 $\mu$l of each fraction with 40 $\mu$l of water and 10 $\mu$l gel dye). Fractions containing DNA of the correct size (see step 8) should be pooled and the sucrose removed by passing the pooled fractions over a Centricon 30 concentrator (Amicon, Beverly, MA) and washing three times with an equal volume of TE (pH 8.0). The final volume should be 50–100 $\mu$l.

8. If the DNA is fractionated on a sucrose gradient, it is preferable to select two size ranges of pooled fractions containing DNA since the average insert size in a library can often be significantly smaller than the average size of the pooled insert DNA used to make it. For example, when we used fragments from a sucrose gradient with an average size of 13 kb, the library contained an average insert size of 6–7 kb. This bias toward smaller size inserts in these libraries may occur because of incomplete size fractionation on sucrose gradients (smaller fragments are more easily ligated to the vector) and because the pPL2.7 plasmid used for the library construction does not readily accept large fragments (greater than 10 kb). This bias is minimized if gel fractionated DNA is used.

9. Check the concentration of pooled insert DNA; it should be between 0.1 and 0.5 $\mu$g/$\mu$l. Recheck the size ranges of the pooled fractions by running approximately 0.5 $\mu$g of each pooled sample on a 0.6% agarose gel.

*Plasmid DNA Preparation*

1. Incubate approximately 60 $\mu$g of plasmid DNA with 100 U of *Bam*HI in a total volume of 120 $\mu$l for 2 hr at 37°. Check 0.5 $\mu$l of the reaction

mixture on an agarose gel to confirm that the plasmid was completely cleaved. Retain 2 μl for later use in control ligations.

2. Add 1 U of shrimp alkaline phosphatase (United States Biochemical/Amersham, Cleveland, OH) to the restriction digest mixture and incubate for 45 min at 37°.

3. Extract the reaction mixture once with phenol/chloroform, followed by chloroform only, and then precipitate the DNA with ethanol. Wash the DNA pellet with 70% ethanol and resuspend it in 40 μl TE (pH 8). Check the DNA concentration and adjust it to 0.5 μg/μl with TE (pH 8).

4. Set up vector only test ligations, using standard ligation conditions, to check the digestion, dephosphorylation, and ligation efficiencies. Use 0.5 μg of vector DNA in 10 μl of total ligation volume. These test ligations should include uncut DNA/no ligase, BamHI cut DNA/no ligase (restriction digest efficiency), BamHI cut DNA/with ligase (ligase efficiency), and BamHI cut DNA/dephosphorylated/with ligase (dephosphorylation efficiency). The BamHI cut DNA/with ligase control should produce at least 95% of the colony forming units (cfus) obtained from the uncut DNA control. The BamHI cut DNA/dephosphorylated/with ligase control should result in at least 50 times fewer cfus compared to the BamHI cut DNA/with ligase control and should have approximately the same number of CFU as the BamHI cut DNA/no ligase control.

## Ligation and Transformation

1. Some optimization of the ligation conditions may be necessary depending on the vector that is used to make the library. For 13-kb average size *F. diplosiphon* genomic DNA ligated into the shuttle vector pPL2.7, the following conditions were used:

| | |
|---|---|
| Sau3AI cleaved genomic DNA (0.4 μg/μl) | 8.2 μl |
| BamHI cut pPL2.7 vector DNA (0.5 μg/μl) | 5.0 μl |
| 10× One-Phor-All Buffer *Plus* (Pharmacia) | 1.8 μl |
| 10 mM ATP (pH 7.0) | 1.0 μl |
| T4 DNA ligase (1 U/μl) | 2.0 μl |
| | 18.0 μl |

2. This ligation contains a molar ratio of approximately 0.5:1 of insert to vector and a final DNA concentration in the ligation mix of about 300 ng/μl. In constructing our library, we set up five separate ligation mixtures to increase the total number of recombinants obtained. Allow the ligations to proceed overnight at 16°.

3. Add 1 μl of the ligation mix to 4 μl of TE (pH 8) and use 1 μl of this dilution to transform 100 μl of *E. coli* DH5α competent cells. The competent cells should have a transformation efficiency of at least $5 \times 10^7$ cfu/μg of DNA. Spread one-tenth of the final transformation volume onto

plates with the appropriate drug selection. An average of approximately 70 cfu per plate should be obtained using these conditions. Thus for each ligation tube the total number of cfu should be 70 cfu/$\mu$l $\times$ 50 (dilution factor) = 3.5 $\times$ 10$^3$ cfu/$\mu$l $\times$ 18 $\mu$l = 6.3 $\times$ 10$^4$ cfu. Five separate ligation reactions give a total library complexity of 63,000 cfu/tube $\times$ 5 tubes = approximately 3 $\times$ 10$^5$ cfu. Assuming an average insert size of 6.5 kb in the library and a genome size of 1.2 $\times$ 10$^7$ bp for *Calothrix* spp. PCC 7601 and *F. diplosiphon*,[21] only 8.5 $\times$ 10$^3$ recombinants would be needed to attain a 99% probability of having any given sequence represented in the library.[20]

4. Check the plasmids from a small number of colonies on the transformation efficiency test plate to determine the average insert size and the percentage of plasmids containing inserts. Approximately 90% of the plasmids in our library contained detectable inserts.

*Library Amplification*

1. A single round of amplification of the library should be conducted to increase the total amount of recombinant molecules. A combination of plate and liquid growth should be used to minimize the bias introduced during the amplification.

2. For our library, 150-mm TB plates containing 25 $\mu$g/ml of kanamycin were used and transformants were plated to a density of approximately 2 $\times$ 10$^4$ cfu/plate.

3. Incubate the plates at 37° overnight (about 15 hr). For each plate, scrape off the colonies with a spatula into 100 ml of prewarmed M9 medium containing 25 $\mu$g/ml of kanamycin. Rinse the plate with 5 ml more of M9 and pool the media. Continue to grow the cells in the liquid medium at 37° for 1.5 hr, then pellet the cells by centrifugation, isolate the DNA by alkaline lysis, and purify the plasmid DNA on a CsCl gradient containing 0.8 mg/ml of ethidium bromide.[20]

4. Butanol extract the negatively supercoiled plasmid DNA three times to remove the ethidium bromide completely, then ethanol precipitate the DNA. Wash the pellet with 70% ethanol and resuspend the DNA in 100 $\mu$l of TE (pH 8). The final concentration should be approximately 1 $\mu$g/$\mu$l. If 10 plates are used in the amplification, the final yield of the library will be approximately 1 mg of plasmid library DNA.

5. Check the DNA on a 0.6% agarose TAE gel. The DNA should appear as a diffuse band of the size of the plasmid plus the average size insert, but should also be spread above and below this band; these are plasmids with larger and smaller than average size inserts, respectively.

[21] M. Herdman, *in* "The Biology of Cyanobacteria" (N. G. Carr and B. A. Whitton, Eds.), Vol. 19, pp. 263. University of California Press, Berkeley, 1982.

Transformation / Complementation of Mutants

Both conjugation and electroporation have been used successfully to complement CCA mutants.[7,15,16,19] Until recently, both of these approaches resulted in low transformation rates (25–250 transformants/$\mu$g DNA).[19] Several changes in the electroporation methodology have improved the transformation frequency to approximately $3 \times 10^3$ transformants/$\mu$g DNA. In conjunction with the use of the plasmid library described above, the complementation of *F. diplosiphon* has become routine.

In the following protocol, the amount and duration of light received by the cells both before and after electroporation is critical for high-frequency transformation. *Fremyella diplosiphon* cells grown in moderate white light (cool white fluorescent bulbs, 60 $\mu$mol photons m$^{-2}$ s$^{-1}$) tend to adhere to each other and form large clumps of cells, whereas cells grown in the same light of a lower intensity (20 $\mu$mol photons m$^{-2}$ s$^{-1}$) do not aggregate. Filamentous cyanobacteria often produce extracellular sheaths of polysaccharide, and we hypothesized that this material was physically blocking the passage of DNA into the cells during electroporation. This raised the possibility that incubating the cells in darkness for several days would starve the cells for carbon and cause the depletion of the extracellular polysaccharides, making the cell membranes more accessible to exogenous DNA. The amount of light that the cells receive after the electroporation treatment is important as well, because the transformed cells require sufficient energy to grow and survive during the antibiotic selection.

*Growth and Preparation of Cells*

1. All of this work should be done under sterile conditions, since the presence of contaminating bacteria will allow nontransformed cyanobacterial cells to grow on plates containing antibiotics.

2. Inoculate 50 ml of BG-11[22] (except that 0.012 g/l of ferric ammonium citrate is used) in culture tubes set up for bubbling with the strain to be transformed. Grow the cultures under approximately 20 $\mu$mol photons m$^{-2}$ s$^{-1}$ of continuous cool white fluorescent light (LiCor Quantum Meter, Lincoln, Nebraska, Li-185A with a "flat" detector) until the A$_{750}$ is between 0.8 and 1.0. The transformation efficiencies will decrease if the cells are grown beyond this cell density. If the light intensity used is above 30–40 $\mu$mol photons m$^{-2}$ s$^{-1}$, the abundant polysaccharides that accumulate will not be depleted during the subsequent dark treatment and transformation efficiencies will be low. The cultures should be grown at 30–32° and bubbled

[22] R. Rippka, J. Deruelles, J. B. Waterbury, M. Herdman, and R. Y. Stanier, *J. Gen. Microbiol.* **111**, 1 (1979).

with a light air stream supplemented with 3–4% $CO_2$. It is best to buffer the BG-11 with 10 m$M$ HEPES (pH 8.0).

3. Put the cultures into darkness for 48 hr. This is usually achieved by wrapping the culture tubes with aluminum foil. The bubbling of the cultures should continue during this time. The $A_{750}$ does not change significantly during these 2 days.

4. Harvest the cells by centrifuging for 10 min at 4000–5000$g$ at room temperature. The cells can be collected in a volume of approximately 5 ml by drawing the loose pellet off of the bottom with a 10-ml pipette.

5. Transfer the cells to a fresh tube containing 40–45 ml of sterile distilled water and resuspend completely. Recentrifuge and wash the pellet in distilled water two more times as above. Because it is filamentous, *F. diplosiphon* does not form a tight pellet. Therefore, after the final wash the cells are contained in approximately 5 ml of distilled water. To concentrate the cells further, transfer them to a smaller tube (such as a 15-ml conical polypropylene tube) and recentrifuge once more as above. Remove the supernatant until a final volume of approximately 1 ml is reached. Resuspend the cells completely in the remaining supernatant and use them promptly for electroporation.

*Electroporation of Cells*

1. The plasmids used for transformations are routinely grown in the *E. coli* strain DH5$\alpha$. The plasmid DNA should be very pure and predominantly (>90%) negatively supercoiled; it is best to use DNA that has been purified on a CsCl gradient[20] or a Qiagen plasmid kit. The DNA should be a concentration of 2–4 $\mu$g/ml in sterile distilled water. Be careful to remove the salt completely from the DNA or it will significantly reduce the electroporation efficiencies. Use 6 $\mu$g of DNA per transformation.

2. For each electroporation, place 40 $\mu$l of cells into a 500-$\mu$l microfuge tube and add the appropriate volume of DNA, mixing well. Keep the cells on ice, although this does not appear to be absolutely critical for good transformation efficiency.

3. Our electroporations are conducted using a Gene Pulser (Bio-Rad, Hercules, CA). The following parameters are used: capacitance = 25 $\mu$F, resistance = 200 $\Omega$, field strength = 10 kV/cm of electrode gap. These parameters may need to be varied somewhat depending on the species being used and the size of the cells. We use custom-made cuvettes that accept small volumes and sterilize them with ethanol before each use. The face of each of the electrodes at the gap is 5 × 5 mm and the gap between the electrodes is set at 1 mm. This design allows the cells to be loaded into and removed from the cuvette easily. Biorad cuvettes with a gap of 1 mm can also be used.

4. Place the cells in the cuvette and electroporate. The time constants obtained under these conditions should be between 4.5 and 5.0 ms. If they are below this (which will reduce the transformation efficiency), it may be due to inefficient removal of salt either from the DNA sample or the cells.

5. As quickly as possible after each electroporation, place the electroporated solution on ice in a microfuge tube. After 20 min, transfer the electroporated cells to a bubbling tube containing 10 ml of BG-11 and bubble under the conditions described earlier for 8–16 hr.

## Selection and Screening of Transformants

1. Centrifuge the cells for 10 min at 4000–5000$g$ and remove the supernatant, leaving the cells in only a few hundred microliters of medium. Resuspend the cells in the remaining medium and plate them directly onto BG-11 plates containing the appropriate antibiotic. When transforming with the plasmid pPL2.7, plates containing 25 $\mu$g/ml of kanamycin are used.

2. Place the plates at 25–30° in approximately 40 $\mu$mol photons m$^{-2}$ s$^{-1}$ of continuous cool white, red, or green fluorescent light. Transformants should be visible within 2 weeks.

3. Screening for complemented CCA mutants can be accomplished by examining the colonies for correct CCA under the appropriate light quality. For example, CCA mutants that are always red in color can be transformed with the plasmid library containing wild-type genomic DNA and placed under red light. Any colony containing a DNA fragment that complements the lesion in the mutant will chromatically adapt and therefore grow as a blue-green, rather than a red, colony.

4. We were concerned that a high rate of mutant generation by electroporation could result in the creation of an overwhelming number of "false positives" that would make it difficult to identify mutants that were actually complemented (secondary CCA mutations can occur, such as a red mutant converting to a green mutant). The average frequency of secondary mutant generation after electroporation was determined for several types of CCA mutants and was found to be generally 10-fold below the calculated frequency for complementation of any mutant phenotype with the plasmid library.

## Acknowledgments

The authors would like to thank a number of people who have contributed to the development of these methods. In particular, Drs. Brigitte Bruns, Gisela Chiang, and Michael Schaefer for establishing the basic electroporation conditions; Drs. Elena Casey, Nicole Tandeau de Marsac, and Jean Houmard for helpful discussions and comments on the electroporation protocol; and Dr. Rakefet Schwarz for thoughtful suggestions on the library construction protocol and for critical review of this paper. We also thank Dr. Michael Schaefer for sharing unpublished data and Kathi Bump for assistance in preparing the manuscript. The development of the protocols presented here was in part supported by a NSF Postdoctoral Fellowship in Plant Biology to D.M.K. and by NSF award MCB 9513576 to ARG. This is Carnegie Institution of Washington publication number 1329.

# Section V

## Photosynthetic Mutants: Construction and Biological/Biochemical/Biophysical Analysis

## [20] Gene Modifications and Mutation Mapping to Study the Function of Photosystem II

*By* Wim F. J. Vermaas

### Introduction

Gene modifications and other molecular genetic approaches have truly revolutionized the ways in which protein function and cofactor–polypeptide interactions can be analyzed. Gene modifications can be introduced at specific sites in appropriate organisms or randomly. Both specific and random modifications of photosynthesis-related genes are covered here. Specific modifications of wild-type genes can be introduced effectively only in systems with homologous recombination mechanisms. Such modifications include interruption or deletion of the whole gene, introduction of site-directed mutations, deletion of domains within a gene, and gene replacement by genes from other organisms. Apart from gene inactivation studies, introduction of specific modifications is generally most useful if some background knowledge exists regarding the function of specific residues or domains of the gene product. This background knowledge is not a prerequisite for the introduction of random mutations into either a particular gene or gene domain or anywhere in the genome. However, for identification of desired random mutants, a strong selection for a particular phenotype is generally required. Once mutants with a desired phenotype have been identified, the precise location of the mutation can be easily determined by PCR (polymerase chain reaction) and DNA sequencing if random mutations had been targeted to a particular gene or domain. However, a major drawback of random mutations that may have been introduced anywhere in the genome of the organism is that identification of the locus of the mutation is often tedious. However, as described in this article, for organisms with a known genome sequence (and therefore with known restriction maps) it is possible to identify the region carrying a random mutation by functional complementation with size-separated restriction fragment pools.

Several molecular genetic methods that are useful for mutant studies in photosynthetic organisms are described, and applications for the study of Photosystem II (PS II) are highlighted. Particular emphasis will be placed on the use of the spontaneously transformable cyanobacterium *Synechocystis* sp. PCC 6803 (hereafter to be referred to as *Synechocystis* 6803), which readily integrates introduced DNA into its genome by homologous recombination, and whose genome has been sequenced in its entirety. Other

METHODS IN ENZYMOLOGY, VOL. 297

photosynthetic prokaryotes, including the cyanobacterium *Synechococcus* sp. PCC 7002, and chloroplasts of photosynthetic eukaryotes (including the green alga *Chlamydomonas reinhardtii*) also show homologous recombination and are therefore suitable for many of these studies. Apart from molecular genetic characteristics, a crucial feature is that the organism of choice can grow in a number of different conditions, ranging from photoautotrophically to essentially heterotrophically, and that the presence of functional photosystems is not absolutely required for growth. This is of importance considering that many gene replacement strategies require the use of mutants in which a particular gene has been deleted. If this gene deletion leads to a loss of photoautotrophic growth, then the mutant must be maintained under (photo)heterotrophic conditions.

Transformation Strategies

*Synechocystis* 6803 spontaneously takes up foreign DNA[1] and can integrate this DNA into its genome via homologous double recombination. Basic *Synechocystis* 6803 transformation methods have been described[2]; here some additional information is provided. For *Synechocystis* 6803, efficient homologous double recombination between the genomic DNA of the organism and introduced DNA requires at least 100- to 400-bp sequence identity in both recombination domains. However, if complete identity exists in one recombination domain, significant homology (but not identity) has been shown to be sufficient for reasonably efficient recombination in the other domain.[3,4] *Synechocystis* 6803 can be transformed with linear as well as circular DNA in either single-stranded or double-stranded form. In our group little consistent difference in transformation efficiency has been observed between single- or double-stranded circular DNA, but DNA has been suggested to enter the cell in single-stranded form.[5] Transformation with linear DNA (for example, with PCR products) requires larger flanking regions in which recombination can occur (optimally ≥500 bp), possibly as a consequence of exonuclease activity inside or outside the cell. Methylation of linear DNA by means of *SssI* methylase increased the transformation efficiency significantly (H. Kless and W. Vermaas, unpublished).

[1] G. Grigorieva and S. Shestakov, *FEMS Microbiol. Lett.* **13**, 367 (1982).
[2] J. G. K. Williams, *Methods Enzymol.* **167**, 766 (1988).
[3] S. D. Carpenter, I. Ohad, and W. F. J. Vermaas, *Biochim. Biophys. Acta* **1144**, 204 (1993).
[4] W. F. J. Vermaas, G. Shen, and I. Ohad, *Photosynth. Res.* **48**, 147 (1996).
[5] R. Barten and H. Lill, *FEMS Microbiol. Lett.* **129**, 83 (1995).

Deletion Mutagenesis

To introduce site- or region-directed mutations, preferably the wild-type genesis deleted from the genome prior to introduction of the modified gene copy. In this way, the contribution of any remaining wild-type gene copies to the mutant's phenotype can be excluded. To make a gene deletion, a plasmid is constructed that replicates in *Escherichia coli* but not in *Synechocystis* 6803 (pUC or pBR derivatives are suitable) and that contains *Synechocystis* 6803 sequences from upstream and downstream of the gene to be deleted, with the gene itself replaced by a selectable marker (see Fig. 1). Suitable selectable markers are indicated in Table I.

Transformation of *Synechocystis* sp. PCC 6803 is very simple and has been described by Williams[2]; it entails addition of DNA to 0.1- to 0.5-ml volume of culture in BG11[6] medium, followed by incubation for 1–6 hr and plating out. After 20–24 hr selective conditions can be applied. Colonies of transformants come up in about 1 week if Photosystem I (PS I) is active, and can then be restreaked on plates with increasingly higher concentration of the antibiotic for which a resistance marker has been introduced (initially, the antibiotic concentration can be doubled at every other restreak, until growth becomes visibly impaired).

*Synechocystis* 6803 appears to carry multiple genome copies per cell, a property shared with a number of other cyanobacteria.[7–9] To obtain a pure mutant phenotype, all wild-type genome copies need to be replaced. Two factors are important to readily obtain segregation: (1) a gradual increase in antibiotic selection pressure and (2) selection of growth conditions under which the mutant phenotype has a competitive advantage or is not very much impaired in comparison with the wild type. In this respect it is important to consider what might help the growth of the mutant as well as what would impede the growth of the background strain (but not of the mutant). For example, when using a *Synechocystis* 6803 wild-type strain to delete a gene for a PS II component that is critical for photoautotrophic growth, segregation of mutant and wild-type genomes is carried out best under obligate photoheterotrophic conditions under which PS II activity does not contribute to the growth of the organism (in the presence of 5–15 m$M$ glucose and 20–50 $\mu M$ of the PS II electron transport inhibitors diuron or atrazine). An example of positive selection against the background strain is the deletion of *apcE*, the gene encoding the anchor protein connecting

[6] R. Rippka, J. Deruelles, J. B. Waterbury, M. Herdman, and R. Y. Stanier, *J. Gen. Microbiol.* **111**, 1 (1979).
[7] N. Mann and N. G. Carr, *J. Gen. Microbiol.* **83**, 399 (1974).
[8] B. J. Binder and S. W. Chisholm, *J. Bacteriol.* **172**, 2313 (1990).
[9] B. J. Binder and S. W. Chisholm, *Appl. Environm. Microbiol.* **61**, 708 (1995).

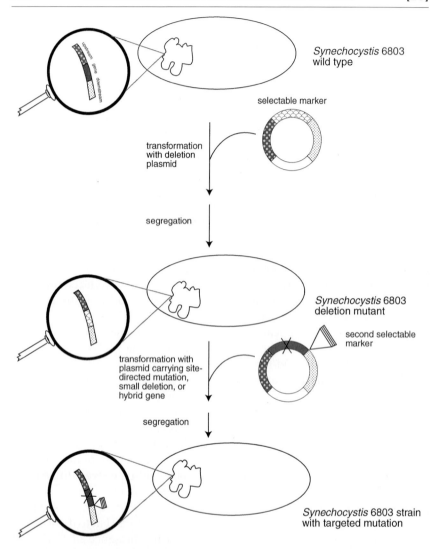

Fig. 1. Schematic representation of targeted gene deletion and replacement in the cyano-bacterium *Synechocystis* sp. PCC 6803. The upper part of the diagram reflects deletion of a wild-type *Synechocystis* sp. PCC 6803 gene from the genome through transformation with plasmid carrying regions upstream and downstream of the gene to be deleted, with the gene itself replaced by a selectable marker (see Table I). After segregation of wild-type and mutant genomes in a *Synechocystis* sp. PCC 6803 transformant, a pure deletion mutant has been created. Modified gene copies carrying site-directed mutations, small deletions, etc., can be reintroduced in the deletion mutant by transformation with a plasmid carrying the desired mutation(s) and generally also a second selectable marker for transformant identification.

TABLE I

SELECTABLE MARKERS (ANTIBIOTICS RESISTANCE MARKERS) SUITABLE FOR USE
IN *SYNECHOCYSTIS* SP. PCC 6803[a]

| Antibiotic to which resistance is achieved | Plasmid from which cassette was derived | Antibiotic concentration ($\mu$g/ml) suitable for selection |
| --- | --- | --- |
| Chloramphenicol | pBR325 | 5–150 |
| Erythromycin | pRL425 | 5–300 |
| Kanamycin | pUC4K (*Tn* 903) | 5–500 |
| Spectinomycin | pHP45$\Omega$ | 3–250 |

[a] On initial selection, the lowest antibiotic concentration listed should be used. Concentrations should be increased gradually during propagation of transformants.

phycobilisomes (the peripheral light-harvesting complex of cyanobacteria) to thylakoids, in the PS I-less *Synechocystis* 6803 strain. This strain is sensitive to moderate light intensity (50 $\mu$E $\cdot$ m$^{-2}$ $\cdot$ s$^{-1}$) in the presence of active PS II[10] presumably because of an overreduction of thylakoid components by PS II-generated electrons. The resulting PS I-less/*apcE*$^-$ transformants are more resistant to moderate light intensity as the antenna size of PS II has been reduced,[10] and these mutants can be segregated best at somewhat higher light intensity (20–40 $\mu$E $\cdot$ m$^{-2}$ $\cdot$ s$^{-1}$) where cells without intact *apcE* have an advantage over the PS I-less cells with this gene.

Verification of Segregation

A convenient and rapid method to screen for segregation of wild-type and mutant genotypes is to prepare DNA from propagated transformants and to amplify the region of the mutation by PCR. DNA can be isolated from cells on plates (from large single colonies or, preferably, from streaks) by the following procedure: Wash the cells twice in 1 ml TE (10 m*M* Tris-HCl, pH 8.0, 0.1 m*M* EDTA), resuspend in 200 $\mu$l TE, and transfer to a 0.5-ml microcentrifuge tube; add about 150 $\mu$l glass beads (0.1 mm in diameter), and shake for 30 sec using a MiniBeadbeater (Biospec Products, Bartlesville, OK). Spin the samples for 10 sec in a microcentrifuge, and transfer the supernatant to a new microcentrifuge tube. Add 200 $\mu$l of a phenol/chloroform mix [the phenol is equilibrated with TES (TE with 50 m*M* NaCl), and the chloroform contains 4% isoamyl alcohol] and vortex for 30 sec. Spin in a microcentrifuge, and remove the upper phase to a new tube. If desired, an additional extraction with chloroform can be carried out to remove traces of phenol. Add 0.5 volume of 7.5 *M* ammonium

[10] G. Shen, S. Boussiba, and W. F. J. Vermaas, *Plant Cell* **5**, 1863 (1993).

acetate, and two volumes of 100% ethanol. DNA precipitates overnight on incubation at $-20°$. When starting from sufficient cell material, the purity of DNA can be improved (at the expense of the quantity of DNA recovered) by precipitating at $-20°$ after addition of 0.5 volumes 7.5 $M$ ammonium acetate and one (rather than two) volume of 100% ethanol. The precipitate is collected by centrifugation, washed with 70% ethanol, and dried. The dry pellet is then resuspended into 50 $\mu$l TE and 1–2 $\mu$l of this suspension is adequate for use as a template for PCR.

If a PCR product corresponding to the wild-type size can still be detected, segregation obviously is not yet complete. However, small PCR products tend to be amplified more readily than larger ones, and thus the absence of a wild-type band is not necessarily indicative of a pure mutant if the size of the mutant PCR product is smaller than that of the wild type. Southern analysis is valuable to confirm the genetic homogeneity of a transformant strain. If an apparently segregated transformant grows significantly slower than the corresponding background strain under particular conditions, final proof of the genetic purity of this transformant strain is that it can be propagated under these conditions in the absence of antibiotics for many generations without generating any evidence of the genotype of the background strain.

### Genetic Modification of Photosystem II Genes

Genetically accessible cyanobacteria serve as good model systems for PS II studies because the cofactors and core subunits of cyanobacterial PS II are comparable to those of PS II found in plants. However, there are several potentially important differences in PS II from the two systems. For example, particular subunits of plant PS II are not obviously present in cyanobacteria and vice versa.[11] With the completion of the *Synechocystis* sp. PCC 6803 genomic sequence,[12–15] a comparison between plant and

[11] W. F. J. Vermaas, S. Styring, W. P. Schröder, and B. Andersson, *Photosynth. Res.* **38,** 249 (1993).

[12] T. Kaneko, A. Tanaka, S. Sato, H. Kotani, T. Sazuka, N. Miyajima, M. Sugiura, and S. Tabata, *DNA Res.* **2,** 153 (1995).

[13] T. Kaneko, A. Tanaka, S. Sato, H. Kotani, T. Sazuka, N. Miyajima, M. Sugiura, and S. Tabata, *DNA Res.* **2,** 191 (1995).

[14] T. Kaneko, S. Sato, H. Kotani, A. Tanaka, E. Asamizu, Y. Nakamura, N. Miyajima, M. Hirosawa, M. Sugiura, S. Sasamoto, T. Kimura, T. Hosouchi, A. Matsuno, A. Muraki, N. Nakazaki, K. Naruo, S. Okumura, S. Shimpo, C. Takeuchi, T. Wada, A. Watanabe, M. Yamada, M. Yasuda, and S. Tabata, *DNA Res.* **3,** 109 (1996).

[15] T. Kaneko, S. Sato, H. Kotani, A. Tanaka, E. Asamizu, Y. Nakamura, N. Miyajima, M. Hirosawa, M. Sugiura, S. Sasamoto, T. Kimura, T. Hosouchi, A. Matsuno, A. Muraki, N. Nakazaki, K. Naruo, S. Okumura, S. Shimpo, C. Takeuchi, T. Wada, A. Watanabe, M. Yamada, M. Yasuda, and S. Tabata, *DNA Res.* **3,** 185 (1996).

cyanobacterial PS II genes (Table II) has been facilitated in that no cyano-bacterial genes with limited but significant homology to PS II genes from higher plants can be overlooked. The large core components of the PS II complex (the reaction center proteins D1 and D2, the chlorophyll-binding proteins CP47 and CP43, and the extrinsic PS II-O protein) have high homology between cyanobacterial and higher plant systems. Most of the small (4–10 kDa) PS II polypeptides with one putative transmembrane span (cytochrome $b_{559}$, PS II-H, PS II-I, PS II-J, PS II-K, PS II-L, PS II-M, PS II-N, PS II-T, PS II-X) are also highly conserved between cyanobacterial and plant systems, particularly in the hydrophobic region. However, homologs of the nuclearly encoded extrinsic proteins PS II-P, PS II-Q, and PS II-R and of the light-harvesting-type protein PS II-S of higher plants are not found in the cyanobacterial genome, and for at least two cyanobacterial PS II proteins (PS II-U and the $psbV$-encoded cyto-chrome $c_{550}$) no higher plant homologs have been found.

Deletion mutagenesis has been applied to most PS II genes of *Synecho-cystis* 6803. As has been reviewed,[16,17] genetic deletion of the D1, D2, CP43, CP47, or cytochrome $b_{559}$ proteins leads to a loss of photoautotrophic growth and to a significant depletion of other PS II core components, whereas a loss of one of the other PS II gene products in *Synechocystis* 6803 generally slows photoautotrophic growth but does not abolish it. The reason that cyanobacterial mutants that lack any of the small nonessential subunits grow slower photoautotrophically is often related to a decreased amount of PS II, whereas remaining PS II centers appear to behave rather normally. The magnitude of the effect of deletion of a small subunit in PS II depends on the experimental organism. In *Chlamydomonas rein-hardtii*, deletion of some of the small subunits leads to a loss of PS II activity,[16,17] suggesting that these subunits have a more prominent role in assembly or stabilization of the PS II complex than in cyanobacterial systems.

Deletion mutagenesis of entire genes does not provide much informa-tion regarding the precise function of specific PS II components: if deletion of a subunit results in an essential loss of the PS II complex, this implies a critical role in structure or function of this subunit but does not indicate which domains or residues are important. To further study this issue, several other approaches to genetic modifications of PS II need to be applied. These are covered in the next sections. However, a useful application of deletion mutagenesis that is relevant in this context is to tailor genetically the experimental organism that is investigated. If one is interested in analy-

[16] W. Vermaas, *Annu. Rev. Plant Physiol. Plant Mol. Biol.* **44,** 457 (1993).
[17] H. B. Pakrasi, *Annu. Rev. Genet.* **29,** 755 (1995).

TABLE II

Photosystem II Genes in *Synechocystis* sp. PCC 6803 and in Higher Plants[a]

| Gene | Product | Present in | | Comments |
|------|---------|-----------|--------|----------|
| | | S. 6803 | Plants | |
| *psbA* | D1 | Yes | Yes (cp) | Three copies in *S.* 6803; one silent |
| *psbB* | CP47 | Yes | Yes (cp) | Monocistronic in *S.* 6803; *psbB/T–petBD* operon in plants |
| *psbC* | CP43 | Yes | Yes (cp) | *psbDC* Operon in *S.* 6803 and plants |
| *psbD* | D2 | Yes | Yes (cp) | Two copies in *S.* 6803, one in plants; *psbDII* of *S.* 6803 monocistronic |
| *psbEF* | Cyt $b_{559}$ | Yes | Yes (cp) | *psbEFLJ* Operon in both *S.* 6803 and plants |
| *psbH* | PS II-H | Yes | Yes (cp) | Phosphorylation near N terminus in plants (lacking in *S.* 6803) |
| *psbI* | PS II-I | Yes | Yes (cp) | Polypeptide present in PS II reaction center preparations |
| *psbJ* | PS II-J | Yes | Yes (cp) | |
| *psbK* | PS II-K | Yes | Yes (cp) | |
| *psbL* | PS II-L | Yes | Yes (cp) | |
| *psbM* | PS II-M | Yes | Yes (cp) | |
| *psbN* | PS II-N | Yes | Yes (cp) | |
| *psbO* | PS II-O | Yes | Yes (n) | Lumenal protein; OEE1 or MSP (manganese-stabilizing protein) |
| *psbP* | PS II-P | No | Yes (n) | Lumenal protein; OEE2 |
| *psbQ* | PS II-Q | No | Yes (n) | Lumenal protein; OEE3 |
| *psbR* | PS II-R | No | Yes (n) | |
| *psbS* | PS II-S | No | Yes (n) | Homology with light-harvesting chl *a/b*-binding (*cab*) proteins |
| *psbT* | PS II-T | Yes | Yes (n, cp) | Nuclear in green algae and plants; chloroplast encoded in red algae |
| *psbU* | PS II-U | Yes | No | May be functionally equivalent to PS II-Q |
| *psbV* | Cyt $c_{550}$ | Yes | No | May be functionally equivalent to PS II-P |
| *psbW* | PS II-W | No | Yes (n) | Varying amounts present in PS II reaction center preparations |
| *psbX* | PS II-X | Yes | Yes? | Chloroplast-encoded in red algae; not yet detected in green systems |

[a] Genes in higher plants may be located in the chloroplast (cp) or the nucleus (n).

sis of PS II in cyanobacterial mutants, it is desirable to delete genetically PS I because this photosystem is abundant in cyanobacterial systems and contributes EPR (electron paramagnetic resonance) and optical signals that interfere with PS II analysis. Functional deletion of PS I can be obtained by deletion of the *psaAB* operon,[10,18] which codes for PS I-A and PS I-B, the two core proteins of PS I. *Synechocystis* 6803 deletion mutants lacking both PS I (*psaAB*) and one of the PS II components (such as D2 encoded by *psbDIC* and *psbDII*) have been generated.[19]

## Site-Directed Mutagenesis

This term is generally used to indicate the introduction of a mutation in a specific residue of a protein. This approach therefore is useful to establish the function of one particular amino acid residue in a specific protein. Procedures for site-directed mutagenesis in DNA of plasmids or of the bacteriophage M13 are numerous and will not be covered here because they have become relatively standard.[20,21] To introduce such plasmid/M13-based mutations into the cyanobacterial genome, it is important to first create a *Synechocystis* 6803 deletion mutation in which the gene of interest has been replaced by a selectable marker (Fig. 1). This deletion strain can then be transformed with a plasmid construct with the *Synechocystis* gene carrying the site-directed mutation, with another selectable marker downstream. If a *Synechocystis* 6803 strain carrying the wild-type gene was transformed with a plasmid carrying the mutated gene and a selectable marker downstream of the coding region, crossover between the site-directed mutation and the marker could occur, thus uncoupling the marker from the introduction of the mutation.

A number of *Synechocystis* 6803 strains carrying deletions in PS II genes have been created.[22–25] Plasmids carrying site-directed mutations in a particular PS II gene linked to a selectable marker can be introduced

[18] L. B. Smart, S. L. Anderson, and L. McIntosh, *EMBO J.* **10**, 3289 (1991).
[19] S. Ermakova-Gerdes, S. Shestakov, and W. Vermaas, *in* "Photosynthesis: from Light to Biosphere" (P. Mathis, ed.), Vol. I, pp. 483–486, Kluwer Academic Publishers, Dordrecht, 1995.
[20] R. Wu, L. Grossman, and K. Moldave, eds., "Recombinant DNA Methodology." Academic Press, San Diego, California, 1989.
[21] R. Wu, ed., "Recombinant DNA Methodology II." Academic Press, San Diego, California, 1995.
[22] W. Vermaas, J. Charité, and B. Eggers, *in* "Current Research in Photosynthesis" (M. Baltscheffsky, ed.), Vol. I, pp. 231–238. Kluwer Academic Publishers, Dordrecht, 1990.
[23] P. Nixon, J. T. Trost, and B. A. Diner, *Biochemistry* **31**, 10859 (1992).
[24] J. J. Eaton-Rye and W. F. J. Vermaas, *Plant Mol. Biol.* **17**, 1165 (1991).
[25] H. B. Pakrasi, P. DeCiechi, and J. Whitmarsh, *EMBO J.* **10**, 1619 (1991).

into a *Synechocystis* deletion mutant in which this particular PS II gene had been replaced by a different marker. Resulting PS II mutants are then selected for through antibiotic resistance screening (Fig. 1).

Domain Deletion Mutagenesis

If a gene deletion leads to a loss of the entire PS II complex and insufficient information is available to rationally target specific amino acid residues for mutations, one of the options is to make smaller deletions in the corresponding PS II polypeptide. If a domain in a polypeptide can be deleted without major functional consequences, the residues in this region are apparently dispensable and are either not functionally active or can be replaced by nearby residues. If the deletion greatly affects the strain's phenotype, the role of the region that was deleted can be investigated further. Introduction of a short deletion mutation can be accomplished either by introduction of a stop codon in the gene (premature termination) using methods identical to those described for site-directed mutagenesis, or by an in-frame deletion of a relatively small part of a gene (internal deletion). The latter can be accomplished using plasmid or M13 constructs similar to those used for site-directed mutagenesis, with the mutation induced through a mutagenic primer carrying the desired deletion.

Interestingly, selected sequences of up to 11 amino acid residues in conserved loop regions in the D1, D2, and CP47 proteins were found to be dispensable in that they could be deleted or replaced without major effects.[23,26–30] In other cases deletions of similar length in D1, D2, CP43, or CP47 were found to lead to a loss of photoautotrophic growth, which often was accompanied by a physical loss of the PS II reaction center complex.[27–29,31] If a loss of function is observed, this suggests that the deleted region is of functional and/or structural relevance; the region may simply play a "connector" role in that two domains adjacent to this region need to be a certain distance apart, or the region may contain residues of functional relevance. To discriminate between these two possibilities, a simple option is to introduce random DNA sequences of a length identical to the sequence that was deleted, and determine which residue combinations can function-

[26] P. Mäenpää, T. Kallio, P. Mulo, G. Salih, E.-M. Aro, E. Tyystjärvi, and C. Jansson, *Plant Mol. Biol.* **22**, 1 (1993).

[27] B. Eggers and W. Vermaas, *Biochemistry* **32**, 11419 (1993).

[28] E. Haag, J. J. Eaton-Rye, G. Renger, and W. F. J. Vermaas, *Biochemistry* **32**, 4444 (1993).

[29] H. Kless, M. Oren-Shamir, S. Malkin, L. McIntosh, and M. Edelman, *Biochemistry* **33**, 10501 (1994).

[30] H. Kless, W. Vermaas, and M. Edelman, *Biochemistry* **31**, 11065 (1992).

[31] M. G. Kuhn and W. F. J. Vermaas, *Plant Mol. Biol.* **23**, 123 (1993).

ally substitute the wild-type sequence that had been deleted by analyzing the relevant sequence of photosynthetically viable transformants.[32] This is covered in more detail in a subsequent section.

## Introduction of Gene Chimaeras

To determine whether a gene from a different source is able to functionally complement a deleted *Synechocystis* gene, the homologous gene from another organism can be introduced into a *Synechocystis* mutant lacking the appropriate native gene. For this purpose, the appropriate *Synechocystis* deletion mutant is transformed with a plasmid construct carrying *Synechocystis* DNA regions that flank the site of the deletion, with at the site of the deletion the heterologous gene next to a selectable marker. In the case of the CP43 or CP47 genes, a corresponding higher plant gene could not functionally replace the *Synechocystis* one: *Synechocystis* mutants carrying a higher plant *psbB* or *psbC* were obligate photoheterotrophs and did not accumulate PS II in their thylakoids.[3,4] However, in the case of *psbA* a higher plant gene can successfully function in the cyanobacterial system.[33] The reason for this difference is not understood, but may be related to the fact that D1 is more highly conserved between higher plants and cyanobacteria than is CP47 or CP43.

## Combinatorial Mutagenesis

Because 19 different mutations can be introduced at any particular amino acid residue, in a domain of just a few amino acid residues the number of possible amino acid sequence combinations is enormous. Yet only one or several sequence combinations may lead to a functional polypeptide. To determine expediently which sequence(s) can be functionally accommodated in a certain domain, combinatorial mutagenesis can be applied. In this approach, essentially all different possible sequence combinations in this domain are presented to a (usually obligate photoheterotrophic) host strain lacking this region in the appropriate gene, and combinations leading to a functional phenotype (or another positively selectable phenotype) are selected for. Several practical approaches exist for the production of DNA strands that have sequence degeneracy in a defined region and that can be used for combinatorial mutagenesis. One approach utilizes two parallel PCRs and has been described by Kless and Vermaas.[32] For one of the PCR reactions, one of the primers is synthesized as a mix

---

[32] H. Kless and W. Vermaas, *J. Mol. Biol.* **246,** 120 (1995).
[33] P. J. Nixon, M. Rögner, and B. A. Diner, *Plant Cell* **3,** 383 (1991).

carrying a degenerate sequence at the desired domain. The other PCR primer for this PCR reaction recognizes a domain closeby. The PCR template (either plasmid or *Synechocystis* DNA) carries a deletion at the site of the degeneracy in order to not favor particular degenerate primer sequences for hybridization. The resulting PCR product has the degenerate region close to one of the ends, which is unfavorable for effective integration of degenerate sequences into the *Synechocystis* genome. To eventually generate a PCR product with sufficiently long flanking regions on both sides of the degenerate domain, the same PCR template is used for a second PCR reaction, in which one primer hybridizes to a region not covered by the first PCR reaction and at about 500–1000 bp from the site at which degeneracy has been introduced in the first PCR reaction; the other primer hybridizes to the template inside the region that was amplified by the first PCR reaction. The products of the two PCR reactions are combined, melted and hybridized, and extended to obtain DNA carrying either the deletion or the degenerate region in the middle. This DNA can be used for transformation of the *Synechocystis* strain carrying an appropriate deletion. Transformants with desired characteristics (for example, restored photoautotrophic growth) can be selected and analyzed in terms of their sequence and functional properties.[32] This method has been applied to determine D1 sequences that are located in a hydrophilic loop at the cytoplasmic side of PS II near the binding site of herbicides and the secondary quinone $Q_B$, and that can support PS II electron transport. Surprisingly, a four-residue D1 region that is fully conserved throughout evolution could be replaced by a large number of other sequences without causing major functional effects on PS II activity and survival in the laboratory.[32] Two out of the four D1 residues were shown to have a much larger tendency to be conserved in combinatorial mutants as compared to the other two, presumably indicating functional relevance of the former two residues.[32]

Combinatorial mutagenesis is an appropriate method to learn about functional relevance of particular residues in a sequence of four to five amino acids. Combinatorial mutagenesis involving a larger number of codons becomes less and less practical in that, for example, 64 million different amino acid combinations of six residues are expected, and therefore a large number of transformants may need to be generated before finding any functionally active (photoautotrophic) transformants. A derivative of this combinatorial mutagenesis method, incorporating multiple degenerate primers, has been described as well.[34] In this experiment, the question was asked whether electron transfer pathways could be created that did not involve the redox-active Tyr residue in D1 ($Y_Z$; Y161) between the water-

[34] H. Kless and W. Vermaas, *Biochemistry* **35**, 16458 (1996).

splitting system and the primary donor P680. In all photoautotrophic strains resulting from transformation of an obligate photoheterotrophic strain having Trp at position 161 of D1 with a mix of *psbA* derivatives that was combinatorially mutagenized around $Y_Z$, a Tyr codon at position 161 was restored in the *psbA* gene. This suggested that Y161 is not easily replaced functionally by redox-active residues at nearby positions.[34] Combinatorial mutagenesis therefore is a useful approach to rather easily determine whether other sequence combinations are functionally acceptable.

## Targeted Random Mutagenesis

The method described above is used to introduce random mutations into a very small region (several codons) of a particular gene. This is appropriate if the correspondingly small region of the polypeptide is of particular interest. However, during the initial stages of a study the precise regions of interest in a polypeptide may not yet be well defined and a first step may be to identify codons in a larger gene region or even in the entire gene that upon mutation lead to a particular phenotype. For this purpose, *targeted random mutagenesis* may be applied. This type of mutagenesis involves the introduction of random mutations into a (region of a) particular gene without modification of other genes in the organism. Several options exist to introduce random mutations into a targeted region. One of them is PCR under suboptimal conditions, leading to frequent misincorporations. However, the drawback of this method is that *Synechocystis* 6803 transformations require significant flanking regions, thus making the region into which mutations will be introduced rather long. A more elegant approach is to make use of the property of sodium bisulfite to introduce mutations ($C \rightarrow U$) preferentially into single-stranded (rather than double-stranded) DNA.[35] This property can be capitalized on by constructing two plasmids that are identical except that in one plasmid the region where random mutations should be introduced has been deleted. After cutting each plasmid with a different unique restriction enzyme, the plasmids are mixed in a 1 : 1 ratio, heat-denatured, and slowly annealed.[36] In this process both homoduplex and heteroduplex DNA molecules are formed. The latter are nicked double-stranded circles because of the use of restriction enzymes with different sites for the two plasmids. In heteroduplexes the DNA region corresponding to the sequence deleted in the one plasmid is single-stranded and is thereby susceptible to chemical mutagenesis by sodium bisulfite.[37]

---

[35] R. Shapiro, B. Braverman, J. B. Louis, and R. E. Servis, *J. Biol. Chem.* **248,** 4060 (1973).
[36] R. Pine and P. C. Huang, *Methods Enzymol.* **154,** 415 (1987).
[37] K. W. Peden and D. Nathans, *Proc. Natl. Acad. Sci. U.S.A.* **79,** 7214 (1982).

Reaction conditions with sodium bisulfite need to be determined empirically, and should introduce an appropriate number (usually one to two) of mutations per plasmid. Plasmids can then be introduced into uracil-DNA glycosylase deficient strains of *E. coli*, and transformants will have plasmids carrying the deletion or the mutagenized loop-out region. *Escherichia coli* transformants with plasmids carrying the loop-out region often can be selected by examining the size of the *E. coli* colony: as plasmids with *Synechocystis* 6803 genes coding for membrane proteins often lead to a decreased rate of growth of *E. coli*, picking small colonies generally is sufficient to greatly enrich for *E. coli* transformants with a full-length gene of interest. Plasmid DNA can then be used for transformation of an appropriate *Synechocystis* 6803 strain, and transformants with desired properties can be selected and analyzed further. This method has been applied successfully to a region of the *psbDI* gene of *Synechocystis* 6803.[38] Ser residues in the first lumenally located hydrophilic loop of the D2 protein were identified that on mutation to Phe gave rise to an obligate photoheterotrophic phenotype. Even though the precise function of these Ser residues has not yet been elucidated, this shows the use of this approach to identify functionally important residues in a protein domain. More detailed information on the procedure and on mutant characterization can be found in Ermakova-Gerdes *et al.*[38]

The drawback of this method is that sodium bisulfite causes only deamination of cytosine to uracil, leading to only C → T and G → A mutations in the coding strand (depending on whether deamination occurred in the coding or noncoding strand of the heteroduplex). Therefore, codons without C or G cannot be mutagenized by this method whereas Gln and Trp codons can be modified only to stop codons.

## Pseudorevertant Analysis

In previous sections, mutagenesis approaches have been described to modify particular PS II-related genes, domains, or residues. In some cases mutant strains will be obligate photoheterotrophs or will have impaired photoautotrophic growth. An important question to be asked in such cases is whether spontaneous secondary modifications can be introduced that reestablish photoautotrophic growth. Such events are simple to select for (restoration of photoautotrophic capacity) and may involve secondary changes in the gene carrying the original mutation or in other genes. In the case of site-directed mutants (with nucleotide changes limited to one codon) photoautotrophic strains may have been derived from additional

[38] S. Ermakova-Gerdes, S. Shestakov, and W. Vermaas, *Plant Mol. Biol.* **30**, 243 (1996).

mutations in the same codon. Of more interest are mutations elsewhere in the gene or secondary mutations in other genes that will lead to a restoration of photoautotrophic function. Such pseudorevertants (second-site mutants with essentially restored function) can be analyzed in terms of sequence and function to determine proteins and interactions that are generally close to the site of the primary mutation.

Photoautotrophic pseudorevertants have been observed after introduction of short deletions into PS II genes that impaired photoautotrophic growth. An important question is the location of the secondary mutation. Whether or not the site of pseudoreversion is in the gene carrying the primary mutation can be determined simply by PCR amplification of this gene from the pseudorevertant. If this PCR product can functionally complement the original mutant to photoautotrophic growth, the site of the pseudoreversion is in the gene carrying the original deletion. In cases where the secondary mutation site can be mapped to the gene carrying the primary deletion, the pseudoreversion often has been found to involve an in-frame duplication of a nearby sequence in the gene.[33,39] The primary sequence of the modified gene in the pseudorevertant is different from that in either the original deletion mutant or the wild type, and the length of the duplication can be very significant, up to almost 100 nucleotides.[33] The precise mechanism for the partial gene duplication events has not yet been elucidated, but a model explaining the observations has been presented.[39]

## Mapping of Unknown Genes Influencing Photosystem II Function

Until recently, if the site of pseudoreversion was not found to be in the gene carrying the primary mutation, identification of the pseudoreversion locus generally was time consuming because it involved making a partial library of pseudorevertant DNA. However, with the sequencing of the entire *Synechocystis* 6803 genome pseudorevertant mapping has been greatly facilitated. The locus of the secondary mutation in pseudorevertants can be mapped simply by a complementation experiment as shown in Fig. 2 (S. Ermakova-Gerdes and W. Vermaas, unpublished observations). After isolation of DNA from the pseudorevertant, it is cut with any 5–10 of the 16 restriction enzymes (*Afl*II, *Bam*HI, *Bgl*II, *Bsa*BI, *Bsa*HI, *Bsp*DI, *Bsp*EI, *Eco*RI, *Eco*RV, *Kpn*I, *Nhe*I, *Pst*I, *Sca*I, *Sma*I, *Stu*I, and *Xba*I) for which we have constructed a restriction map of the *Synechocystis* 6803 genome, and then DNA samples digested with one of these enzymes are size-separated on a gel. The lanes of the gel are cut into 20–25 fractions, each corresponding to a size category between about 1 and 20 kb pairs, and

[39] H. Kless and W. Vermaas, *J. Biol. Chem.* **270**, 16536 (1995).

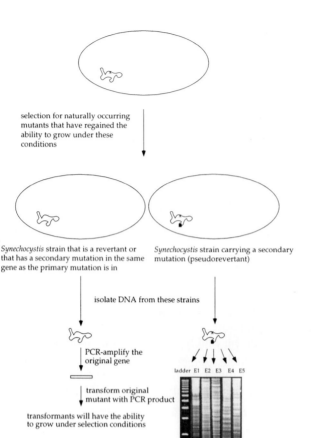

selection for naturally occurring
mutants that have regained the
ability to grow under these
conditions

*Synechocystis* strain that is a revertant or
that has a secondary mutation in the same
gene as the primary mutation is in

*Synechocystis* strain carrying a secondary
mutation (pseudorevertant)

isolate DNA from these strains

PCR-amplify the
original gene

transform original
mutant with PCR product

transformants will have the ability
to grow under selection conditions

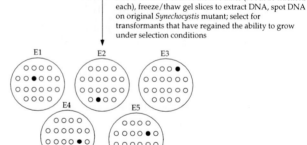

ladder E1 E2 E3 E4 E5

size-fractionate each of the five lanes (about 20 fractions
each), freeze/thaw gel slices to extract DNA, spot DNA
on original *Synechocystis* mutant; select for
transformants that have regained the ability to grow
under selection conditions

E1

E2

E3

E4

E5

Mutant *Synechocystis* plates under selection conditions and spotted with DNA size
fractions (small circles) from each restriction digest. Areas with complementing DNA
are visualized by dark circles

Identify the site of the secondary mutation as the unique region
in the cyanobacterial genome with this restriction pattern

each fraction is put into an Eppendorf tube. The gel pieces are frozen and thawed, and then spun down in a microcentrifuge. The supernatant liquid contains a significant amount of DNA, which can be precipitated or used directly for transformation of the original mutant strain plated on medium that will support growth of the pseudorevertant but not of the original mutant (for example, on BG11 without glucose). The size range of the collection of restriction fragments that leads to functional complementation of the mutant can be determined for each restriction enzyme. We have made a list of genome locations of restriction sites and of the corresponding restriction fragment sizes (ordered by size) for the 16 restriction enzymes mentioned earlier. By determining which genome location(s) are present in all restriction fragment collections that complement, a unique solution of a 0.5–3 kb pair region carrying the pseudoreversion locus usually can be found when using only 5 different restriction enzymes. The Kaleidagraph files with the 16 *Synechocystis* 6803 restriction maps and size-sorted restriction fragment tables are available from the author upon request.

The region that was identified to carry the pseudoreversion locus can then be amplified as a whole and/or in part by means of PCR, and PCR products can be used for transformation. Based on which PCR products can lead to functional complementation, the site of the secondary mutation can be narrowed down. Once a sufficiently small region has been identified, this region can be sequenced in both the original mutant and in the pseudorevertant. Because a large number of open reading frames of unknown function have been identified,[14,15] the probability that these pseudoreversion studies help to identify "new" genes involved with the structure, function, and assembly of PS II is considerable. Indeed, by this method pseudorevertants of an obligate photoheterotrophic D2 mutant have been shown to

---

FIG. 2. Flow scheme for revertant and pseudorevertant (secondary mutation) mapping in *Synechocystis* sp. PCC 6803. The left part of the scheme represents the scenario that the secondary mutation has occurred in the same codon or at least in the same gene as the primary mutation. After PCR amplification of this gene in the secondary mutant, the PCR product will be able to functionally complement the original mutant. DNA analysis can then be applied to identify the intactness of the original mutation and/or the precise site of the secondary mutation. The right part of the figure represents mapping of the site of pseudoreversion making use of a genomic restriction map. Genomic DNA from the pseudorevertant is digested with five restriction enzymes (E1–E5) and fragments are separated by size using gel electrophoresis. DNA corresponding to distinct size classes is cut from the gel and used for functional complementation of the original mutant. Using the restriction maps for all enzymes used, from the complementation pattern a single DNA region (≤3 kb pairs) is likely to be identifiable that harbors the pseudoreversion locus. If no single region can be identified as the pseudoreversion locus, additional restriction enzymes should be used for genomic digestion of DNA from the pseudorevertant followed by functional complementation of the original mutant.

have a secondary mutation in a *Synechocystis* 6803 open-reading frame that has homology to an unidentified open reading frame in the chloroplast genome of red algae (S. Ermakova-Gerdes and W. Vermaas, unpublished).

The selection criteria for pseudorevertants are not necessarily limited to reversion to photoautotrophic growth. Any positive selection method can be utilized. For example, a reduction in PS II activity can be selected for by screening for resistance to moderate light intensity of PS I-less strains that usually are light sensitive due to PS II-generated electrons.

Molecular genetic approaches have proven to provide powerful insight into PS II function and structure. With time, these molecular genetic tools may become perfected further and other useful techniques may be developed. However, it is important to realize that these approaches are most fruitful if combined with techniques from other disciplines such as biochemistry. Nonetheless, molecular genetic tools have become such important contributors to progress in photosynthesis research that the amenability to gene modification should be a very prominent criterion in the selection of experimental systems. With a known genome sequence and a highly convenient transformation system, it is likely that *Synechocystis* 6803 will become an even more popular system for the study of photosynthesis.

Acknowledgments

Work from the Vermaas group that was presented in this chapter has been funded primarily by grant MCB-9316857 from the National Science Foundation and grant GM-51556 from the National Institutes of Health.

## [21] Specific Mutagenesis of Reaction Center Proteins by Chloroplast Transformation of *Chlamydomonas reinhardtii*

*By* Hyeonmoo Lee, Scott E. Bingham, and Andrew N. Webber

Introduction

The initial charge separation events that convert light energy into chemical energy occur in specialized membrane protein complexes called reaction centers. Application of genetic engineering to reaction center proteins has led to a significant advancement in our understanding of primary electron transfer events and the role of the protein environment in modulating these processes. Much of this work has been restricted to purple bacteria and cyanobacteria in which manipulation of the genes encoding reaction centers is fairly straightforward. In eukaryotes, genes encoding the PS I and PS II reaction center proteins are located in the chloroplast genome. The development of biolistic transformation procedures for introducing DNA into the chloroplast[1-3] and suitable systems for expression of antibiotic resistance markers,[4] together with the presence of an active homologous recombination system in chloroplasts, now make it quite routine to generate mutants in chloroplast reaction center proteins.

## Principle of the Method

Despite the presence of an active homologous recombination system, transformation of the chloroplast genome has several inherent complications.[1,5,6] First, plants and many green algae are unable to grow in the absence of photosynthesis. Second, most eukaryotes contain several to hundreds of chloroplasts with each organelle containing 80–100 copies of chloroplast (cp) DNA. Third, three cell membranes and a cell wall act as a barrier to exogenously added DNA. Several of these limitations are overcome by using the green alga *Chlamydomonas reinhardtii,* which contains only a single chloroplast and grows heterotrophically when supplemented with acetate. *Chlamydomonas* has, therefore, served as a model organism for the development of chloroplast transformation procedures and the study of photosynthetic mutants generated using this method.[6]

Exogenous cloned cpDNA can be introduced into the chloroplast by use of the gene gun.[1,2] The transforming DNA is precipitated onto micron-sized gold or tungsten particles and propelled into cells using a gas pressure wave to accelerate the particles. Inside the chloroplast the exogenous DNA recombines with complementary regions of the cpDNA.[1,2,7] Some of the earliest transformation experiments used rDNA with mutations conferring resistance to the antibiotics spectinomycin, streptomycin, or erythromy-

[1] J. E. Boynton, N. W. Gillham, E. H. Harris, J. P. Hosler, A. M. Johnson, A. R. Jones, B. L. Randolph-Anderson, D. Robertson, T. M. Klein, K. B. Shark, and J. C. Sanford, *Science* **240,** 1534 (1988).
[2] A. D. Blowers, L. Bogorad, K. B. Shark, and J. C. Sanford, *Plant Cell* **1,** 123 (1989).
[3] K. L. Kindle, K. L. Richards, and D. B. Stern, *Proc. Natl. Acad. Sci. U.S.A.* **88,** 1721 (1991).
[4] M. Goldschmidt-Clermont, *Nucleic Acids Res.* **19,** 4083 (1996).
[5] J. M. Erickson, *in* "Oxygenic Photosynthesis: The Light Reactions" (D. R. Ort and C. F. Yocum, eds.), p. 589. Kluwer Academic Publishers, The Netherlands (1998).
[6] A. N. Webber, S. E. Bingham, and H. Lee, *Photosynth. Res.* **44,** 191 (1995).
[7] S. M. Newman, J. E. Boynton, N. W. Gillham, B. L. Randolph-Anderson, A. M. Johnson, and E. H. Harris, *Genetics* **126,** 875 (1990).

cin.[7,8] Transformants with resistance to the antibiotics were found to contain rDNA harboring the mutation. Following several rounds of single colony selection all copies of cpDNA contained the introduced mutation.[7,8]

Two other procedures have been useful for selection of successful transformation. Nonphotosynthetic mutants containing small deletions or frame shifts in the *atpB, psbA,* or *psaB* gene have been complemented with wild-type copies that restore photosynthetic growth capability to the transformed cells.[1,5,9–11] This selection procedure has been used to introduce nonphotosynthesis-lethal mutations into reaction center proteins.

The most versatile selection procedure, however, is to use chimeric constructs of the bacterial *aadA* gene, encoding aminoglycoside adenyltransferase, which renders transformed cells resistant to spectinomycin and streptomycin.[4] This article focuses on procedures for constructing and using chimeric constructs for chloroplast transformation that can be applied to the study of any chloroplast gene.

Protocols

*Construction of Chimeric Genes Conferring Antibiotic Resistance*

For high-level expression of a foreign gene in chloroplasts it is necessary to engineer chimeric constructs that effectively drive transcription and translation.[4] In many cases 3′ flanking sequences from a chloroplast gene are also included in the construct to guide transcription termination and enhance mRNA stability.[4] The chimeric genes are made by cloning polymerase chain reaction (PCR) amplified fragments of cpDNA that contain transcription and translation initiation signals and linking these to the antibiotic resistance gene in the correct context. Incorporation of restriction sites in the primers used for amplification allows for rapid construction of the chimeric constructs with expression driven by a promoter of choice. The first report by Goldschmidt-Claremont[4] used the *atpA* promoter and 5′ flanking sequences to drive transcription and translation of *aadA*. In this construct 3′ flanking sequences from *rbcL* were cloned downstream of *aadA*. We have found that the 3′ sequences are not required for normal accumulation of the *aadA* mRNA.[12]

[8] S. M. Newman, N. W. Gillham, E. H. Harris, A. M. Johnson, and J. E. Boynton, *Mol. Gen. Genet.* **230,** 65 (1991).
[9] S. E. Bingham, R. Xu, and A. N. Webber, *FEBS Lett.* **292,** 137 (1991).
[10] A. N. Webber, P. B. Gibbs, J. B. Ward, and S. E. Bingham, *J. Biol. Chem.* **268,** 12990 (1993).
[11] J. Minagawa and A. R. Crofts, *Photosynth. Res.* **42,** 121 (1994).
[12] S. E. Bingham and A. N. Webber, *J. App. Phycol.* **6,** 239 (1994).

A scheme for constructing the chimeric *atpA–aadA* construct by PCR, based on procedures described in Ref. 4, is shown in Fig. 1 with the necessary primer sequences provided in Table I. The chloroplast promoter and the *aadA* gene are first amplified by PCR using primers that contain convenient restriction enzyme sites (Table I and Fig. 1) that aid with cloning. The 5' primer for amplification of the *atpA* sequences contains *Eco*RI or *Cla*I, whereas the 3' primer has an *Nco*I site. The 5' and 3' primers for amplification of *aadA* have an *Nco*I and *Sal*I site, respectively (Table I). The *atpA* fragment is amplified from a preparation of total chloroplast DNA (protocol provided in a later section) using standard PCR conditions.[13] The *aadA* gene is amplified using a plasmid, PH45-Ω,[14] as the DNA template. The PCR-amplified DNA from each reaction is then separately cloned into the TA vector (Invitrogen, Carlsbad, CA) and sequenced to ensure that no base changes have occurred during amplification. This cloned *aadA* gene can be saved and used for construction of chimeric *aadA* genes driven by other chloroplast promoter regions.

The cloned fragments are next digested with the restriction enzymes shown (Fig. 1, step 2), gel purified by acrylamide gel electrophoresis and then ligated together with the pUC19 plasmid that has been linearized by digestion with *Eco*RI and *Sal*I (Fig. 1). The resulting product has the *atpA* and *aadA* genes linked at a unique *Nco*I site, which forms a direct fusion between the first several amino acids of AtpA and the entire AadA protein.[4] This serves as the initial chimeric construct, which is later inserted into the desired cpDNA clone to be used as the donor plasmid for chloroplast transformation.

For chloroplast transformation the *aadA* cassette is placed in the non-coding flanking sequences either 5' or 3' to the gene of interest, being sure to avoid disruption of any polycistronic mRNAs. We have typically placed the chimeric construct in the 3' flanking sequences of either *psaA* or *psaB,* a few hundred bases downstream of the inverted repeat sequence.[15] Since the inverted repeats play a significant role in mRNA stability, transcription termination, and processing in chloroplasts, the chimeric *atpA–aadA* should be placed far enough downstream so as not to interfere with these processes. As an example (Fig. 1, step 3), the *atpA–aadA* cassette was inserted, by blunt-ended ligation, at an *Eco*RI restriction site downstream of a portion of *psaA* exon 3 (an *Avr*II-*Bam*HI fragment from chloroplast DNA restriction

[13] J. Sambrook, E. F. Fritsch, and T. Maniatis, pp. 1.21–1.52. "Molecular Cloning." Cold Spring Harbor Laboratory Press, New York, 1989.
[14] P. Prentki and H. M. Krisch, *Gene* **29,** 303 (1984).
[15] A. N. Webber, H. Su, S. E. Bingham, H. Käss, L. Krabben, M. Kuhn, E. Schlodder, and W. Lubitz, *Biochemistry* **39,** 12857 (1996).

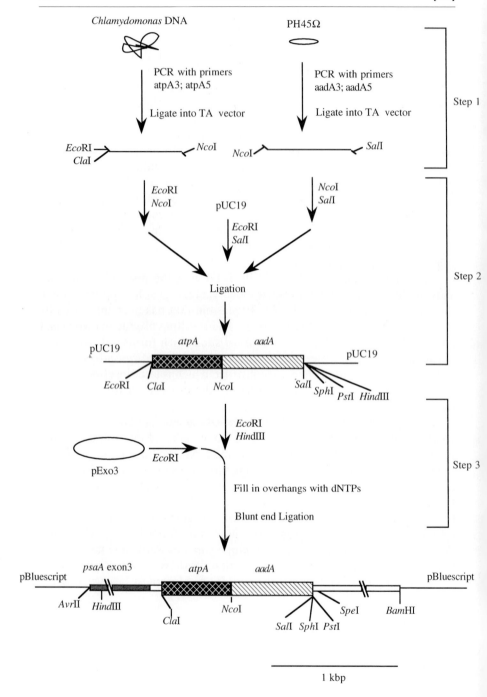

1 kbp

TABLE I

OLIGONUCLEOTIDES USED TO CONSTRUCT THE CHIMERIC *aadA* CASSETTE FOR
TRANSFORMATION OF *CHLAMYDOMONAS* CHLOROPLAST

| Primer name | Primer sequence[a] | Restriction sites introduced |
|---|---|---|
| AAD5 | cgca**ccatggc**tcgtGAAGCGGTtATCGCCGAAG | *Nco*I |
| AAD3 | gcac**gtcgac**TTATTTGCCaACTACCTTaGTGATC | *Sal*I |
| ATPA5 | gtactc**gaattcatcgat**GACTTTATTAGAGGCAGTG | *Cla*I, *Eco*RI |
| ATPA3 | gtcatcggat**ccatgg**aCATTTTCACTTCTGGAGTG | *Nco*I |
| ATPB5 | attcgt**atcga**TCCAAGACATTGTATGCTTA | *Cla*I3 |
| ATPB3 | ttcactc**catg**GATATTTTAACTTATTTTACTTAAAT | *Nco*I |
| PSBD5 | gcta**atcgat**GAGTCATATGAAATTAAATG | *Cla*I |
| PSBD3 | gcta**ccatgg**CGTGTATCTCCAAAATAAA | *Nco*I |

[a] Each oligonucleotide is written 5' to 3'. Lowercase letters do not match sequence. Bold letters indicate restriction endonuclease recognition sites.

enzyme fragment *Bam*11). The orientation of the antibiotic resistance cassette to the *psaA* exon 3 is important in this case. The parallel orientation to exon 3 gives good transformation frequency but not the opposite orientation. In our research we have been interested in making mutations in regions of *psaA* that encoded in the final construct shown in Fig. 3. This final construct is small enough that we can use this plasmid directly to make site-directed mutations in *psaA* using commercially available kits.

Homologous recombination between direct repeats in the chloroplast genome occurs very actively. Recombination can occur between a chimeric cassette and complementary cpDNA sequences when placed close to a gene from which the promoter was also used in the chimeric gene. This

FIG. 1. Outline of the procedure used to construct the *atpA–aadA* chimeric construct and subsequent cloning adjacent to the *psaA* gene. Details of the procedure are provided in the text. Step 1: PCR is used to amplify a portion of the 5' end of *atpA* from *Chlamydomonas* total cellular DNA using primers ATPA3 and ATPA5 (Table I). In a separate PCR reaction the *aadA* gene is amplified from the vector PH45Ω using primers AADA3 and AADA5 (Table I). Each product is digested with restriction enzymes indicated ligated into the TA vector and sequenced. Step 2: The two PCR products in the TA vector are digested with the restriction enzymes indicated. The products are purified by agarose gel electrophoresis and then ligated with pUC19 (previously digested with *Eco*RI and *Sal*I) to give the *atpA–aadA* plasmid construct. This construct can be cloned adjacent to any gene of interest. In this example the *aadA* chimeric construct is inserted in a unique *Eco*RI site located 3' of *psaA* as described in the next step. Step 3: The *aadA* construct is digested from pUC19 using *Eco*RI and *Hin*dIII and the chimeric construct purified by agarose gel electrophoresis. In a separate reaction pExo3 containing a portion of *psaA* exon 3 (see text) is digested with *Eco*RI. The ends of the DNA fragments are filled in using the Klenow enzyme and ligated.

can result in deletion of the region between the promoters (one from the chimeric gene and the other from the native gene) in a portion of the cpDNA copies. This deletion may significantly affect expression of the gene of interest and so care must be taken when the promoter for the chimeric construct is chosen. We have used promoter regions from different chloroplast genes, such as *atpB* and *psbD* (primers provided in Table I), that are comparable to the *atpA* sequences in their ability to drive transcription and translation of *aadA*.

### Transformation Procedures

Several procedures for biolistic transformation of *Chlamydomonas* chloroplasts have been published.[5,16] Efficiency of transformation is very sensitive to the conditions used. We have found that the following three factors considerably affect efficiency of chloroplast transformation: (1) growth stage of the recipient cells, (2) preparation of tungsten particles, and (3) conditions of bombardment.

*Growth Media.* In our laboratory we have been using a medium developed by the late Dr. John C. Cox, which he named CC. CC contains the following components in 1 liter: glacial acetic acid, 1 ml; Tris base, 2.5 g; $NH_4NO_3$, 0.5 g; $MgSO_4 \cdot 7H_2O$, 0.1 g; $CaCl_2 \cdot 2H_2O$, 0.02 g; $KH_2PO_4$, 0.05 g; KCl, 0.1 g; and 1 ml Hunter's mineral salts.[17] Yeast extract (0.1%) may added to liquid medium at a final concentration of 0.1%. All components are autoclaved together. A convenient 5× stock solution can be prepared and stored without autoclaving for approximately 2 weeks at 4°. Recipes for other commonly used media are provided in Ref. 17.

*Recipient Strains.* The *Chlamydomonas* Culture Collection maintained at Duke University provides a source for wild-type and mutant strains. Many mutants are available that lack various thylakoid membrane protein complexes and so careful selection of the recipient strain can make it easy to screen the transformants and to characterize resulting mutants by spectroscopy.[5,6] As an example, we have recently been using the CC2696 strain as a recipient for site-directed mutants of the *psaA* or *psaB* genes. CC2696 contains a deletion of the *psbA* gene and a nuclear lesion that blocks synthesis of chlorophyll *b* resulting in a mutant strain that lacks both photosystem II and chlorophyll *a/b* containing light-harvesting complexes in the thylakoid membrane.

*Growth Conditions.* Recipient cells are typically grown in CC liquid media containing 0.1% yeast extract to a density of $1-1.5 \times 10^6$ cells/ml. When cells are grown over $1.5 \times 10^6$ cells/ml the transformation efficiency

[16] J. E. Boynton and N. W. Gillham, *Methods Enzymol.* **217**, 510 (1993).
[17] E. H. Harris, "The *Chlamydomonas* Sourcebook." Academic Press, San Diego, 1989.

is significantly reduced. The cells are concentrated to $5 \times 10^7$ cells/ml, and $2 \times 10^7$ cells are spread on plates containing spectinomycin (100 $\mu$g ml$^{-1}$). The cells should cover at least four-fifths of the surface of the plates. It is not necessary to first plate cells on nonselective media for bombardment followed by transfer to the selective media.

*Particle Bombardment Procedure*

SOLUTIONS

   2.5 $M$ CaCl$_2$

   0.1 $M$ Spermidine free base

   100% Ethanol [high-performance liquid chromatography (HPLC) grade]

   M10 Tungsten (Bio-Rad, Richmond, CA)

PREPARATION OF M10 TUNGSTEN PARTICLES

1. Tungsten particles (60 mg) are suspended in 2 ml of 100% ethanol.
2. The suspension is sonicated three times for 20 sec at full power using a microtip in order to deaggregate the particles. Sonication works significantly better than conventional vortexing.
3. Collect the tungsten particles in a 1.5-ml microfuge tube by centrifugation.
4. Wash the pellet once with 1 ml sterile deionized water followed by centrifugation.
5. Resuspend the particles in 1 ml sterile deionized water by vortexing.

*Precipitation of DNA onto Tungsten Particles.* Quality of DNA is a parameter affecting transformation efficiency. In our laboratory, CsCl purified DNA described by Sambrook *et al.*[13] is used for chloroplast transformation.

1. After vortexing, 60 $\mu$l of the particle suspension is transferred to a microfuge tube.
2. Add 5$\mu$g of DNA and continue to vortex for 30 sec.
3. Add 60 $\mu$l of 2.5 $M$ CaCl$_2$.
4. Add 12 $\mu$l of 0.1 $M$ spermidine and continue to vortex for 1 min.
5. Leave at room temperature for 15 min.
6. Centrifuge for 5 sec and discard the supernatant.
7. Carefully suspend the particles in 250 $\mu$l of 100% ethanol by repeated pipetting until there is no visible aggregate.
8. Vortex for 2 min.
9. Centrifuge for 5 sec and discard the supernatant.
10. Resuspend the particle in 60 $\mu$l of 100% ethanol by repeated pipetting.

Steps 2, 3, and 4 should be done while vortexing. If available, 100% ethanol should be used; if not available, 95% ethanol containing isopropanol and methanol (HPLC grade from Fisher, Pittsburgh, PA) can be used.

*Particle Bombardment using PDS-1000/He*

1. 60 $\mu$l of DNA precipitated tungsten particles is loaded on the thin macrocarrier.
2. Allow the particles to air-dry.
3. Place the macrocarrier five-eighths in from the rupture disk (1350 psi).
4. Plates are placed on the bottom shelf of the chamber.
5. The chamber is evacuated to at least 28.5 in. Hg immediately prior to bombardment.

Using these conditions, transformation efficiencies between $1 \times 10^{-5}$ and $1 \times 10^{-4}$ (between several hundred to a thousand colonies per plate) can be reached. The most important variables are the distance between the rupture disk and the macrocarrier, and the vacuum pressure in the bombardment chamber. A small change can dramatically reduce the efficiency of the transformation.

*Selection and Screening for Homoplasmic Transformants*

*PCR Analysis.* Bombarded cells are incubated on the selective plates under dim light for 7–10 days until single colonies appear. Single colonies are then streaked onto new plates and incubated under dim light. At this stage it is important to select for transformant cells in which every copy of cpDNA contains the introduced mutation. PCR can be used to screen for homoplasmy of each colony if the donor plasmid introduces a unique restriction site in the region amplified. Primers are designed to flank the site of the introduced mutation and then used to amplify the region of interest. The PCR product is then digested with the restriction enzyme and size fractionated by agarose gel electrophoresis (Fig. 2). PCR analysis can be carried out from minipreparation DNA from a patch of plated cells as described next. Generally, two or three rounds of single colony selection are sufficient to obtain completely homoplasmic transformants.

*Minipreparation of DNA from Chlamydomonas Cells for PCR Analysis*

SOLUTIONS
  TEN buffer (10 m$M$ Tris-HCl, p$H$ 8.0, 10 m$M$ EDTA, 50 m$M$ NaCl)
  20 mg/ml pronase or 1 mg/ml proteinase K
  10% Sodium dodecyl sulfate (SDS)

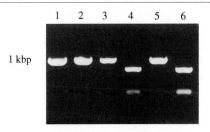

FIG. 2. Analysis of PCR products from wild-type and mutant cells. A 1-kb pair fragment of *psaB* was amplified from total cellular DNA using primers that flank the site of the silent mutation in the donor plasmid that generates a *StuI* site. The product was either loaded directly onto the 2% agarose gel (lanes 1, 3, and 5) or digested with *StuI* before electrophoresis (lanes 2, 4, and 6). Complete digestion with *StuI* indicates that all copies of the wild-type *psaB* gene have been replaced with the mutant copy. Wild-type, lanes 1 and 2; mutants containing the *StuI* site, lanes 3, 4, 5, and 6. Note that the PCR product from wild-type cells is not digested with *StuI*.

1. Collect a patch of cells ($0.5$ cm$^{-2}$) from plates to a microfuge tube containing 400 $\mu$l TEN buffer.
2. Suspend cells by pipetting or vortexing.
3. Add 50 $\mu$l of pronase or proteinase K and 10% SDS, and incubate for 1 hr at 50°.
4. Extract once with phenol/chloroform/IAA (25 : 24 : 1).
5. Precipitate DNA with two volumes of 100% ethanol.
6. Pellet DNA by centrifugation, wash once with 70% ethanol, air-dry.
7. Dissolve DNA in 30 $\mu$l of sterile deionized water.
8. Use 1–2 $\mu$l of this DNA for the PCR reaction (final volume: 100 $\mu$l).

Following amplification, approximately 20 $\mu$l of the PCR reaction mixture can be digested with restriction enzyme and size fractionated by agarose gel electrophoresis. Figure 2 shows an example of PCR amplified DNA from wild-type cells (Fig. 2, lanes 1 and 2) and two mutants (lanes 2, 3, 4, and 5) that have a unique *StuI* site introduced in the transforming plasmid. As shown in Fig. 2, the DNA from wild-type cells does not cut with *StuI* (Fig. 2, lane 1), whereas DNA from the mutants cuts completely (lanes 4 and 6). If the mutants are not heteroplasmic then the DNA would not cut completely with *StuI*. This assay is a very sensitive screen for heteroplasmy. Erickson[5] reports that PCR can be used to detect one wild-type copy at a dilution of 1 in 20,000.

*Southern Analysis.* Southern analysis can also be used for screening the homoplasmic transformants. This protocol will provide DNA of adequate quality and quantity for Southern analysis on a small scale. The amount of DNA obtained using this procedure would be sufficient for at least four restriction enzyme digestions.

Solutions

Lysis buffer (37.5 m$M$ sodium citrate, pH 7.0, 6 $M$ sodium thyocyanate)
10% L-Lauryl sacrosine
Phenol equilibrated with 0.5 $M$ Tris-HCl, pH 8.0
Chloroform/isoamylalcohol (24:1)

1. Grow cells to 1 × 10$^6$ cells/ml.
2. Harvest the cells and transfer to a 2-ml microfuge and wash once with TEN buffer.
3. Suspend cells with sterile water and make the total volume up to 200 $\mu$l.
4. Add 400 $\mu$l of lysis buffer and 60 $\mu$l of 10% L-lauryl sacrosine. (Do not use SDS, which will be precipitated.)
5. Add the same volume (600 $\mu$l) of phenol and mix gently using a pipette.
6. Add one-quarter of chloroform/IAA (150 $\mu$l) and mix by hands.
7. Centrifuge at room temperature for 5 min.
8. The supernatant should be clear. If not, repeat steps 5, 6, and 7.
9. Transfer the supernatent to a new tube and precipitate DNA by adding the same volume of isopropanol.
10. Centrifuge for 5 min and wash the DNA pellet with 70% ethanol, and briefly air-dry (do not dry completely).
11. Dissolve the pellet in TE buffer and extract once with phenol/chloroform/IAA (25:24:1).
12. Reprecipitate DNA in the presence of 0.3 $M$ sodium acetate with two volumes of ethanol. Pellet and air-dry briefly.
13. Dissolve DNA in appropriate volume of TE (10 m$M$ Tris-HCl, pH 8, 1 m$M$ EDTA) buffer with 1 $\mu$g/ml RNase.

# [22] Directed Mutagenesis in Photosystem II: Analysis of the CP 47 Protein

By Terry M. Bricker, Cindy Putnam-Evans, and Jituo Wu

## Introduction

Directed mutagenesis is a powerful tool for the investigation of protein structure–function relationships. Site-directed mutagenesis has been used extensively in the study of Photosystem II (PS II) with mutations being produced in the D1 and D2 proteins,[1,2] CP 47 and CP 43,[3,4] cytochrome $b_{559}$,[5] the extrinsic 33-kDa protein,[6] and others. The majority of these alterations have been engineered into the cyanobacterium *Synechocystis* 6803, although some investigators have utilized chloroplast transformation systems to introduce mutations into the unicellular green alga *Chlamydomonas reinhardtii*.[7] Additionally, other investigators have used heterologous bacterial expression systems to produce mutagenized extrinsic PS II proteins in *Escherichia coli* followed by their purification and reconstitution of extrinsic protein-depleted PS II membranes.[8,9]

In this article, we describe the techniques we have used in our laboratory to introduce directed mutations into the chlorophyll-protein CP 47 (Fig. 1) of the cyanobacterium *Synechocystis* 6803. This protein is an integral membrane component of the proximal chlorophyll *a* antenna of PS II, which also interacts with the oxygen-evolving site.[10] We wished to investigate the roles of the conserved, charged residues located principally in the large extrinsic loop E of CP 47, which appears to be lumenally exposed. This domain has been examined using a variety of biochemical techniques and there exists a strong body of evidence indicating that this portion of CP 47 interacts with components required for oxygen evolution.[10,11] Site-directed mutations were introduced into this region of CP 47 using the method of Kunkel.[12] Random mutations directed against the large extrinsic loop domain of CP 47 were introduced using the mutator strain of *E. coli* XL-1 Red (Stratagene, La Jolla, CA).[13] An excellent and detailed description of the methods and strategies used for the introduction of site-directed mutations in the D1 protein of *Synechocystis* 6803 was presented in this

[1] R. J. Debus, B. A. Berry, I. Sithole, G. T. Babcock, and L. McIntosh, *Biochemistry* **27,** 9071 (1988).
[2] R. J. Debus, B. A. Berry, G. T. Babcock, and L. McIntosh, *Proc. Natl. Acad. Sci. U.S.A.* **85,** 427 (1988).
[3] C. Putnan-Evans and T. M. Bricker, *Biochemistry* **31,** 11482 (1992).
[4] M. G. Kuhn and W. F. J. Vermaas, *Plant Mol. Biol.* **23,** 123 (1993).
[5] H. B. Pakrasi, P. D. Ciechi, and J. Whitmarsh, *EMBO J.* **10,** 1619 (1991).
[6] R. L. Burnap, M. Quian, J.-R. Shen, Y. Inoue, and L. A. Sherman, *Biochemistry* **33,** 13712 (1994).
[7] A. Roffey, D. M. Kramer, Govindjee, and R. T. Sayre, *Biochim. Biophys. Acta* **1185,** 257 (1994).
[8] A. Seidler and H. Michel, *EMBO J.* **9,** 1743 (1990).
[9] S. D. Betts, J. R. Ross, E. Pichersky, and C. F. Yocum, *Biochemistry* **35,** 6302 (1996).
[10] T. M. Bricker, *Photosynth. Res.* **24,** 1 (1990).
[11] T. M. Bricker and D. F. Ghanotakis, *in* "Oxygenic Photosynthesis: The Light Reactions" (D. R. Ort and C. F. Yocum, eds.), p. 113. Kluwer Academic Publishers, Dordrecht, 1996.
[12] T. A. Kunkel, *Proc. Natl. Acad. Sci. U.S.A.* **82,** 448 (1985).
[13] A. Greener and M. Callahan, *Strategies* **7,** 32 (1994).

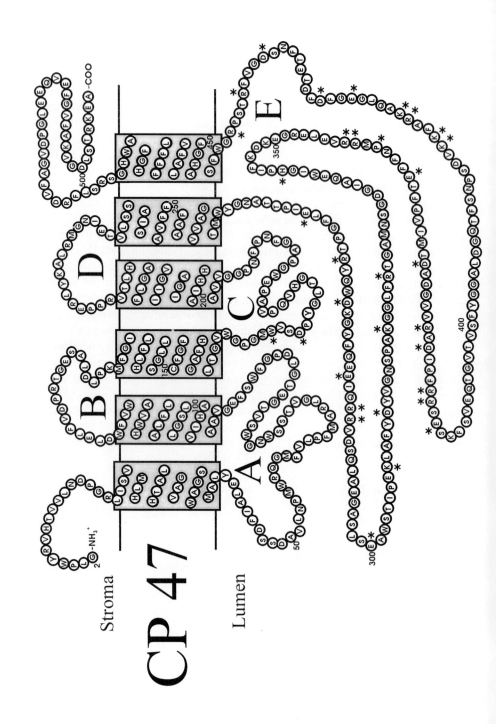

CP 47

Stroma

Lumen

series by Williams,[14] which should be consulted prior to the initiation of any directed mutagenesis studies in this organism.

### Why Directed Mutagenesis?

The principal problem we face in the study of the structure and function of PS II is the absence of high-resolution structural information. Although several groups have reported the crystallization of various PS II preparations,[15,16] no crystals have been obtained that diffract beyond about 10 Å. Consequently, we have had to rely on less direct methods to identify important domains and residues within the various protein subunits of the photosystem. In *Synechocystis* 6803, directed mutagenesis studies are generally performed as follows: Within a given protein, amino acid residues (site-directed mutagenesis) or protein domains (directed random mutagenesis) are identified for modification, the appropriate changes are introduced *in vitro* into plasmid DNA bearing the gene of interest, the mutated DNA is then introduced into a host deletion strain where it integrates into the host's chromosomal DNA by homologous recombination, mutants expressing the targeted alteration are identified, and the phenotype of the mutant is evaluated. Numerous methods are currently available for the introduction of mutations at specific sites in target DNA and a number of methods are currently available for the localized introduction of random mutations.[13,17,18] The use of directed random mutagenesis, however, has been underutilized in the study of PS II structure and function.[19]

Unfortunately, a number of problems are associated with the interpretation of any directed mutagenesis results within PS II. In the ideal situation, these studies are carried out on proteins for which high-resolution structures *are* available for both the wild-type *and* mutant proteins. An excellent example of this type of study has been performed in *Rhodobacter sphae-*

[14] J. G. K. Williams, *Methods Enzymol.* **167,** 766 (1988).
[15] N. Adir, M. Y. Okamura, and G. Feher, *Biophys. J.* **61,** A101 (1992).
[16] C. Fotinou, M. Kokkinidis, G. Fritzsch, W. Haase, E. Michel, and D. F. Ghanotakis, *Photosynth. Res.* **37,** 41 (1993).
[17] R. Pine and P. C. Huang, *Methods Enzymol.* **154,** 415 (1987).
[18] M. J. Zoller and M. Smith, *Methods Enzymol.* **100,** 468 (1983).
[19] S. Ermakova-Gerdes, S. Shestakov, and W. F. J. Vermaas, *Plant Mol. Biol.* **30,** 243 (1996).

FIG. 1. Predicted protein topology of CP 47 and the location of residues modified by site-directed mutagenesis. Shown is the CP 47 protein from *Synechocystis* 6803. The extrinsic loops are identified (A–E). The location of residues that have been modified by site-directed mutagenesis in our laboratory are indicated by asterisks.

*roides.*[20] In the absence of a high-resolution structure of both the wild-type and mutant proteins, it is extremely difficult to differentiate between the direct perturbation of a protein's function brought about by a directed alteration and an indirect perturbation induced by more global structural rearrangements within the target protein induced by the directed mutation. Thus, careful and thorough analysis of the physiological, biochemical, and biophysical phenotype of the resultant mutants is an essential component of directed mutagenesis studies within PS II.

Note that a strong motivation for performing directed random mutagenesis studies is the difficulty in productively choosing appropriate sites for the introduction of site-directed mutations. There is an implicit assumption upon choosing a particular residue for site-directed alteration that we understand the general structural and functional correlates embodied by particular amino acid residues and/or sequences of residues within the secondary and tertiary structural context of the protein. This is certainly not the case. Our understanding of these correlates is rudimentary at best and misleading at worst. Pragmatically, residues are generally targeted for mutagenesis which are highly conserved in all organisms for which sequence information is available for a given gene of interest. The rationale is that if a residue has not changed through the protein's evolutionary history, then it must be important for the structure and/or function of that protein. In reality, alteration of conserved residues usually results in no *apparent* phenotypic change for most proteins examined. In our studies, we have introduced site-directed mutations at 33 residues in CP 47; only 5 of these sites yielded mutants with an apparent phenotype.[21] Note that for a phenotype to be *apparent,* a significant and consistent deviation from the control phenotype, for any given parameter being measured, must be present. Given the rather large variability observed in control *Synechocystis* strains for parameters such as growth rate, oxygen evolution, quantum yield for oxygen evolution, and fluorescence yield, our laboratory requires a minimum deviation of 25–50% from the control values to be considered significant. Of course, from an evolutionary viewpoint, a change that results in a deviation of a fraction of a percent would, if deleterious, be selected strongly against and eliminated from the population.

### Advantages in the Use of *Synechocystis* 6803

As noted earlier, most of the directed mutagenesis studies performed in PS II have utilized *Synechocystis* 6803. This is because the organism

---

[20] A. C. Chirino, E. J. Lous, J. P. Allen, C. C. Schenck, G. Feher, and D. C. Rees, *Biochemistry* **33,** 4584 (1994).

[21] C. Putnam-Evans, J. Wu, and T. M. Bricker, unpublished observations (1996).

exhibits a number of attractive features that permit the relatively simple genetic analysis and manipulation of the photosystem. First, cyanobacterial PS II is very similar to that of higher plants and algae. The major intrinsic proteins of the photosystem and the 33-kDa extrinsic protein are highly homologous to their higher plant counterparts. Additionally, the inorganic and organic cofactors for oxygenic photosynthesis appear to be very similar, if not identical, in both systems. The signature period four oxygen evolution pattern observed with saturating light flashes as well as the characteristic EPR signals associated with PS II (the "multiline" signal, the "g = 4.1" signal, $Y_Z$ and $Y_D$) all appear to be very similar in higher plants and cyanobacteria.[11] Nevertheless, some differences do exist. No homologs for the 24- and 17-kDa extrinsic proteins present in higher plants are found in *Synechocystis*. Two analogous proteins, however, cytochrome $c_{550}$ and a 12-kDa protein, appear to functionally replace the 24- and 17-kDa proteins in cyanobacteria.[22] Additionally, in higher organisms, the light-harvesting array for PS II is the intrinsic light-harvesting chlorophyll–protein complex, whereas in the cyanobacteria the extrinsic phycobilisomes perform this critical function. Finally, the assembly of PS II appears to differ in the two systems. Deletion of the *psbO* gene, which encodes the 33-kDa extrinsic protein in *Chlamydomonas*, leads to a PS II-minus phenotype with only a small amount of assembled PS II being present in the thylakoid membranes.[23] The analogous deletion in *Synechocystis* leads to a decrease in PS II activity but the assembly of 80–90% of functional PS II reaction centers.[24]

Second, *Synechocystis* can grow either photoautotrophically or photoheterotrophically (on glucose in the absence of PS II). This allows for the facile deletion of PS II genes, a prerequisite for the efficient production of directed mutants. Additionally, Photosystem I (PS I)-minus/phycobilisome-minus strains are available.[25] These allow the characterization of mutants within PS II in the absence of PS I. Because most of the chlorophyll present in cyanobacterial thylakoids (85%) is associated with PS I, removal of this photosystem can be beneficial during some types of analysis of PS II parameters.

Finally, *Synechocystis* is naturally trasformable and readily takes up exogenous DNA. If this exogenous DNA contains cyanobacterial sequences it can then be efficiently integrated into the host chromosome by homologous recombination. These features allow the investigator to manipulate

---

[22] J.-R. Shen and Y. Inoue, *Biochemistry* **32,** 1825 (1993).
[23] S. P. Mayfield, P. Bennoun, and J.-D. Rochaix, *EMBO J.* **6,** 313 (1987).
[24] R. L. Burnap and L. A. Sherman, *Biochemistry* **30,** 440 (1991).
[25] G. Shen, S. Boussiba, and W. F. J. Vermaas, *Plant Cell* **5,** 1853 (1993).

cloned *Synechocystis* genes *in vitro,* to easily integrate the altered genes into the *Synechocystis* genome, and to observe the effects of the genetic alterations *in vivo.*

## Bacterial Strains and their Maintenance

A glucose-tolerant strain of *Synechocystis* 6803 is used in these studies.[14] This is maintained on BG-11 media[26] supplemented with 10 m$M$ Tes-KOH, pH 8.2. Glucose (5 m$M$) is added for mixotrophic and photoheterotrophic growth. Solid media is additionally supplemented with 1.5% Difco agar and 3 g/liter sodium thiosulfate. When necessary, kanamycin and spectinomycin are added at concentrations of 10 $\mu$g/ml. Liquid cultures and plates are maintained at a temperature of 30° at 20–50 $\mu$mol photons $(m^2)^{-1}$ sec$^{-1}$ of continuous white light. Liquid cultures are continuously bubbled with sterile, humidified air as previously described.[14] Cultures of *Synechocystis* can be grown autotrophically (with $CO_2$ as the sole carbon source, BG-11 media), mixotrophically (with $CO_2$ and glucose available as carbon sources, BG-11/glucose media), or photoheterotrophically (with glucose as a sole carbon source, BG-11/glucose/dichloromethylurea [DCMU] media). Cultures that are grown photoheterotrophically are supplemented with 10 $\mu$M DCMU to block PS II electron transport. For long-term culture, mutant strains are generally maintained photoheterotrophically on solid media. During transfers we also regularly streak cells onto BG-11/glucose plates to observe mixotrophic growth. The mixotrophically grown cultures are then examined using either an inverted or dissecting microscope for the appearance of revertants and other visible mutations.[27] For long-term storage, liquid cultures are supplemented with 15% glycerol (added from an 80% sterile stock solution), distributed in 1.0-ml aliquots and immediately frozen at −80°.

The *E. coli* strains CJ236, MV1190, InVαF', and DH5α are maintained on Luria broth, which is supplemented with the appropriate antibiotics. CJ236 and MV1190 cells are used during site-directed mutagenesis. CJ236 is *dut⁻ ung⁻*, which allows the incorporation of uracil into DNA. MV1190 is *dut⁺ ung⁺* and is used to remove parental strand uracil-containing DNA

---

[26] M. M. Allen and R. Y. Stanier, *J. Gen. Microbiol.* **51,** 203 (1968).

[27] It is very important to observe the cultures under mixotrophic growth conditions. Revertants and secondary mutations occur often and must be rigorously excluded during mutant characterization. Both growth-rate revertants and color secondary mutants have been observed. Even though the maintenance of mutant cultures under photoheterotrophic conditons prevents the selection and preferential growth of revertants, nevertheless, reversions do occur spontaneously in these cultures.

during plasmid replication. InV$\alpha$F' is used during the cloning of PCR products into pCR-1000 while DH5$\alpha$ serves as a general cloning host strain.

### Construction of pTZ18K3 and pTZ18K9

The primary parental plasmid used for mutagenesis, pTZ18K3, consists of the *Kpn*I/*Kpn*I fragment of the *psb*B gene cloned into the phagemid pTZ18U. The *Kpn*I/*Kpn*I fragment spans nucleotides 1122–2584 of the *psb*B gene, supplied to us in the plasmid pUC*psb*B (kind gift of Dr. Vim Vermaas, Arizona State University). To aid in screening and maintenance of potential mutant phenotypes, a kanamycin resistance gene is inserted into pUC*psb*B at an *Nco*I site located 369 base pairs downstream of the *psb*B coding region. Following digestion of pUC*psb*B with the restriction enzyme *Nco*I, the DNA is ethanol-precipitated, resuspended in a minimal volume, and treated with calf intestinal alkaline phosphatase to prevent resealing of ends during ligation with the kanamycin cartridge. The kanamycin cartridge is prepared by digesting the plasmid pUC4K (Pharmacia Biotech, Inc., Piscataway, NJ) with the restriction enzyme *Eco*RI. Following electrophoresis in a 1% agarose gel, the desired 1.2-kb kanamycin resistance gene is extracted from the gel using GLASSMILK® (Bio101). The overhanging ends generated by *Nco*I are filled in using the Klenow fragment of DNA polymerase. The kanamycin cartridge is then blunt-end ligated to *Nco*I linkers (United States Biochemical). Following digestion of ligation products with *Nco*I and purification from an agarose gel as above, the kanamycin cartridge is ligated into the *Nco*I-digested pUC*psb*B vector. Ligation products are used to transform competent *E. coli* DH5$\alpha$ cells and transformants are selected on LB plates containing 50 $\mu$g/ml kanamycin. Plasmid DNA (designated pUC*psb*B-Kan) is isolated from overnight cultures of kanamycin-resistant colonies using anion exchange columns (Qiagen, Chatsworth, CA), and digested with the restriction enzyme *Kpn*I. The desired 2.5-kb *psb*B-Kan fragment is gel-purified as above, then ligated into *Kpn*I digested pTZ18U. Again, the ligation products are introduced into competent *E. coli* DH5$\alpha$ cells and transformants selected on LB plates containing 50 $\mu$g/ml kanamycin. Plasmid DNA (designated pTZ18K3) is isolated as above and restriction mapped to determine the orientation of the insert. A partial restriction map of pTZ18K3 is shown in Fig. 2.

The second parental plasmid used for mutagenesis, pTZ18K9, consists of a 1.8-kb fragment of the *psb*B gene cloned into pTZ18U. Genomic DNA is isolated from wild-type *Synechocystis* with minor modifications of the procedure outlined by Williams[14] (see later discussion). PCR primers are designed to amplify a 1.8-kb fragment of the *psb*B gene corresponding to

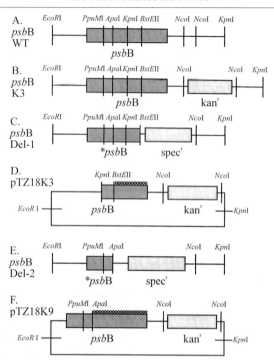

FIG. 2. Partial restriction maps of the *Synechocystis* 6803 and phagemids used in these studies. (A) The genomic *psb*B gene of wild type. (B) The genomic *psb*B gene from the control strain K3. (C) The genomic truncated *psb*B gene of the host strain Del-1. (D) The phagemid pTZ18K3. (E) The genomic truncated *psb*B gene of the host strain Del-2. (F) The phagemid pTZ18K9. The *psb*B genes are darkly shaded; the antibiotic resistance cassettes are lightly shaded. The truncated *psb*B genes are labeled *psbB. The domains on pTZ18K3 and pTZ18K9, which are useful for the introduction of directed mutations, are marked with open circles.

nucleotides 693–2537. This fragment contains virtually all of the sequence found in the 1.4-kb *Kpn*I/*Kpn*I fragment described earlier as well as an additional 0.4 kb of sequence immediately 5' to the *Kpn*I/*Kpn*I fragment. PCR is performed as described later. Following amplification, the 1.8-kb fragment is directly cloned into the pCR-1000 vector (Invitrogen, San Diego, CA). The ligation product is then used to transform competent *E. coli* InVαF' cells (Invitrogen) and colonies containing inserts identified by α-complementation on LB/kanamycin plates containing isopropylthio-galactoside [IPTG] and X-gal. PCR products are routinely cloned in this manner in order to maintain a "hard copy" of desired fragments. Plasmid DNA from overnight cultures is isolated as above, then digested with the restriction enzymes *Hind*III and *Eco*RI. The desired 1.8-kb *psb*B fragment is isolated and ligated into *Hind*III and *Eco*RI-digested pTZ18U. Ligation products are used to transform competent *E. coli* DH5α cells and colonies containing inserts are selected on LB plates containing 50 μg/ml ampicillin, IPTG, and X-gal. Plasmid DNA (designated pTZ18-8) is isolated and sequenced to confirm the presence of the desired 1.8-kb *psb*B fragment and to verify the absence of PCR-induced mutations. The kanamycin resistance gene is isolated as above from pUC4K, filled in with the Klenow fragment, and gel purified. The kanamycin resistance cartridge is then blunt-end ligated into the pTZ18-8 that had been digested with *Nco*I and filled in with the Klenow fragment. Ligation products are used to transform competent *E. coli* DH5α. Transformants are selected on LB/kanamycin plates, and plasmid DNA (designated pTZ18K9) is isolated. A partial restriction map of pTZ18K9 is shown in Fig. 2.

### Construction of *Synechocystis* Control Strain K3

The control strain for all of our mutagenesis studies was a strain of *Synechocystis* identical to wild type with the exception that it contained a kanamycin resistance gene downstream of the *psb*B gene. This control strain was designated K3 and was produced by transformation of wild-type *Synechocystis* with the parental plasmid pTZ18K3 according to the procedure of Williams.[14] Briefly, wild-type cells were scraped from a stock plate and resuspended at a concentration of $1 \times 10^8$ cells/ml in sterile BG-11 media. Then 100-μl aliquots were added to sterile 15-ml polystyrene tubes, followed by the addition of 1–3 μg of plasmid DNA. Following incubation at 25° for 4–6 hr at a light intensity of 20 μmol photons $(m^2)^{-1}$ sec$^{-1}$, transformation mixtures were plated onto BG-11 plates containing 5 m*M* glucose and 10 μ*M* DCMU. After 2 days of growth to allow for expression of the kanamycin resistance gene, plates were underlayed[28] with kanamycin to give a final antibiotic concentration of 10 μg/ml agar media. Using an inverted microscope, mutant colonies at very early stages of growth (10–25 cells) could be observed after only 2–3 days. Visible colonies appeared after approximately 1 week of growth. Individual colonies were selected and restreaked a minimum of three times on BG-11/glucose/DCMU/kanamycin plates in order to allow for sorting out of the mutations.[14] The proper location of the kanamycin resistance cassette was verified by PCR amplification, cloning into the pCR-1000 vector, and sequenc-

---

[28] This is accomplished by using a flame-sterilized spatula to gently lift the agar media. An appropriate amount of antibiotic is then applied in about 200 μl of sterile water underneath the agar.

ing (see below). The phenotype of the control strain K3, with respect to PS II function, was indistinguishable from wild type.

## Construction of *Synechocystis* 6803 *psbB* Partial Deletion Strains Del-1 and Del-2

For the introduction of site-directed mutations, pTZ18K3 or pTZ18K9 plasmid constructs bearing the mutations were used to transform either the *psbB* partial deletion strain Del-1 (pTZ18K3 constructs) or Del-2 (pTZ18K9 constructs). Del-1 and Del-2 are *Synechocystis* strains in which the portion of the *psbB* gene targeted for mutagenesis, along with the 3' flanking region including the kanamycin resistance gene, was deleted and replaced by a gene conferring resistance to spectinomycin (Fig. 2). Thus, in the selection of mutants, a loss of spectinomycin resistance accompanied by acquisition of kanamycin resistance was the primary screen.

The *psbB* partial deletion strain Del-1 can be used for alteration of bases in the region of the gene encoded by nucleotides 1392–1883. This region includes amino acid residues $^{344}$Pro–$^{450}$Trp, which encode the C-terminal half of the large extrinsic loop E. In addition, this region also includes the sixth membrane-spanning $\alpha$-helix ($^{451}$Phe–$^{470}$Gly) as well as the extreme C terminus of the protein ($^{471}$Ser–$^{507}$Ala). The construction of Del-1 has been summarized by us previously.[3] The parental plasmid pTZ18K3 was digested with the restriction enzymes *Bst*EII and *Nco*I, which removed nucleotides 1392–2385 from the *Kpn*I/*Kpn*I fragment of the *psbB* gene. This included the above-stated portions of coding region as well as 0.37 kb of 3' noncoding region and the kanamycin resistance cartridge. It should be pointed out that after digestion with *Bst*EII and *Nco*I, approximately 250 bp of *psbB* sequence 5' to the *Bst*EII site and 200 bp of 3' flanking sequence downstream of the *Nco*I site remained. These flanking sequences are required for efficient homologous recombination. The overhanging ends generated by *Bst*EII and *Nco*I were filled in using the Klenow fragment. The DNA was then ethanol precipitated, resuspended in a minimal volume, and treated with calf intestinal alkaline phosphatase. Following electrophoresis in a 1% agarose gel, the desired 3.26-kb fragment was extracted from the gel using GLASSMILK®. To prepare the spectinomycin cartridge, the plasmid pBR322$\Omega$ (a kind gift from Dr. Alan Biel, Department of Microbiology, Louisiana State University) was digested with the restriction enzyme *Eco*RI and the DNA fragments filled in with the Klenow fragment. The DNA fragments were separated on a 1% agarose gel and the desired 2.0-kb spectinomycin cartridge excised and gel purified. The spectinomycin cartridge was then blunt-end ligated into the pTZ18K3 vector. *Escherichia coli* DH5$\alpha$ cells were transformed with the ligation products and transformants were selected on LB plates containing 50 $\mu$g/ml spectinomycin. Plasmid DNA (designated pTZDel-1) was isolated from overnight cultures of spectinomycin-resistant colonies using anion exchange column chromatography (Qiagen). DNA was sequenced either manually or on an ABI 373A automated DNA sequencer to confirm the sequence of the construct. pTZDel-1 DNA was then used to transform wild-type *Synechocystis* as described earlier, with the exception that selection was on BG-11/glucose/DCMU plates containing spectinomycin at a final concentration of 10 $\mu$g/ml.

Because the Del-1 strain contains the first half of the coding region for the large extrinsic loop (nucleotides 1131–1391), it was unsuitable for the introduction of mutations in this portion of the *psb*B gene. To introduce mutations into nucleotides 1131–1391, the partial deletion strain Del-2 was constructed. To make the deletion construct, pTZ18K9 DNA was digested with the restriction enzymes *Apa*I and *Nco*I, which removed nucleotides 852–2385 from the 1.8-kb fragment of the *psb*B gene. This region included the entire large extrinsic loop coding region. The spectinomycin resistance gene was blunt-end ligated into the vector as described earlier. Following transformation of competent *E. coli* DH5$\alpha$ cells with ligation products and selection of transformants on LB/spectinomycin plates, plasmid DNA was isolated as above. This DNA, designated pTZDel-2, was used to transform wild-type *Synechocystis* to produce the partial deletion strain Del-2.

### Introduction of Site-Directed Mutations into pTZ18K3

Desired site-directed mutations were introduced into pTZ18K3 (or pTZ18K9) by oligonucleotide-mediated mutagenesis using a commercially available kit (Bio-Rad Laboratories) based on the method of Kunkel *et al.*[12] pTZ18K3 (pTZ18K9) was used to transform competent *E. coli* CJ236 cells, in which the genes encoding dUTPase and uracil *N*-glycosylase were inactive, allowing the incorporation of uracil into nascent DNA. Superinfection of CJ236 with helper phage M13K07 resulted in the packaging of single-stranded pTZ18K3 (pTZ18K9) DNA containing uracil. This single-stranded DNA served as the template for complementary strand synthesis using oligonucleotide primers containing the desired mutations. Introduction of the resulting double-stranded plasmids into *E. coli* MV1190 containing a functional dUTPase and uracil *N*-glycosylase resulted in the excision of the nonmutagenized, uracil-containing strand, and replication of the mutagenized strand.

To produce single-stranded uracil-containing DNA, pTZ18K3 (pTZ18K9) is used to transform CJ236, with transformants being selected on LB plates containing 70 $\mu$g/ml kanamycin. One milliliter of a 5-ml overnight culture grown in LB/kanamycin/chloramphenicol (30 $\mu$g/ml) is used to inoculate 50 ml 2× YT/chloramphenicol. (CJ236 must be grown in the presence of chloramphenicol to maintain its F' episome.) The culture is grown to an OD$_{600}$ of 0.3 and then helper phage added to obtain an MOI of 20 phage particles/cell. After 1 hr of incubation, kanamycin is added to a final concentration of 70 $\mu$g/ml, and the culture is allowed to incubate for an additional 6 hr. The culture is then centrifuged at 17,000$g$ for 15 min. Phagemid particles are transferred to a fresh tube, 150 $\mu$g of RNase A added, and the supernatant incubated at room temperature for 30 min. Phagemids are then precipitated by addition of one-quarter volume of 3.5% ammonium acetate/20% PEG-6000 and incubation on ice for 30 min. Phagemids are harvested by centrifugation as above and resuspension of the pellet in 200 $\mu$l of 300 m$M$ NaCl/100 m$M$ Tris, pH 8.0, 1 m$M$ EDTA. DNA is isolated by extraction of phagemid particles 2× in neutralized phenol, 1× with phenol/chloroform, and several times with chloroform/isoamyl alcohol. DNA is precipitated by addition of 1/10 volume of 7.8 $M$ ammonium acetate and 2.5 volumes of ethanol and incubation at −70° for 30 min. Following centrifugation, the DNA pellets are resuspended in 20 $\mu$l TE (10 m$M$ Tris, pH 8.0, 1 m$M$ EDTA).

Annealing of mutagenic oligonucleotides occurs in a 10-$\mu$l reaction containing 200 ng of single-stranded phagemid DNA and 6–9 pmol of phosphorylated mutagenic primer in annealing buffer (20 m$M$ Tris, pH 7.4, 2 m$M$ MgCl$_2$, 50 m$M$ NaCl). The mutagenic primers are generally 36- to 42-mers with at least 15 fully complementary bases both upstream and downstream from the introduced mutation. These mutations incorporated, where possible, a two-base change to the appropriate mutant DNA codon. Annealing reactions are incubated at 70° for 1 min, then allowed to cool to room temperature. For synthesis of the complementary strand, the following are added to the annealing reactions which had been placed in an ice water bath: 0.4 m$M$ of each dNTP, 0.75 m$M$ adenosine triphosphate (ATP), 17.5 m$M$ Tris, pH 7.4, 3.75 m$M$ MgCl$_2$, 1.5 m$M$ dithiothreitol (DTT), 3 units of T4 DNA ligase, and 0.5 unit of T7 DNA polymerase. Reactions are incubated on ice for 5 min, at 25° for 5 min, and then at 37° for 30 min. Reactions are terminated by the addition of 60 $\mu$l of 10 m$M$ Tris, pH 8.0, 10 m$M$ EDTA followed by freezing of the sample at −20°. Three microliters of the second strand synthesis reactions are used to transform competent E. coli MV1190 and transformants are selected on LB/kanamycin plates. Plasmid DNA is isolated and sequenced to confirm the presence of the intended mutations prior to transformation into one of the two Synechocystis psbB partial deletion strains. The transformation of these partial deletion strains is performed as previously described.

## Verification of Mutants

After several rounds of streaking on BG-11/glucose/DCMU/kanamycin plates to allow sorting out of the multiple chromosomal copies of the *Synechocystis* genomic DNA, putative mutants were identified by the loss of spectinomycin resistance and the acquisition of kanamycin resistance. The kanamycin-resistant colonies were tested for spectinomycin sensitivity on solid media. Genomic DNA was then isolated from the kanamycin-resistant, spectinomycin-sensitive putative mutants with minor modifications of the procedure of Williams[41]: Fifty milliliters of cells of a log-phase culture are spun (11,000g for 5 min) and resuspended in 2 ml saturated NaI. Cells are incubated at 37° for 20 min, then 40 ml of deionized water is added and the cells spun again. The pellet is resuspended in a total of 1 ml of 50 m$M$ Tris, pH 8.5, 50 m$M$ NaCl, and 5 m$M$ EDTA followed by the addition of 0.5 ml lysozyme (80 mg/ml) and 20 $\mu$l of RNase A (10 mg/ml). The mixture is incubated for 10 min at 37°, followed by the addition of 0.5 ml protease K (1 mg/ml) and incubated an additional 20 min. One-half milliliter of a 10% lauroylsarcosine solution is then added and the mixture is incubated an additional 20 min at 37°. The cell lysate is then extracted 2× in an equal volume of phenol:chloroform and 1× with chloroform. Genomic DNA is precipitated with one-tenth volume of sodium acetate and two volumes of ethanol. DNA is resuspended in 100 $\mu$l TE.

Oligonucleotides flanking the *Kpn*I/*Kpn*I fragment of the *psb*B gene are used to amplify this region from the genomic DNA using the polymerase chain reaction. PCR is performed in 100-$\mu$l reactions containing 20 m$M$ Tris, pH 8.4, 50 m$M$ KCl, 20 pmol of each primer, 4 m$M$ MgCl$_2$, 1 m$M$ dNTPs, 2 units of *Taq* DNA polymerase (Gibco-BRL, Gaithersburg, MD), and varied amounts of the isolated genomic DNA (10–50 ng). The thermal cycling routine consists of the following steps: 1 min denaturation at 93°, 45 sec annealing at 65°, and 2 min elongation at 72° for a total of 20 cycles. The PCR products are then directly ligated into the pCR-1000 vector (Invitrogen). The resulting plasmids are pooled and then sequenced to verify that the intended mutation has been introduced and that no PCR-induced mutations are evident.

## Localized Random Mutagenesis with XL-1 Red *E. coli* Strain

While site-directed mutagenesis has proven to be a very powerful tool in identifying structurally or functionally important amino acid residues, only a limited number of residues can be practically targeted in this approach. For a given protein, especially when critical domains or residues have not been identified, prescreening with random mutagenesis is an attractive approach.

Chemical treatment or UV radiation are conventional methods for the production of random mutants. In both cases, however, the mutation frequency is relatively low and is difficult to control. Additionally, these treatments are not entirely random due to mutational hot spots existing within the DNA. Recently, many laboratories have begun using PCR to generate random mutations in genes.[29] This method exploits the inherent infidelity of *Taq* DNA polymerase. However this method is time consuming, expensive, and only a limited number of random mutants can be generated.

In our laboratory, we have used the *E. coli* strain XL-1 Red (Stratagene) to introduce random mutations within DNA. This strain is deficient in three of the primary DNA repair pathways: *mutS* (error-prone mismatch repair), *mutD* (deficient in 3'-5' exonuclease activity of DNA polymerase III), and *mutT* (unable to hydrolyze 8-oxo-dGTP). Because of these defects, this strain exhibits a mutation frequency ~5000-fold higher than that of wild type. The advantage of using this strain is that one can adjust the mutation frequency in the target plasmid by controlling the growth parameters of the host XL-1 Red cells. The longer the period of time that the target plasmid is maintained in XL-1 Red cells, the more mutations accumulate in the plasmid DNA. By introducing the phagemid pTZ18K3 into this strain, we found that the following procedure could give rise to satisfactorily random mutants in the loop E region of CP 47.

Competent *E. coli* XL-1 Red cells were provided by the manufacturer (Stratagene). Because of the inherent high mutation rate of this strain, it is not possible to maintain stock cultures. The frozen competent cells are thawed on ice and 100-$\mu$l aliquots are distributed into prechilled 15-ml Falcon 2059 polypropylene tubes (one tube per transformation). Then 1.7 $\mu$l 2-mercaptoethanol is added to the cells and the tubes gently agitated every 2 min for 10 min. Phagemid DNA (pTZ18K3) (10–50 $\mu$g) is added to each tube. The tubes are swirled gently and incubated on ice for 30 min. The transformation mixture is given a 42° heat pulse in a water bath for 45 sec, then put back on ice for 2 min. Then 0.9 ml of preheated (42°) SOC medium is added and the transformation mixture is incubated at 37° for 1 hr with shaking at 225–250 rpm. Variable amounts of the transformation mixture are plated onto the LB plates containing 70 $\mu$g/ml kanamycin. These are incubated at 37° for 24–30 hr. Note that the XL-1 Red strain grows relatively slowly and additional time is often needed for colonies to appear.

---

[29] See H. Kless and W. F. J. Vermaas, *Biochemistry* **35,** 648 (1996), for an example in PS II.

Two hundred kanamycin-resistant colonies are picked at random from the transformation plates and used to inoculate (*en masse*) 5 ml of LB medium containing 70 μg/ml kanamycin. The cells are then grown overnight at 37° with agitation. This 5-ml culture is then used to inoculate 100 ml LB medium containing 70 μg/ml kanamycin, and this culture is again grown at 37° overnight. About 100 μg of phagemid DNA is isolated from this 100-ml overnight culture using anion exchange column chromatography (Qiagen).

Transformation of the host Del-1 strain of *Synechocystis* 6803 with the randomly mutated pTZ18K3 phagemid DNA is performed as described previously for the site-directed mutagenesis studies. The *Synechocystis* transformation mixture is spread onto BG-11/glucose/DCMU plates for about 20 hr. At this time, the agar is underlayed with 200 μl of kanamycin (1 mg/ml). After about 2 weeks of growth, individual colonies are selected and streaked six times sequentially on BG-11/glucose/DCMU/kanamycin plates to allow sorting out of the chromosomal DNA copies.[14]

To screen for the loss or alteration of photoautotrophic growth, individual colonies are streaked on BG-11 plates that do not contain glucose or DCMU. Colonies that grow normally on BG-11/glucose/DCMU media but exhibited little or no photoautotrophic growth on BG-11 plates are selected for further study.

Since we used the randomly mutated pTZ18K3 to transform the deletion strain Del-1, only random mutations within the partial *psb*B gene should have been introduced into the genomic DNA of the cyanobacterium. About 2000 kanamycin-resistant clones were screened for photoautotrophic growth on BG-11 plates. Of these 24 exhibited aberrant photoautotrophic growth and were further characterized (Table I). To verify that the random mutation(s) was introduced into the appropriate domain of the *psb*B gene, individual *Synechocystis* mutants were transformed, as described above, with wild-type pTZ18K3. If the random mutation lay within the *psb*B sequence encoded on this phagemid, high-frequency reversion (>100 times the background reversion rate) to the wild-type phenotype (i.e., photoautotrophic growth) should be observed. This was the case for all of the random mutants examined. To identify the site of the introduced mutations, genomic DNA was isolated from several mutant cell lines as described earlier and PCR was used to amplify the *Kpn*I/*Kpn*I fragment of the *psb*B gene. After cloning into the pCR-1000 vector and the transformation of competent InVαF' *E. coli*, plasmid DNA was isolated from pooled *E. coli* colonies and sequenced to identify the site(s) of the random mutation(s). The results (Table I) indicate that the XL-1 Red mutator strain can be used effectively to incorporate random mutation into targeted PSII sequences.

TABLE I

PRELIMINARY CHARACTERIZATION OF DIRECTED RANDOM MUTANTS IN THE
psbB GENE OF SYNECHOCYSTIS 6803

| Mutant | Photoautotrophic growth | Oxygen evolution (%) | Transformation with pTZ18K3 | Mutation |
|---|---|---|---|---|
| Control, K3 | +++[a] | 100[b] | na[c] | na |
| RM1 | 0 | 35 | Yes | H455T |
| RM3 | 0 | 0 | Yes | H466R, F432L |
| RM4 | ++ | 50 | nd[d] | nd |
| RM5 | 0 | 0 | nd | nd |
| RM6 | 0 | 0 | nd | nd |
| RM7 | + | 30 | nd | R4485 |
| RM8 | 0 | 0 | Yes | W450Stop |
| RM9 | + | 25 | Yes | H455Y |
| RM10 | 0 | 0 | Yes | nd |
| RM11 | 0 | 0 | Yes | nd |
| RM12 | 0 | 0 | Yes | W302Stop |
| RM13 | 0 | 0 | Yes | Q409Stop |
| RM14 | 0 | 0 | nd | nd |
| RM15 | 0 | 0 | Yes | W302Stop |
| RM16 | ++ | 10 | nd | nd |
| RM17 | 0 | 0 | Yes | nd |
| RM18 | 0 | 0 | Yes | nd |
| RM19 | + | 30 | nd | nd |
| RM20 | 0 | 0 | Yes | nd |
| RM21 | 0 | 30 | Yes | H469Y |
| RM22 | 0 | 0 | Yes | nd |
| RM23 | 0 | 0 | Yes | nd |
| RM24 | 0 | 0 | Yes | nd |

[a] +++, wild-type growth rate; ++, moderately affected; +, strongly affected; 0, no growth.
[b] Control oxygen evolution rate was 400 $\mu$mol photons $(m^2)^{-1}$ $sec^{-1}$.
[c] na, not applicable.
[d] nd, not determined.

## Characterization of Directed Mutations

The methods used to characterize directed mutations in *Synechocystis* 6803 cannot be adequately discussed in the space allotted. In our laboratory, we include characterization of photoautotrophic and photoheterotrophic growth rate, growth under chloride- and calcium-limiting conditions, steady-state oxygen evolution and quantum yield measurements, flash oxygen yield measurements and S-state lifetime determinations, photoinactivation studies, fluorescence yield measurements, [14C] atrazine-binding determinations, and the analysis of the polypeptide composition of the cyanobacterial thylakoid membranes. These studies have allowed us to identify domains involved in the association of CP 47 with the 33-kDa extrinsic protein,[30] residues which appear to be involved in the regulation of the chloride requirement for PS II activity,[31] and to identify regions which may be important in PS II assembly.[32]

## Acknowledgments

This work was supported by National Science Foundation grants to TMB and CPE. Special thanks to Ms. Laurie K. Frankel for her critical reading of this manuscript.

[30] C. Putnam-Evans, J. Wu, R. Burnap, J. Whitmarsh, and T. M. Bricker, *Biochemistry* **35,** 4046 (1996).
[31] C. Putnam-Evans and T. M. Bricker, *Biochemistry* **33,** 10770 (1994).
[32] J. Wu, C. Putnam-Evans, and T. M. Bricker, *Plant Mol. Biol.* **32,** 537 (1996).

# [23] Application of Spectroscopic Techniques to the Study of Photosystem II Mutations Engineered in *Synechocystis* and *Chlamydomonas*

*By* Bruce A. Diner

## Introduction

In recent years, site-directed mutagenesis has been used extensively for the purpose of examining structure–function relationships in the reaction centers of Photosystem II (PSII).[1-4] This effort has been guided and stimulated by a number of developments: (1) by determination of the X-ray crystallographic structure of the reaction centers of the purple nonsulfur photosynthetic bacteria which have substantial homology to PSII[5-10];

[1] J. G. K. Williams, *Methods Enzymol.* **167,** 766 (1988).
[2] S. D. Carpenter and W. F. J. Vermaas, *Physiol. Plant.* **77,** 436 (1989).
[3] P. J. Nixon, D. A. Chisholm, and B. A. Diner, *in* "Plant Protein Engineering" (P. R. Shewry and S. Gutteridge, eds.), p. 93. Cambridge University Press, Cambridge, 1992.
[4] H. P. Pakrasi and W. F. J. Vermaas, *in* "The Photosystems: Structure, Function and Molecular Biology" (J. Barber, ed.), p. 231. Elsevier, Amsterdam, 1992.
[5] H. Michel and J. Deisenhofer, *Biochemistry* **27,** 1 (1988).
[6] J. P. Allen, G. Feher, T. O. Yeates, H. Komiya, and D. C. Rees, *Proc. Natl. Acad. Sci. U.S.A.* **84,** 5730 (1987).
[7] O. El-Kabbani, C. H. Chang, D. Tiede, J. Norris, and M. Schiffer, *Biochemistry* **30,** 5361 (1991).
[8] A. Trebst, *Z. Naturforsch* **41c,** 240 (1986).

(2) by the relative ease of genetic transformation of certain photosynthetic organisms[1,11-13]; (3) by the substantial progress in the techniques of mutant construction, including the engineering of host deletion strains,[14-19] of intron-free copies of reaction center genes,[18,19] and of antibiotic resistance cassettes for selection[3,20]; and (4) the availability of PSII-enriched biochemical preparations.[21-30] These developments have been particularly marked for two organisms, the cyanobacterium, *Synechocystis* PCC 6803, and the green alga, *Chlamydomonas reinhardtii,* and have made them the organisms of choice in the site-directed mutagenesis of the PSII reaction center.

Quite a range of spectroscopic techniques has been applied to the characterization of PSII mutants.[31] These have been applied to whole cells,

[9] S. V. Ruffle, D. Donnelly, T. L. Blundell, and J. H. A. Nugent, *Photosynth. Res.* **34,** 287 (1992).

[10] B. Svensson, C. Etchebest, P. Tuffery, P. van Kan, J. Smith, and S. Styring, *Biochemistry* **35,** 14486 (1996).

[11] G. Grigorieva and S. Shestakov, *FEMS Microbiol. Lett.* **13,** 367 (1982).

[12] J. Labarre, F. Chauvat, and P. Thuriaux, *J. Bacteriol.* **171,** 3449 (1989).

[13] J. Boynton, N. Gillham, E. Harris, J. Hosler, A. Johnson, A. Jones, B. Randolph-Anderson, D. Robertson, T. Klein, K. Shark, and J. Sanford, *Science* **240,** 1534 (1988).

[14] R. J. Debus, A. P. Nguyen, and A. B. Conway, *in* "Current Research in Photosynthesis" (M. Baltscheffsky, ed.), Vol. I, p. 829. Kluwer Academic Publishers, Dordrecht, 1990.

[15] W. Vermaas, J. Charité, and B. Egger, *in* "Current Research in Photosynthesis" (M. Baltscheffsky, ed.), Vol. I, p. 231. Kluwer Academic Publishers, Dordrecht, 1990.

[16] P. J. Nixon, J. T. Trost, and B. A. Diner, *Biochemistry* **31,** 10859 (1992).

[17] X.-S. Tang, D. A. Chisholm, G. C. Dismukes, G. W. Brudvig, and B. A. Diner, *Biochemistry* **32,** 13742 (1993).

[18] U. Johanningmeier and S. Heiss, *Plant Mol. Biol.* **22,** 91 (1993).

[19] J. Minagawa and A. R. Crofts, *Photosynth. Res.* **42,** 121 (1994).

[20] M. Goldschmidt-Clermont, *Nucleic Acids Res.* **19,** 4083 (1991).

[21] R. L. Burnap, H. Koike, G. Sotiropoulou, L. A. Sherman, and Y. Inoue, *Photosynth. Res.* **22,** 123 (1989).

[22] G. H. Noren, R. J. Boerner, and B. A. Barry, *Biochemistry* **30,** 3943 (1991).

[23] D. I. Kirilovsky, A. G. P. Boussac, F. J. E. van Mieghem, J.-M. R. C. Ducruet, P. R. Sétif, J. Yu, W. F. J. Vermaas, and A. W. Rutherford, *Biochemistry* **31,** 2099 (1992).

[24] M. Rögner, P. J. Nixon, and B. A. Diner, *J. Biol. Chem.* **265,** 6189 (1990).

[25] X.-S. Tang and B. A. Diner, *Biochemistry* **33,** 4594 (1994).

[26] H. Shim, J. Cao, Govindjee, and P. G. Debrunner, *Photosynth. Res.* **26,** 223 (1990).

[27] B. A. Diner and F.-A. Wollman, *Eur. J. Biochemistry* **110,** 521 (1980).

[28] L. B. Giorgi, P. J. Nixon, S. A. P. Merry, D. M. Joseph, J. R. Durrant, J. De Las Rivas, J. Barber, G. Porter, and D. R. Klug, *J. Biol. Chem.* **271,** 2093 (1996).

[29] L. B. Giorgi, J. R. Durrant, S. Alizadeh, P. J. Nixon, D. M. Joseph, T. Rech, J. Barber, G. Porter, and D. Klug, *Biochim. Biophys. Acta* **1186,** 247 (1994).

[30] S. Alizdeh, P. J. Nixon, A. Telfer, and J. Barber, *Photosynth. Res.* **43,** 165 (1995).

[31] J. Amesz and A. J. Hoff, *in* "Biophysical Techniques in Photosynthesis," Kluwer Academic Publishers, Dordrecht, 1995.

thylakoids, core complexes, and reaction centers. The techniques discussed here will emphasize those that have most contributed to mutant characterization. They include measurement of the variation of the chlorophyll fluorescence yield, thermoluminescence, ultraviolet/visible (UV/Vis) optical spectroscopy, Fourier transform infrared spectroscopy (FTIR), and the magnetic resonance techniques of electron paramagnetic resonance (EPR), electron spin echo envelope modulation (ESEEM) and cw and pulsed electron nuclear double resonance (ENDOR).

To facilitate the discussion that follows we will refer, by their symbols, to the components that comprise the electron donor and electron acceptor sides of the PSII reaction center. These are shown in their respective positions within the electron transport chain, with the arrows indicating the direction of electron flow:

$$\text{electron acceptors}$$

$$\text{OEC (Mn}_4) \rightarrow Y_Z \rightarrow P680 \xrightarrow{h\nu} \text{Pheo} \rightarrow Q_A \rightarrow Q_B$$

$$\text{electron donors}$$

where OEC ($Mn_4$) is the tetranuclear manganese cluster that comprises the oxygen evolving complex; $Y_Z$ is a redox active tyrosine (D1-Tyr161); P680 is the primary electron donor chlorophyll(s); Pheo is the primary electron acceptor pheophytin, and $Q_A$ and $Q_B$ are, respectively, the primary and secondary plastoquinone electron acceptors.

## Fluorescence Kinetics

The measurement of the increase of the chlorophyll fluorescence yield with time upon continuous illumination (fluorescence induction) is one of the easiest and most noninvasive of the analytical techniques of photosynthesis and is easily applied to whole cells.[32–34] The primary phenomenon responsible for the variation in the fluorescence yield is the photochemical reduction of the primary quinone electron acceptor, $Q_A$, with the fluorescence yield increasing from a minimum, $F_0$, where $Q_A$ is oxidized, to a maximum, $F_m$, where $Q_A$ is reduced. Other redox species that impact the chlorophyll fluorescence yield of Photosystem II and that are indicators of lesions in PSII electron transfer are the primary electron donor, P680,[35,36]

[32] G. H. Krause and E. Weis, *Annu. Rev. Plant Physiol. Plant Mol. Biol.* **42**, 313 (1991).
[33] J. Lavergne and H.-W. Trissl, *Biophys. J.* **68**, 2474 (1995).
[34] L. N. M. Duysens and H. E. Sweers, *in* "Studies in Microalgae and Photosynthetic Bacteria" (S. Miyachi, ed.), p. 353. University of Tokyo Press, Japan, 1963.
[35] D. Mauzerall, *Proc. Natl. Acad. Sci. U.S.A.* **69**, 1358 (1972).
[36] W. L. Butler, J. W. M. Visser, and H. L. Symons, *Biochim. Biophys. Acta* **292**, 140 (1973).

and the plastoquinone pool,[37] both of which quench fluorescence in their cation radical and quinone forms, respectively. The kinetics of increase of the fluorescence induction curve are also determined by such nonredox factors as the connectivity of the light-harvesting antenna to the PSII reaction center[38] (the optical cross-section[39]) and energy transfer between centers.[33,40] Where there is energy transfer between PSII centers, the fluorescence yield increases nonlinearly with the reduction of $Q_A$, as originally described by Joliot and Joliot.[40] Where there is no energy transfer between PSII centers, the fluorescence yield is a linear function of $[Q_A^-]$. More typically, in whole cells and thylakoids, there is energy transfer producing a nonlinear dependence on $[Q_A^-]$. For a detailed discussion of the relationship between reaction center/antenna energetics and fluorescence, see reviews by Renger,[41] van Grondelle et al.,[42] Diner and Babcock,[43] and Sauer and Debreczeny.[44] For a discussion of the non-photochemical quenching factors that affect the fluorescence yield of PSII, see Krause and Weis[32] and Horton.[45]

In its simplest form, a combination excitation/detection beam of <680 nm is used to excite chlorophyll fluorescence. The chlorophyll fluorescence emission is monitored at >680 nm and often at right angles to the light source using a photomultiplier or photodiode on which is mounted a suitable high bandpass filter to block the actinic excitation. The right-angle geometry minimizes direct illumination of the blocking filter, reducing filter fluorescence. The combination excitation/detection beam can be gated on and off in the case of a continuous beam by using a shutter (transient time, ~1 ms) or by switching on and off a cluster of light-emitting diodes (transient time, ≤1 ms). The variation with time of the variable fluorescence is monitored using a digital or analog data storage device.

An alternative to this method is that provided by the "pulsed amplitude

[37] C. Vernotte, A.-L. Etienne, and J.-M. Briantais, *Biochim. Biophys. Acta* **545,** 519 (1979).
[38] C. Bonaventura and J. Myers, *Biochim. Biophys. Acta* **189,** 366 (1969).
[39] A. C. Ley and D. C. Mauzerall, *Biochim. Biophys. Acta* **680,** 95 (1982).
[40] A. Joliot and P. Joliot, *Comp. Rend. Acad. Sci. Paris* **258,** 4622 (1968).
[41] G. Renger, *in* "The Photosystems: Structure Function and Molecular Biology" (J. Barber, ed.), p. 45. Elsevier, Amsterdam, 1992.
[42] R. van Grondelle, J. P. Dekker, T. Gilbro, and V. Sundstrom, *Biochim. Biophys. Acta* **1187,** 1 (1994).
[43] B. A. Diner and G. T. Babcock, *in* "Oxygenic Photosynthesis: The Light Reactions" (D. R. Ort and C. F. Yocum, eds.), p. 213. Kluwer Academic Publishers, Dordrecht, 1996.
[44] K. Sauer and M. Debreczeny, *in* "Biophysical Techniques in Photosynthesis" (J. Amesz and A. J. Hoff, eds.), p. 41. Kluwer Academic Publishers, Dordrecht, 1996.
[45] P. Horton, *in* "Light as an Energy Source and Information Carrier in Plant Physiology" (R. C. Jennings, G. Zucchelli, F. Ghetti, and G. Colombetti, eds.), pp. 99. Plenum Press, New York, 1966.

modulation" (PAM)[46] technique where the detection is provided by a rapidly modulated collection of light-emitting diodes (LEDs) and the actinic illumination is provided by a continuous source or by a light flash. Modulation-sensitive detection makes for selective detection of the detecting beam even in the presence of intense actinic illumination.

The fluorescence yield monitors a complex array of donor and acceptor side electron transfer rates that modulate the redox state of $Q_A$, P680, and the plastoquinone pool and which can be distinguished by using light of different intensities as well as by the addition of electron transport inhibitors and electron donors. Use of this methodology as a mutant screening technique has been described by Bennoun and Delepelaire[47] in which individual colonies on petri plates could be excited using a finely focused shuttered beam of continuous light (e.g., using a microscope condenser). A further refinement has been recently developed by Fenton and Crofts[48] and by Bennoun and Béal[49] in which a CCD camera simultaneously images an entire plate, uniformly illuminated by a quartz-halogen lamp and by an array of LEDs, respectively. The fluorescence induction of all of the colonies is measured and stored simultaneously and can be individually or collectively viewed on a computer monitor. In this method, the sampling rate is typically of the order of 30 ms.

The fluorescence induction kinetics in the absence of an electron transport inhibitor is capable of revealing rate limitations on the electron donor or acceptor side of PSII. In the absence of any change in the size or connectivity of the light-harvesting antenna, a slowed rise of the fluorescence induction curve and a lowered steady-state fluorescence yield compared to wild type is an indication of a slowing of electron transfer on the donor side of the reaction center. An accelerated rise and an elevated steady-state fluorescence yield is an indication of a slowing of $Q_A^-$ oxidation on the acceptor side of the reaction center. In the presence of an inhibitor of $Q_A^-$ oxidation (e.g., diuron, atrazine), the fluorescence induction kinetics are a measure of the size of the PSII light-harvesting antenna. This type of analysis is easier to do in *Chlamydomonas* than in *Synechocystis*, owing to the substantially greater $F_v/F_m$ of the former than the latter. $F_v$ is the difference between $F_m$ and $F_0$, and $F_v/F_m$ is smaller in *Synechocystis* because of PSI fluorescence quenching of $F_m$ in an organism where PSI/PSII is 3- to 5-fold higher than in *Chlamydomonas*.

[46] U. Schreiber, *Photosynth. Res.* **9**, 261 (1986).
[47] P. Bennoun and P. Delepelaire, *in* "Methods in Chloroplast Molecular Biology" (M. Edelman, R. B. Hallick, and N.-H. Chua, eds.), p. 25. Elsevier Biomedical Press, Amsterdam, 1982.
[48] J. M. Fenton and A. R. Crofts, *Photosynth. Res.* **26**, 59 (1990).
[49] P. Bennoun and D. Béal, *Photosynth. Res.* **51**, 161 (1997).

Examples of mutants that show low steady-state fluorescence yields as an indicator of impaired donor side function are the LF-1 mutant of *Scenedesmus*[50] with a mutation in its D1 processing protease[50a] and site-directed mutants of *Chlamydomonas* (at D1-Asp170).[51] These lack, respectively, all and to varying degrees a functional manganese cluster responsible for water oxidation. The lowered fluorescence yield caused by the electron donor side limitation is a reflection of an enhanced concentration of P680$^+$.[36] An example of a mutant that shows an accelerated rise in the fluorescence induction curve as an indicator of impaired acceptor side function is the herbicide-resistant strain of *Chlamydomonas,* Br24 (D1-Gly256Asp), with slowed electron transfer from $Q_A^-$ to $Q_B$.[52] The elevated fluorescence yield caused by the rate limitation on the electron acceptor side is a reflection of an enhanced concentration of $Q_A^-$. Centers containing $Q_A^-$ are unable to discharge, except by charge recombination, the primary charge pair, P680$^+$Pheo$^-$, thought to be within ~70 meV of P680*,[53] the lowest excited singlet state of P680. P680* can thus be reformed by charge recombination, which in turn is nearly isoenergetic with the excited state chlorophyll (chl*) of the PSII antenna. The consequent enhanced lifetime of chl* in the antenna/reaction center complex is reflected in the enhancement of the overall fluorescence yield.[32,41-44]

Kinetic analysis of PSII secondary electron transfer rates can be readily obtained using a pump/probe technique. Here a saturating single turnover actinic light flash is followed by low-intensity xenon flashes[54-58a] or pulsed photodiodes to detect the fluorescence relaxation.[46,59] Algal cells are dark adapted in the presence of 0.3 m$M$ *p*-benzoquinone (purified by sublimation) plus 0.3 m$M$ K$_3$Fe(CN)$_6$, which together ensure the oxidation of the plastoquinone pool and a synchronization of the electron acceptor side of the reaction center in the $Q_AQ_B$ state.[58,58a] After a 10-min incubation, the

[50] J. G. Metz, T. M. Bricker, and M. Seibert, *FEBS Lett.* **185,** 191 (1985).

[50a] J. T. Trost, D. A. Chisholm, D. B. Jordan, and B. A. Diner, *J. Biol. Chem.* **272,** 29348 (1997).

[51] J. P. Whitelegge, D. Koo, B. A. Diner, I. Domian, and J. M. Erickson, *J. Biol. Chem.* **270,** 225 (1995).

[52] J. M. Erickson, K. Pfister, M. Rahire, R. K. Togasaki, L. Mets, and J.-D. Rochaix, *Plant Cell* **1,** 361 (1989).

[53] V. V. Klimov and A. A. Krasnovskii, *Photosynthetica* **15,** 592 (1981).

[54] R. Delosme, *in* "Proc. 2nd Intl. Congress Photosynthesis Research" (G. Forti, M. Avron, and A. Melandri, eds.), p. 187. Dr. W. Junk, The Hague, 1971.

[55] A. Joliot, *Biochim. Biophys. Acta* **357,** 439 (1974).

[56] H. H. Robinson and A. R. Crofts, *FEBS Lett.* **153,** 221 (1983).

[57] D. M. Kramer, H. R. Robinson, and A. R. Crofts, *Photosynth. Res.* **26,** 181 (1990).

[58] F. A. Wollman, *Biochim. Biophys. Acta* **503,** 263 (1978).

[58a] P. J. Nixon and B. A. Diner, *Biochemistry* **31,** 942 (1992).

[59] H.-A. Chu, A. P. Nguyen, and R. J. Debus, *Biochemistry* **33,** 6137 (1994).

cells are subjected to a train of saturating microsecond-duration blue light flashes, typically provided at a frequency of 1–20 Hz. This kind of analysis can also be applied to isolated chloroplasts or thylakoids containing low concentrations of $K_3Fe(CN)_6$.[56,60] The evolution of the fluorescence yield is then recorded following each of the actinic flashes. The rate of fluorescence relaxation, corrected for energy transfer, reflects the rate of oxidation of $Q_A^-$.

Equation (1) shows the reaction following the odd-numbered flashes, while Eq. (2) shows that following the even-numbered flashes. In the case of *Chlamydomonas,* this relaxation rate needs to be corrected for energy transfer to obtain the true rate of electron transfer, whereas in the case of *Synechocystis,* where energy transfer between centers is much less marked, the fluorescence relaxation is a more direct reflection of the kinetics of $Q_A^-$ oxidation.

$$Q_A^- + PQ \underset{kq_{off}}{\overset{kq_{on}}{\rightleftharpoons}} Q_A^- Q_B \underset{kb_{AB(1)}}{\overset{kf_{AB(1)}}{\rightleftharpoons}} Q_A Q_B^- \tag{1}$$

$$Q_A^- Q_B^- + 2H^+ \underset{kb_{AB(2)}}{\overset{kf_{AB(2)}}{\rightleftharpoons}} Q_A Q_B H_2 \tag{2}$$

$$Q_A Q_B H_2 \underset{kgh_{2_{on}}}{\overset{kgh_{2_{off}}}{\rightleftharpoons}} Q_A + PQH_2 \tag{3}$$

$$K_{AB(1)} = \frac{kf_{AB(1)}}{kb_{AB(1)}} \tag{4}$$

$$K_{AB(2)} = \frac{kf_{AB(2)}}{kb_{AB(2)}} \tag{5}$$

$$K_q = \frac{kq_{off}}{kq_{on}} \tag{6}$$

$$K_{qh_2} = \frac{kqh_{2_{off}}}{kqh_{2_{on}}} \tag{7}$$

$$K'_{AB(1)} = \frac{K_{AB(1)}}{1 + K_q/[PQ]} \tag{8}$$

$$K'_{AB(2)} = K_{AB(2)}(1 + K_{qh_2}/[PQH_2]) \tag{9}$$

[60] S. Taoka and A. R. Crofts, *in* "Current Research in Photosynthesis" (M. Baltscheffsky, ed.), Vol. I, p. 547. Kluwer Academic Publishers, Dordrecht, 1990.

The rate $kf_{AB(1)}$ ($t_{1/2}$ = 100–200 $\mu$s) is normally substantially faster than that of $kf_{AB(2)}$ ($t_{1/2}$ = 300–500 $\mu$s).[56] Mutations that affect the delivery of protons to the quinol form of $Q_B$ are likely to have a much more marked effect on $k_{AB(2)}$ than on $k_{AB(1)}$ (Ref. 61 and references therein). Mutations that lower the affinity of quinone for the $Q_B$ site (larger $K_q$) are likely to show composite kinetics with rate constants $kf_{AB(1)}$ and $kq_{on}$[61a] as well as to lower the apparent equilibrium constant, $K'_{AB(1)}$ for Eq. (1). Many of the mutants that demonstrate resistance to herbicides acting at the $Q_B$ binding site fall into this category, showing slowed $Q_A^-$ to $Q_B$ electron transfer.[52,60–62] One extreme example of a mutation perturbing acceptor side electron transfer function is a mutation at D1-His252 (His252Leu in *Synechocystis*[61]). Mutations at this site greatly slow $k_{AB(1)}$ either by lowering $K_{AB(1)}$ (e.g., destabilization of $Q_B^-$) or by increasing the dissociation constant of $Q_B$ (larger Kq) at its binding site in the PSII reaction center.

The observation of a periodicity of two with flash number of the rate of oxidation of $Q_A^-$ (fast on odd flashes, slow on even)[56] is not only an indicator of reaction center acceptor side function, but an indicator that the donor side is turning over as well, as the OEC is required to supply electrons to the acceptor side two-electron gate. Another indicator of function on the electron donor side of PSII is the observation of a periodicity of four in the fluorescence yield measured several hundred milliseconds after the actinic flash.[63] This oscillation is marked by an elevated fluorescence yield following the first and second flashes and a diminished yield following the third and fourth flashes. The fluorescence yield follows the sum of the S2 plus S3 states of the OEC with a minimum on the fourth flash. These oscillations of period four can be observed in *Synechocystis* and *Chlamydomonas* cells at 5 ms after the actinic flash[16,51,58a] and at earlier times if the acceptor side is not synchronized by pretreatment with oxidant. Where there is a rate limitation on the donor side or the presence of centers that are inactive for oxygen evolution, the flash train results in a progressive decrease in the fluorescence yield measured at $\leq$50 $\mu$s after the actinic flash.[58a,59] This quenching is caused by an enhanced concentration of P680$^+$ at short times after the actinic excitation and reflects the inability to stabilize oxidizing equivalents on the OEC.

Another type of fluorescence analysis using flash excitation involves measurement of the rate of charge recombination following a single flash

[61] B. A. Diner, V. Petrouleas, and J. J. Wendoloski, *Physiol. Plant.* **81,** 423 (1991).
[61a] A. R. Crofts, I. Baroli, D. Kramer, and S. Taoka, *Z. Naturforsch.* **48c,** 259 (1993).
[62] A. L. Etienne, J.-M. Ducruet, G. Ajlani, and C. Vernotte, *Biochim. Biophys. Acta* **1015,** 435 (1990).
[63] G. Barbieri, R. Delosme, and P. Joliot, *Photochem. Photobiol.* **12,** 197 (1970).

in the presence of $10 \mu M$ DCMU [3-(3,4-dichlorophenyl)-1,1-dimethylurea]. The recombination rate is much slower ($t_{1/2} \geq 1$ ms) than the rate at which the oxidizing equivalent is equilibrated on the donor side of the reaction center ($t_{1/2} < 100 \mu s$). Because charge recombination occurs predominantly between $Q_A^-$ and $P680^+$, the rate of recombination between $Q_A^-$ and the electron donor side is a reflection of the equilibrium concentration of $P680^+$ following equilibration with the S2 state and $Y_Z$. The lower the equilibrium constant ($K_{zp}K_{sz}$) governing the overall equilibrium of Eq. (10), the higher the concentration of $P680^+$ and the higher the observed rate ($k_{obs}$) of charge recombination, expressed in Eq. (13) in terms of the donor side equilibrium constants ($K_{zp}$ and $K_{sz}$) and the intrinsic rate constant for charge recombination ($k_{in}$).

$$S1Y_ZP680^+ \underset{kb_{zp}}{\overset{kf_{zp}}{\rightleftarrows}} S1Y_Z + P680 \underset{kb_{sz}}{\overset{kf_{sz}}{\rightleftarrows}} S2Y_ZP680 \qquad (10)$$

$$K_{zp} = kf_{zp}/kb_{zp} \qquad (11)$$

$$K_{sz} = kf_{sz}/kb_{sz} \qquad (12)$$

$$k_{obs} = \frac{k_{in}}{K_{zp}(1 + K_{sz}) + 1} \qquad (13)$$

where $k_{in}$ is the rate of charge recombination between $P680^+$ and $Q_A^-$ in the absence of equilibration with the secondary electron donors, S1 and $Y_Z$, and the secondary electron acceptor, $Q_B$. This rate (700 $sec^{-1}$) has been measured in PSII core complexes isolated from a $Y_Z$-less mutant (D1-Tyr161Phe) of *Synechocystis* 6803.[64] Mutations that affect the reduction potential or the presence of a donor side component or that affect $k_{in}$ are likely to be observable through a modification of $k_{obs}$. A considerable number of such mutations have been reported in both *Synechocystis*[16,58–59,65,66] and *Chlamydomonas*.[51,67]

A variation on this technique involves giving continuous light for up to 15 sec rather than flash illumination in the presence of DCMU.[59] Such centers can store only one oxidizing equivalent. Mutant strains that fail to store the oxidizing equivalent on the Mn cluster, either because of failure to assemble the cluster or for thermodynamic reasons, can show a stabiliza-

[64] J. G. Metz, P. J. Nixon, M. Rögner, G. W. Brudvig, and B. A. Diner, *Biochemistry* **28,** 6960 (1989).
[65] H.-A. Chu, A. P. Nguyen, and R. J. Debus, *Biochemistry* **34,** 5839 (1995).
[66] H.-A. Chu, A. P. Nguyen, and R. J. Debus, *Biochemistry* **34,** 5859 (1995).
[67] R. Roffey, D. M. Kramer, Govindjee, and R. T. Sayre, *Biochim. Biophys. Acta* **1185,** 257 (1994).

tion of $Q_A^-$ that is associated with the oxidation of $Mn^{2+}$ or of cytochrome $b_{559}$, respectively.

Another analytical technique involves the addition of 10–20 m$M$ $NH_2OH$ to cells preoxidized with $p$-benzoquinone and $K_3Fe(CN)_6$ and incubated with DCMU as above. A train of saturating light flashes (12–18 Hz) followed or not by continuous illumination given within 30 sec of the addition of hydroxylamine results in the accumulation of $Q_A^-$ with a reduced donor side as reflected in an increase in the fluorescence intensity with each flash until $Q_A^-$ is entirely reduced ($F_{max}$).[59,64] The higher the flash yield for the reduction of $Q_A$, the more effective each flash is in the approach to $F_m$. Several types of information can be obtained from such a measurement: (1) Assuming the antenna size to be a constant, then the amplitude of the variable fluorescence ($F_m$–$F_0$) is proportional to the concentration of active centers. Such a correlation has been established, in our own laboratory, by a comparison of such an amplitude with the relative concentration of active centers extracted on a per cell basis and by a comparison with the concentration of DCMU binding sites on a per cell basis.[59] (2) $NH_2OH$ does not donate elections to $P680^+$ or only very poorly.[64] Consequently, the flash yield for the reduction of $Q_A$ is a reflection of the competition between the rate of reduction of $P680^+$ by $Y_Z$ versus the rate of charge recombination between $Q_A^-$ and $P680^+$. The most extreme example is that of the mutant D1-Tyr161Phe, which does not have a functional $Y_Z$. In this case, the flash yield for $Q_A$ reduction is small owing to the more rapid reduction of $P680^+$ by $Q_A^-$ than by $NH_2OH$.[59,64] A close second are mutations at D1-His190 that slow the oxidation of $Y_Z$ 200- to 1000-fold, to the point where charge recombination between $P680^+$ and $Q_A^-$ is 2- to 3-fold faster than reduction of $P680^+$ by $Y_Z$.[59,68,69,69a]

## Thermoluminescence

Thermoluminescence arises from the thermal activation of recombination of electrons and holes trapped at low temperature (e.g. $-20°$) following charge separation (see Ref. 70 for review). One of the pathways of recombination repopulates the lowest singlet excited state of P680 giving rise to a luminescence emission. In the case of PSII, different trapped redox states give rise to characteristic luminescence emission components as the temper-

[68] B. A. Diner, P. J. Nixon, and J. W. Farchaus, *Curr. Opin. Struct. Biol.* **1,** 546 (1991).
[69] R. Roffey, K. van Wijk, R. Sayre, and S. Styring, *J. Biol. Chem.* **269,** 5115 (1994).
[69a] A. M. A. Hays, I. R. Vasiliev, J. H. Golbeck, and R. J. Debus, *Biochemistry,* in press (1998).
[70] Y. Inoue, *in* "Biophysical Techniques in Photosynthesis" (J. Amesz and A. J. Hoff, eds.), p. 93. Kluwer Academic Publishers, Dordrecht, 1996.

ature is increased. These are $S3Q_A^-$ (A-band, $\sim -15°$), $S2Q_A^-$ (Q-band, $\sim +5°$), $S3Q_B^-$ (B1-band, $\sim +20°$), $S2Q_B^-$ (B2-band), $\sim +30°$) and $Y_D \cdot Q_A^-$ (C-band, $\sim +50°$).[70] Demeter,[71] Etienne,[62] and Gleiter[72] and their collaborators (the latter two include *Chlamydomonas* and *Synechococcus* strains) have reported a downshift of the B-band in atrazine-resistant mutants, D1-Ser264Ala and Gly, attributed to a decrease in $E_m$ of $Q_B/Q_B^-$ relative to $Q_A/Q_A^-$. Little effect was observed in another mutation at D1-Phe255Tyr with resistance to phenylureas.[72]

The $A_T$ thermoluminescence band (oxidized donor-$Q_A^-$ recombination), observed at $\sim -16°$ in Tris- or hydroxylamine-washed thylakoid membranes, has been examined in site-directed mutants of *Chlamydomonas* at D1-His190 and His195.[73] A mutation at His195 (D1-His195Asp) decreased $K_{zp}$ 50-fold[67] but did not shift the temperature of the $A_T$ band.[73] It was concluded, based on these studies, that the oxidizing equivalent responsible for charge recombination could not be $Y_Z$. This observation is in contrast with optical spectroscopic evidence that indicates that $Y_Z$ is oxidized in PSII core complexes under conditions similar to those used for thermoluminescence.[74] A distinction, however, should probably be made, in the interpretation of the thermoluminescence emission, between the activation energy required for charge recombination and the free-energy difference between $Y_Z P680^+$ and $Y_Z \cdot P680$. The activation energy could be unchanged despite the alteration of $K_{zp}$ or the thermally activated process may be completely independent of electron transfer of $Y_Z + P680 \leftrightarrow Y_Z P680^+$.

## Ultraviolet/Visible Optical Spectroscopy

The UV/vis difference spectra for most of the redox components of PSII have been well characterized, though in some cases there is not complete agreement on all details of the spectra. These components include the S-state transitions (Sn–Sn-1),[75–77] $Y_Z \cdot - Y_Z$,[78–80] $Y_D \cdot - Y_D$,[74] $P680^+ -$

[71] S. Demeter, I. Vass, E. Hideg, and A. Sallai, *Biochim. Biophys. Acta* **806,** 16 (1985).

[72] H. Gleiter, N. Ohad, H. Koike, J. Hirschberg, G. Renger, and Y. Inoue, *Biochim. Biophys. Acta* **1140,** 135 (1992).

[73] D. M. Kramer, R. A. Roffey, Govindjee, and R. T. Sayre, *Biochim. Biophys. Acta* **1185,** 228 (1994).

[74] B. A. Diner, X.-S. Tang, M. Zheng, G. C. Dismukes, D. A. Force, D. W. Randall, and R. D. Britt, *in* "Photosynthesis: From Light to Biosphere" (P. Mathis, ed.), Vol. II, p. 229. Kluwer Academic Publishers, Dordrecht, 1995.

[75] J. P. Dekker, H. J. van Gorkom, J. Wensink, and L. Ouwehand, *Biochim. Biophys. Acta* **767,** 1 (1984).

[76] J. Lavergne, *Biochim. Biophys. Acta* **894,** 91 (1987).

[77] H. Kretschmann, J. P. Dekker, O. Saygin, and H. T. Witt, *Biochim. Biophys. Acta* **932,** 358 (1988).

P680,[64,81] Pheo$^-$–Pheo,[82,83] $Q_A^-$–$Q_A$[64,80,84,85] and $Q_B^-$–$Q_B$.[85] Such measurements are difficult to carry out in whole cells owing to the induction by light of absorbance changes in both photosystems and the low concentration of reaction centers on a per chlorophyll basis, making for a low signal-to-noise ratio ($S/N$). Aiding such measurements has been a flash detection spectrophotometer first described by Joliot and collaborators.[86] This instrument has figured prominently among the optical techniques used owing to its excellent $S/N$. Also contributing to this work, particularly as regards the S-state absorption changes carried out by Lavergne,[76] is a mutant of *Chlorella sorokiniana* lacking PSI and much of the light-harvesting chlorophyll. However, the inability to introduce site-directed mutations into this organism limits its utility for the establishment of structure–function relationships. In contrast, the ability in *Chlamydomonas* to produce a similar background strain through genetic crosses combined with site-directed mutations in plastid genes should allow this organism to be used for the detection *in vivo* of flash-induced absorbance changes. The recent isolation[87] in *Synechocystis* of a *psa*A$^-$B$^-$, *apc*E$^-$ strain lacking most of the light-harvesting phycobilin should render this organism useful as well for measuring, *in vivo,* flash-induced absorbance changes in site-directed PSII mutants.

Most of the UV/vis optical work has been done on PSII core complexes[25,27] and reaction centers[28–30] with site-directed mutations inducing alterations in difference spectra and in rates of electron transfer. To date there are not many examples of such changes in the optical difference spectra, a few of which follow. Shifts in the Qx band of the active side pheophytin as well as alterations in the quantum yield of primary charge separation have been observed as a result of mutations introduced at D1-Gln130 in *Synechocystis*.[28] This residue is the homolog of L-Glu104 of the reaction centers of the purple photosynthetic bacteria implicated in

[78] J. P. Dekker, H. J. van Gorkom, M. Brok, and L. Ouwehand, *Biochim. Biophys. Acta* **764,** 301 (1984).

[79] B. A. Diner and C. de Vitry, *in* "Advances in Photosynthesis Research" (C. Sybesma, ed.), Vol. I, p. 407. Martinus Nijhoff/Dr. W. Junk, The Hague, 1984.

[79a] G. Renger and W. Weiss, *Biochim. Biophys. Acta* **850,** 184 (1986).

[80] S. Gerken, K. Brettel, E. Schlodder, and H. T. Witt, *FEBS Lett.* **237,** 69 (1988).

[81] S. Gerken, K. Brettel, E. Schlodder, and H. T. Witt, *Biochim. Biophys. Acta* **977,** 52 (1989).

[82] V. V. Klimov, E. Dolan, and B. Ke, *FEBS Lett.* **112,** 97 (1980).

[83] B. A. Diner and R. Delosme, *Biochim. Biophys. Acta* **722,** 452 (1983).

[84] H. J. van Gorkom, *Biochim. Biophys. Acta* **347,** 439 (1974).

[85] G. H. Schatz and H. J. van Gorkom, *Biochim. Biophys. Acta* **810,** 283 (1985).

[86] P. Joliot, D. Béal, and B. Frilley, *J. Chim. Phys.* **77,** 209 (1980).

[87] H.-A. Chu, A. P. Nguyen, and R. J. Debus, *in* "Photosynthesis: From Light to Biosphere" (P. Mathis, ed.), Vol. II, p. 439. Kluwer Academic Publishers, Dordrecht, 1995.

hydrogen bonding to the keto group of ring V of the active side pheophy-tin.[5,9,10] Mutations at D1-His198 produce blue shifts in the soret of the absorbance difference spectrum of P680⁺–P680.[88] This residue is the homo-log of L-His173 of the purple bacterial reaction centers, which is responsible for the coordination of one of the members of the special pair primary electron donor Bchlorophylls,[5,8–10] implicating D1-His198 as well in the coordination of P680. Mutations at this same site also alter the optical difference spectrum of $Y_Z\bullet - Y_Z$ in the blue region of the spectrum[89] that has been attributed to electrochromism.[90] This displacement of the band shift implicates P680 as being part of the electrochromic response.[89]

Much more common consequences of site-directed mutations are changes in the kinetics of electron transfer associated with alterations in the reduction potential of a component, perturbation of the coupling of proton transfer with component oxidation and reduction, or the disappear-ance of the redox component entirely. The mutations mentioned above at D1-His198 produce shifts in the reduction potential of P680⁺/P680 as large as 150 mV as judged by measurements of the rates of charge recombination between $Q_A^-$ and the donor side.[88] The rate of charge recombination can be followed in core complexes lacking $Q_B$ (or in the presence of DCMU) at 325 nm, a peak in the difference spectrum of $Q_A^- - Q_A$.[58a,64,84] Here, unlike for the measurement of the fluorescence yield, there is no correction necessary for energy transfer. Equation (13) expresses the dependence of the observed rate constant for recombination on the donor side equilibrium constants. In core complexes, with the OEC intact, the rate of recombination depends on both $K_{zp}$ and $K_{sz}$. $K_{zp}$ can be examined alone following extrac-tion of Mn by $NH_2OH$ treatment,[91] Tris wash,[22] or high concentration phosphate treatment.[25,92] However, some caution is required because the treatments themselves increase $K_{zp}$ over and above any effects induced by mutation.[93] Knowing $K_{zp}$ [Eq. (11)], measurement of the rate of reduction of P680⁺ by $Y_Z$ ($kf_{zp}$, at 432.5 nm, a minimum of P680⁺–P680 in *Synechocystis*) allows the calculation of the back reaction, $kb_{zp}$.

Measurements of $kf_{zp}$ have been recorded following mutation at D1-His190. These mutations slow the oxidation of $Y_Z$ by a factor of 200 ($t_{1/2} \approx 2$ ms). The flash yield for the oxidation of $Y_Z$ is therefore markedly lowered (to ~1/3) compared to WT, because the oxidation of YZ is two

[88] P. J. Nixon, B. A. Diner, W. J. Coleman, D. A. Chisholm, and W. F. J. Vermaas, in preparation (1998).
[89] J. Lavergne, P. J. Nixon, and B. A. Diner, in preparation (1998).
[90] M. Haumann and W. Junge, in "Oxygenic Photosynthesis: The Light Reactions" (D. R. Ort and C. F. Yocum, eds.), p. 165. Kluwer Academic Publishers, Dordrecht, 1996.
[91] G. M. MacDonald and B. A. Barry, *Biochemistry* **31**, 9848 (1992).
[92] M. Rögner, P. J. Nixon, and B. A. Diner, *J. Biol. Chem.* **265**, 6189 (1990).
[93] C. T. Yerkes, G. T. Babcock, and A. R. Crofts, *FEBS Lett.* **158**, 359 (1983).

times slower than charge recombination between $P680^+$ and $Q_A^-$.[65,68,69] This slowing is interpreted as arising from the loss of the proton acceptor, the protonation of which is coupled to the oxidation of $Y_Z$ to the neutral radical. These results therefore implicate D1-His190 in a proton transfer pathway associated with the oxidation of $Y_Z$.

Mutations that result in the complete loss of a redox component are sometimes difficult to examine spectroscopically as they often result in a complete loss of the assembled reaction center. Examples are D1-His198 Leu and D2-His197Leu, which appear to interfere with chromophore binding,[94] unlike their bacterial homologs, which incorporate Bpheophytin instead.[95] Also Vermaas and collaborators[96] have reported that the mutation, D2-Trp253Leu, results in a complete loss of reaction center, most likely through the inability to bind $Q_A$. Exceptions are the replacement of D1-Tyr161 with Phe.[64,97] D1-Tyr161 is the tyrosine residue responsible for $Y_Z$. While it is not a cofactor, it is nonetheless a redox component. Its replacement with Phe results in a loss of donor side reduction of $P680^+$, as demonstrated by the measurements of identical rates of disappearance of $P680^+$ (432 nm, minimum of $P680^+$–P680) and of $Q_A^-$ (325 nm, maximum of $Q_A^-$–$Q_A$) indicating that both disappear through charge recombination.[64] Also mutations at D1-Asp170,[51,58a,59,65,98–101] His332,[16,66,101] His342,[16,66,101] Ala344Stop,[16,99,101] and Ser345Pro[16,99,101] and other D1 C-terminal processing mutants[16,101] impair assembly or turnover of the Mn cluster in assembled reaction centers. Optical measurements have been employed in the study of some of these (see below)[16,100,101] and fluorescence relaxation kinetics in all of them.

While all of the above measurements involve internal electron transfer, it is possible to monitor net electron transport or to follow turnover of endogenous redox components using externally added electron acceptors. Common acceptors are DCPIP (reduction followed at 574 or 600 nm),

[94] W. J. Coleman, P. J. Nixon, W. F. J. Vermaas, and B. A. Diner, in "Photosynthesis: From Light to Biosphere" (P. Mathis, ed.), Vol. I, p. 779. Kluwer Academic Publishers, Dordrecht, 1995.

[95] C. C. Schenck, D. Gaul, M. Steffen, S. G. Boxer, L. McDowell, C. Kirmaier, and D. Holten, in "Reaction Centers of Photosynthetic Bacteria," Feldafing II Meeting (M. E. Michel-Beyerle, ed.), p. 229. Springer Verlag, Berlin 1990.

[96] W. F. J. Vermaas, J. Charité, and G. Shen, Z. Naturforsch. 45c, 359 (1990).

[97] R. J. Debus, B. A. Barry, I. Sithole, G. T. Babcock, and L. McIntosh, Biochemistry 27, 9071 (1988).

[98] R. J. Boerner, A. P. Nguyen, B. A. Barry, and R. J. Debus, Biochemistry 31, 6660 (1992).

[99] H.-A. Chu, A. P. Nguyen, and R. J. Debus, Biochemistry 33, 6150 (1994).

[100] B. A. Diner and P. J. Nixon, Biochim. Biophys. Acta 1101, 134 (1992).

[101] P. J. Nixon and B. A. Diner, Biochem. Soc. Trans. 22, 338 (1994).

frequently used in chloroplasts, thylakoid membranes,[102] and membrane fragments[103]; $K_3Fe(CN)_6$ (reduction followed at 400 nm or in the UV)[104]; and 2,6 or 2,5-dichlorobenzoquinone alone or in association with $K_3Fe(CN)_6$[105] in BBY membranes and PSII core complexes. Silicomolybdate has been used in PSII core complexes[105] and isolated reaction centers.[106]

DCPIP and silicomolybdate in Mn-depleted PSII membrane fragments and core complexes, respectively, have been used to monitor competition between diphenylcarbazide and $Mn^{2+}$ for electron donation. This technique has been used by Seibert and collaborators[103] to examine Mn binding sites in *Scenedesmus* wild-type and mutant LF-1.

The $K_m$ of $Mn^{2+}$ binding and oxidation has also been characterized[58a,100] by measuring as a function of the $Mn^{2+}$ concentration, the fraction of centers with stabilized $Q_A^-$ (detected at 325 nm) following a light flash. Mutations at D1-His170 result in an up to 60-fold increase in the $K_m$ of binding and oxidation of $Mn^{2+}$ and implicate this site in the first step in the assembly of the Mn cluster.


EPR, ESEEM, and ENDOR

All of the redox components of PSII have at least one redox state that allows them to be detected by magnetic resonance techniques. Miller and Brudvig[107] have written a review of such PSII signals and the conditions for detecting them. One particular advantage to *Synechocystis* in this respect is the ability to grow the cells in the presence of isotopically labeled amino acids that are incorporated into reaction center proteins. This technique provides a means of not only verifying the identity of an EPR active species where an amino acid radical is concerned,[108–110] but allows the identification of amino acid residues that are implicated in either the coordination of metals or in hydrogen bonding (see below). To this end, *Synechocystis* has been labeled with a number of amino acids containing stable magnetic nuclei

[102] W. Auslânder and W. Junge, *Biochim. Biophys. Acta* **357**, 285 (1974).

[103] M. Seibert, N. Tamura, and Y. Inoue, *Biochim. Biophys. Acta* **974**, 185 (1989).

[104] C. F. Fowler and B. Kok, *Biochim. Biophys. Acta* **423**, 510 (1976).

[105] F. Rappaport and J. Lavergne, *Biochemistry* **30**, 10004 (1991).

[106] J. Barber, D. J. Chapman, and A. Telfer, *FEBS Lett.* **220**, 67 (1987).

[107] A.-F. Miller and G. W. Brudvig, *Biochim. Biophys. Acta* **1056**, 1 (1991).

[108] B. A. Barry and G. T. Babcock, *Proc. Natl. Acad. Sci. U.S.A.* **84**, 7099 (1987).

[109] B. A. Barry, M. K. El-Deeb, P. O. Sandusky, and G. T. Babcock, *J. Biol. Chem.* **265**, 20139 (1990).

[110] X.-S. Tang, D. W. Randall, D. A. Force, B. A. Diner, and R. D. Britt, *J. Am. Chem. Soc.* **118**, 7638 (1996).

useful for both magnetic resonance and FTIR spectroscopies ($^2$H,[108,109,111,112] $^{13}$C,[111–113] $^{15}$N,[114,115] and $^{17}$O[116]). These amino acids include tryosine,[108,109,111] histidine,[114,115] and alanine.[117]

For tyrosine labeling,[108] cells can be grown photoautotrophically in the presence of 0.5 m$M$ L-phenylalanine, 0.25 m$M$ L-tryptophan, and 0.25 m$M$ L-tyrosine to shut down biosynthesis of aromatic amino acids, obliging the cells to take up the amino acids present in the growth medium. Consequently any one of these aromatic amino acids could be used to incorporate label.

In the case of histidine, a histidine-tolerant strain of *Synechocystis* was isolated[114] that allowed the cells to be grown photoautotrophically in the presence of 120 $\mu M$ L-histidine. Under conditions of growth in the presence of $^{15}$N-imidazole labeled histidine, 85% of the cell protein had incorporated labeled histidine.

Where 0.5 m$M$ $^{13}$C-alanine was added to the photoautotrophic growth medium, 70% of the alanine in cell protein was similarly labeled (Xiao-Song Tang, personal communication).[117] No preadaptation of cells was required for alanine labeling.

Owing to the instability of Y$_Z$• and the stability of Y$_D$•, it is easy to trap Y$_D$• in the absence of Y$_Z$• in wild-type cells, thylakoids, or core complexes. This generally involves brief illumination of the sample (e.g., ~250 W/m$^2$ white light for 30 sec at 0°), followed by 10–15 min of dark incubation at 0° and freezing the sample in darkness in liquid nitrogen (e.g., Ref. 17). Kodera *et al.*[118] were the first to report trapping of Y$_Z$•. These authors used illumination of PSII membrane fragments of 253 K. Trapping of Y$_Z$• in the absence of Y$_D$• is more difficult. Tang and collaborators[119,120] trapped Y$_Z$• by freezing under illumination in PSII core complexes bearing

[111] G. M. MacDonald, K. A. Bixby, and B. A. Barry, *Proc. Natl. Acad. Sci. U.S.A.* **90**, 11024 (1993).

[112] R. Heinerwadel, A. Boussac, J. Breton, B. A. Diner, and C. Berthomieu, *Biochemistry* **36**, 14712 (1997).

[113] T. Noguchi, Y. Inoue, and X.-S. Tang, *Biochemistry* **36**, 14705 (1997).

[114] X.-S. Tang, B. A. Diner, B. S. Larsen, M. L. Gilchrist, G. A. Lorigan, and R. D. Britt, *Proc. Natl. Acad. Sci. U.S.A.* **91**, 704 (1994).

[115] K. A. Campbell, J. M. Peloquin, B. A. Diner, X.-S. Tang, D. A. Chisholm, and R. D. Britt, *J. Am. Chem. Soc.* **119**, 4787 (1997).

[116] F. Dole, B. A. Diner, C. W. Hoganson, G. T. Babcock, and R. D. Britt, *J. Am. Chem. Soc.* **119**, 11540 (1997).

[117] J. M. Peloquin, X.-S. Tang, G. A. Lorigan, B. A. Diner, and R. D. Britt, in preparation (1998).

[118] Y. Kodera, K. Takura, H. Mino, and A. Kawamori, *in* "Research in Photosynthesis" (N. Murata, ed.), Vol. II, p. 57. Kluwer Academic Publishers, Dordrecht, 1992.

[119] X.-S. Tang, M. Zheng, D. A. Chisholm, G. C. Dismukes, and B. A. Diner, *Biochemistry* **35**, 1475 (1996).

[120] C. Tommos, X.-S. Tang, K. Warncke, C. W. Hoganson, S. Styring, J. McCracken, B. A. Diner, and G. T. Babcock, *J. Am. Chem. Soc.* **117**, 10325 (1995).

a site-directed mutation that eliminates $Y_D\bullet$ (D2-Tyr160Phe). In so doing it was possible to trap $Y_Z\bullet$ free of contamination by $Y_D\bullet$.

By incorporating into PSII reaction centers tyrosine selectively deuterated at specific sites, it has been possible using ESEEM (electron spin echo envelope modulation, see review by Britt[121]) and ENDOR (electron nuclear double resonance, see review by Lubitz and Lendzian[122]) to determine in detail the electronic structure of the tyrosyl radicals $Y_D\bullet$[123–125] and $Y_Z\bullet$.[120] The hyperfine coupling of the $\beta$-methylene protons is strongly dependent on their orientation with respect to the plane of the ring.[126] This orientation is the dominant parameter responsible for the differences in the tyrosyl radical X-band EPR spectra from different sources. Indeed modifications of the EPR spectrum of $Y_D\bullet$ in *Synechocystis* strains with mutations at D2-Pro161Ala and Gln164Leu have been interpreted as arising from such a rotation of the ring plane with respect to the $\beta$-methylene protons.[127]

Site-directed mutations at D2-His189 produce a marked narrowing of the $Y_D\bullet$ EPR spectra[17,128] and a loss in the cwENDOR spectrum[17] of a proton hyperfine component at 3.1 MHz that has been shown to be exchangeable in $D_2O$.[129,130] This component was attributed to a hydrogen bond from D2-His189 to the phenolic oxygen of $Y_D\bullet$.[17,128] The narrowing of the EPR spectrum is probably not solely a consequence of the loss of the hydrogen bond, but may reflect an additional structural change associated with its loss. Mutation of the homologous histidine of the D1 polypeptide, D1-His190, produces a lowered quantum yield of oxidation of $Y_Z$[65,68,69] resulting from a 200-to 1000-fold slowing of $Y_Z \rightarrow P680^+$ electron transfer.[68,69a] While these mutations produce no narrowing of the $Y_Z\bullet$ EPR spectrum,[69] they do implicate D1-His190 in the oxidation of $Y_Z$. Because deuterium kinetic isotope effects, in the absence of the Mn cluster, indicate the rate-limiting step in the oxidation of $Y_Z$ to be the initial proton trans-

[121] R. D. Britt, *in* "Biophysical Techniques in Photosynthesis" (J. Amesz and A. J. Hoff, eds.), p. 235. Kluwer Academic Publishers, Dordrecht, 1996.

[122] W. Lubitz and F. Lendzian, *in* "Biophysical Techniques in Photosynthesis" (J. Amesz and A. J. Hoff, eds.), p. 255. Kluwer Academic Publishers, Dordrecht, 1996.

[123] C. W. Hoganson and G. T. Babcock, *Biochemistry* **31**, 11874 (1992).

[124] S. E. J. Rigby, J. H. A. Nugent, and P. J. O'Malley, *Biochemistry* **33**, 1734 (1994).

[125] K. Warncke, G. T. Babcock, and J. McCracken, *J. Am. Chem. Soc.* **116**, 7332 (1994).

[126] K. Warncke and J. McCracken, *J. Chem. Phys.* **103**, 6829 (1995).

[127] C. Tommos, C. Madsen, S. Styring, and W. Vermaas, *Biochemistry* **33**, 11805 (1994).

[128] C. Tommos, L. Davidsson, B. Svensson, C. Madsen, W. F. J. Vermaas, and S. Styring, *Biochemistry* **32**, 5436 (1993).

[129] I. D. Rodriguez, T. K. Chandrashekar, and G. T. Babcock, *in* "Progress in Photosynthesis Research" (J. Biggins, ed.), Vol. I, p. 471. Matinus Nijhoff Publishers, The Hague, 1987.

[130] R. G. Evelo, A. J. Hoff, S. A. Dikanov, and A. M. Tyryshkin, *Chem. Phys. Lett.* **161**, 479 (1989).

fer,[131] it is likely that this histidine is involved in proton-coupled electron transfer.

Pulsed ENDOR (for review, see Britt[121]) has also been applied to the examination of the hydrogen bonding to the $S = 1/2$ tyrosyl radicals.[132] This technique has the advantage over the cw ENDOR technique of being able to resolve both the axial $A_\perp$ and $A_\parallel$ components of the hydrogen bonded proton, showing the proton in the case of $Y_D\bullet$ to be largely dipolar coupled. In the case of $Y_Z\bullet$, these components are not well resolved because of a marked broadening of the hyperfine coupling that was attributed to a heterogeneity in the positioning of the hydrogen bond(s).[120,132] Despite this broadening, the exchangeable proton hyperfine coupling value in $Y_Z\bullet$ is similar to that of $Y_D\bullet$, indicating approximately equal strength hydrogen bonding in each of the two tyrosyl radicals. Pulsed ENDOR measurements combined with $^{15}N$ isotopic labeling of the imidazole nitrogens of histidine and global $^{15}N$-labeling of mutant D2-His189Gln[133] have definitively shown that the tau nitrogen of the imidazole ring of D2-His189 is the source of the hydrogen bond to $Y_D\bullet$.

Similarly, there have been a number of reports of ESEEM spectra of $Q_A^-$ under conditions in which the semiquinone has become decoupled from the nonheme iron.[134–137] This decoupling which narrows the EPR signal of $Q_A^-$, enhancing the ESEEM spectrum, involves a number of different treatments including extraction of the iron,[134,135] use of high concentrations of $CN^-$,[136,137] and transient exposure to high concentrations of phosphate.[137] Two and possibly three different nitrogens have been detected as being isotropically coupled to $Q_A^-$. Two of these nitrogens have been attributed to an imidazole nitrogen of D2-His214 and to the peptide nitrogen of D2-Ala260, respectively. These assignments were based on homologies between the $Q_A$ binding sites of PSII and the purple bacterial reaction centers and on literature values for the quadrupolar coupling of $^{14}N$ observed using ESEEM in the special condition where the nitrogen hyperfine field exactly cancels the Zeeman field. Direct verification of these assignments by isotope

[131] B. A. Diner, D. A. Force, D. W. Randall, and R. D. Britt, in preparation (1998).

[132] D. A. Force, D. W. Randall, R. D. Britt, X.-S. Tang, and B. A. Diner, *J. Am. Chem. Soc.* **117**, 12643 (1995).

[133] K. A. Campbell, J. M. Peloquin, B. A. Diner, X.-S. Tang, D. A. Chisholm, and R. D. Britt, *J. Am. Chem. Soc.* **119**, 4787 (1997).

[134] A. V. Atashkin, A. Kawamori, Y. Kodera, S. Kuroiwa, and K. Akabori, *J. Chem. Phys.* **102**, 5583 (1995).

[135] F. MacMillan, F. Lendzian, G. Renger, and W. Lubitz, *Biochemistry* **34**, 8144 (1995).

[136] Y. Deligiannakis, A. Boussac, and A. W. Rutherford, *Biochemistry* **34**, 16030 (1995).

[137] X.-S. Tang, J. M. Peloquin, G. A. Lorigan, R. D. Britt, and B. A. Diner, *in* "Photosynthesis: From Light to Biosphere" (P. Mathis, ed.), Vol. I, p. 775. Kluwer Academic Publishers, Dordrecht, 1995.

labeling and by isotope labeling combined with directed mutagenesis is under way.

The consensus now is that there are six chlorophylls bound to the PSII reaction center,[138] two more than in the bacterial reaction centers. It is likely that at least four of the six chlorophylls bound to the reaction center are coordinated by histidine ligands to $Mg^{2+}$. These are D1-His118, D2-His117, D1-His198, and D2-His197.[5,8–10] The latter two are the homologs, respectively, of histidines L-173 and M-200 that coordinate the special pair primary electron donor Bchlorophylls of the purple nonsulfur photosynthetic bacterial reaction centers. Despite this homology, it appears that P680 is quite different from the bacterial primary donor with respect to both the orientation of its spin-polarized triplet[139] and its interaction with other chlorophylls (e.g., weak exciton coupling).[140,141]

The first pair of histidines mentioned above does not have homologs in the bacterial reaction centers, but are likely involved in the coordination of two chlorophylls that have a more peripheral location with respect to P680 and are possibly implicated in a 20- to 30-ps phase of energy transfer.[142] The one coordinated by D1-His118 is responsible (see below, Ref. 161a) for a chlorophyll cation radical ($Chl_Z^+$) that has been implicated in a cyclic pathway of electron transfer involving cytochrome $b_{559}$.[143]

The question is then how to identify ligands to the reaction center chlorophylls. One way is to construct site-directed mutations that induce changes in the properties of the coordinated chlorophylls, as described above for D1-His198. In this case, site-directed mutations produce shifts in the oxidized-minus-reduced absorbance difference spectra and in the reduction potential of the primary donor. Such evidence, while suggestive of involvement of the chlorophyll, coordinated by this histidine, in the ground state absorbance spectrum of P680, has not provided a definitive answer as to the localization of the P680$^+$ cation. Recently, Mac and co-workers[144] found that they could use cw ENDOR to observe $^{15}N$-histidine imidazole coordination to P700$^+$ of PSI. A signal with isotropic hyperfine

[138] C. Eijckelhoff and J. P. Dekker, *Biochim. Biophys. Acta* **1231**, 21 (1995).

[139] F. J. E. van Mieghem, K. Satoh, and A. W. Rutherford, *Biochim. Biophys. Acta* **1058**, 379 (1991).

[140] J. R. Durrant, D. R. Klug, S. L. S. Kwa, R. van Grondelle, G. Porter, and J. P. Dekker, *Proc. Natl. Acad. Sci. U.S.A.* **92**, 4798 (1995).

[141] S. E. J. Rigby, J. H. A. Nugent, and P. J. O'Malley, *Biochemistry* **33**, 10043 (1994).

[142] J. P. M. Schelvis, P. I. van Noort, T. J. Aartsma, and H. J. van Gorkom, *Biochim. Biophys. Acta* **1184**, 242 (1994).

[143] L. K. Thompson and G. W. Brudvig, *Biochemistry* **27**, 6653 (1988).

[144] M. Mac, X.-S. Tang, B. A. Diner, J. McCracken, and G. T. Babcock, *Biochemistry* **35**, 13288 (1996).

coupling of 0.64 MHz appears on replacement of $^{14}N$ with $^{15}N$. This finding now provides a probe for the localization of ligands to chlorophyll cations of PSII. Isotopic labeling with $^{15}N$-histidine imidazole permits one to establish whether histidine is a ligand to the chlorophyll responsible for the cation radical. The combination of such labeling with site-directed replacement of the coordinating histidine with a residue that retains function (e.g., glutamine) should permit the identification of the coordinating histidine.

Histidine has been identified using ESEEM as a ligand to the Mn cluster.[114] Hyperfine coupling from $^{14}N$ of histidine imidazole gives rise to a peak centered near 5 MHz, which disappears upon replacement of $^{14}N$ with $^{15}N$. It is unclear whether this coupling arises from only one or more than one histidine. Potential candidates for histidine coordination of Mn are D1-His190, His332, and His337.[16,65–69,100,101,103] Here too, conservative replacement of histidine coordination by a residue that retains function could result in the loss from the ESEEM spectrum of the $^{14}N$ feature arising from histidine thus identifying the ligand.

High-field (e.g., 245 GHz) EPR is another magnetic resonance technique that has been used in the characterization of site-directed mutants of PSII.[145,146] This technique has been used to resolve the anisotropic components of the g-tensors of $Y_Z\bullet$ and $Y_D\bullet$.[145,146] These components provide information on the chemical environments of the radicals and have been used to examine the hydrogen bonding of the redox active tyrosines of PSII. The gx component of the anisotropic g-tensor is aligned with the C—O axis of the tyrosyl radical and is sensitive to electrostatic interactions with the nonbonding electrons of the phenolic oxygen. Such interactions include hydrogen bonding as well as charge stabilization. Values of gx have been proposed by Un and co-workers[145,146] to vary from 2.0064 for the strongly hydrogen bonded case to 2.0090 for the neutral environment with no hydrogen bond. This technique, in addition to pulsed ENDOR,[132] was used to show that the hydrogen bond strengths for $Y_Z\bullet$ (gx = 2.0075) and $Y_D\bullet$ (gx = 2.0074) are similar, though in the former case there was found to be considerable broadening of the gx component that likely arises from some heterogeneity in the positioning of the phenolic oxygen with respect to one or more hydrogen bonds.[119,120,132] The mutant D2-His189Gln, which had been previously shown by cw ENDOR[17] to lack a hydrogen bond to $Y_D\bullet$, showed a gx component for $Y_D\bullet$ shifted to 2.00832, close to that of the tyrosyl radical

[145] S. Un, M. Atta, M. Fontecave, and A. W. Rutherford, *J. Am. Chem. Soc.* **117,** 10713 (1995).
[146] S. Un, X.-S. Tang, and B. A. Diner, *Biochemistry* **35,** 679 (1996).

of *Escherichia coli* ribonucleotide reductase (gx = 2.00868). The latter radical has been reported to lack a hydrogen bond.[147]

## FTIR and Resonance Raman

FTIR (Fourier transform infrared spectroscopy, for review see Mäntele[148]) is a developing technique in studies of Photosystem II and has been used to record spectra of a considerable number of the redox components associated with this photosystem. These include $Q_A^-/Q_a$,[149,150] $Y_d\bullet$/ $Y_D$,[111,151–153] and $Y_Z\bullet/Y_Z$,[152] $Chl_Z^+/Chl_Z$,[154] $^TP680$,[155] the nonheme $Fe^{3+}$/ $Fe^{2+}$,[156,157] and S2/S1 of the Mn cluster.[113,158,159] The interpretation of these spectra is, however, still in its infancy, though work on model compounds, isotopic labeling, and site-directed mutagenesis are together providing increased confidence in the assignment of vibrational/rotational modes. There has, however, been considerable controversy over mode assignments, particularly those associated with tyrosine oxidation/reduction,[111,151–153] which has arisen from the use of conditions that give mixed spectra, containing more than one redox component.

Site-directed mutations at D1-His190[151] and D2-His189[153] have been used to examine hydrogen bonding to the oxidized and reduced forms of the redox-active tyrosines as well as to try and determine the nature of the proton acceptor on tyrosine oxidation. As mentioned above, both of these residues have been implicated, based on kinetic measurements and magnetic resonance spectroscopy, in proton-coupled electron transfer. D2-His189 has been implicated by FTIR[153] as both a hydrogen bond acceptor

[147] C. J. Bender, M. Sahlin, G. T. Babcock, B. A. Barry, T. K. Chandrashekar, S. P. Salowe, J. Stubbe, B. Lindström, L. Petersson, A. Ehrenberg, and B.-M. Sjöberg, *J. Am. Chem. Soc.* **111**, 8076 (1989).

[148] W. Mäntele, *in* "Biophysical Techniques in Photosynthesis" (J. Amesz and A. J. Hoff, eds.), p. 137. Kluwer Academic Publishers, Dordrecht, 1996.

[149] C. Berthomieu, E. Nebedryk, W. Mäntele, and J. Breton, *FEBS Lett.* **269**, 363 (1990).

[150] T. Noguchi, T.-A. Ono, and Y. Inoue, *Biochemistry* **31**, 5953 (1992).

[151] M. T. Bernard, G. M. MacDonald, A. P. Nguyen, R. J. Debus, and B. A. Barry, *J. Biol. Chem.* **270**, 1589 (1995).

[152] R. Heinerwadel, A. Boussac, J. Breton, and C. Berthomieu, *Biochemistry* **35**, 115447 (1996).

[153] R. Heinerwadel, A. Boussac, J. Breton, B. A. Diner, and C. Berthomieu, *Biochemistry* **36**, 14712 (1997).

[154] T. Noguchi and Y. Inoue, *FEBS Lett.* **370**, 241 (1995).

[155] T. Noguchi, Y. Inoue, and K. Satoh, *Biochemistry* **32**, 7186 (1993).

[156] T. Noguchi and Y. Inoue, *J. Biochemistry* **118**, 9 (1995).

[157] R. Hienerwadel and C. Berthomieu, *Biochemistry* **34**, 16288 (1995).

[158] T. Noguchi, T.-A. Ono, and Y. Inoue, *Biochim. Biophys. Acta* **1228**, 189 (1995).

[159] J. J. Steenhuis and B. A. Barry, *J. Phys. Chem. B* **101**, 6652 (1997).

from the reduced and a hydrogen bond donor to the oxidized forms of $Y_D$, the latter in agreement with results from pulsed ENDOR.[115] It remains unclear, however, whether an imidazolium ion actually forms on $Y_D$ oxidation. No difference was reported between the $Y_Z$• FTIR spectra for wild-type and mutant D1-His190Asp,[151] implying that D1-His190 is not involved in a hydrogen bond to $Y_Z$•. The negative influence of this mutation on the quantum yield of $Y_Z$ oxidation,[59,68,69] however, would be consistent with D1-His190 serving as at least a transient proton acceptor.

Resonance Raman spectroscopy, another technique for the detection of bond vibrational modes (see Robert[160] for a review) has been used to obtain vibrational spectra of PSII reaction center pheophytin[161] and to show that the active side pheophytin is hydrogen bonded by D1-Glu130 in a manner similar to its bacteriopheophytin homolog in the bacterial reaction centers. Unlike FTIR, which can be performed on PSII membrane fragments and core complexes, this technique has traditionally been performed on isolated reaction center complexes containing little more than the D1 and D2 polypeptides. Giorgi and collaborators[28] have done optical spectroscopy on reaction centers containing site-directed mutations at this site. It is likely to be only a matter of time before the resonance Raman spectra appear.

In a further development of this technique, resonance Raman has been recently used with mutant and wild type PSII core complexes to identify the axial ligand to $Chl_Z^+$ (D1-His118).[161a]

## Other Spectroscopic Techniques

Other spectroscopic techniques that have been applied to the study of structure and function of PSII include X-ray absorption spectroscopy [EXAFS (extended X-ray absorption fine structure) and XANES (X-ray absorption near edge structure); see review by Yachandra and Klein[162]], which has been extensively employed by several groups to examine the structure and changes in oxidation state of the Mn cluster during the Kok–Joliot cycle of advance of the S-states.[162a] While samples treated to remove

[160] B. Robert, in "Biophysical Techniques in Photosynthesis" (J. Amesz and A. J. Hoff, eds.), p. 161. Kluwer Academic Publishers, Dordrecht, 1996.

[161] P. Möenne-Loccoz, B. Robert, and M. Lutz, Biochemistry **28**, 3641 (1989).

[161a] D. H. Stewart, A. Cua, D. A. Chisholm, B. A. Diner, D. F. Bocian, and G. W. Brudvig, Biochemistry, in press (1998).

[162] V. Y. Yachandra and M. P. Klein, in "Biophysical Techniques in Photosynthesis" (J. Amesz and A. J. Hoff, eds.), p. 337. Kluwer Academic Publishers, Dordrecht, 1996.

[162a] V. K. Yachandra, V. J. DeRose, M. J. Latimer, I. Mukerji, K. Sauer, and M. P. Klein, Science **260**, 675 (1993).

$Ca^{2+}$ [162b] and $Cl^-$ [162c] or with reductant[162d] have been examined by this technique it has not as yet been used with site-directed mutants. Similarly Mössbauer spectroscopy (see review by Debrunner[163]) has been used to examine extensively the nonheme iron of the iron–quinone complex of PSII. Changes in its oxidation state and structure have been examined following treatments that modify ligands of the first coordination sphere of the metal.[164] This technique requires replacement of the natural abundance $^{56}Fe$ with the Mössbauer isotope $^{57}Fe$ and has been carried out in spinach, *Chlamydomonas,* and *Synechocystis.* It has helped to show that there are two to three coordination sites of the nonheme iron that are exchangeable and that can be occupied by small molecules including $CN^-$, which converts the nonheme iron to the low spin diamagnetic form.[165] Finally, the techniques of spectral hole burning (see review by Johnson *et al.*[166]), optically detected magnetic resonance (ODMR, see review by Hoff[167]), and photoacoustic spectroscopy (see review by Malkin[168]) have all been successfully applied to Photosystem II though little work has been done with mutants. These methods have been used, respectively, to examine weak excitonic interactions of chlorophylls associated with the primary donor P680,[169] to show that the triplet state of P680, resulting from charge recombination in the reaction center, is monomeric at liquid helium temperature,[170] and to measure energy storage and volume changes on charge

[162b] M. J. Latimer, V. J. DeRose, I. Mukerji, V. K. Yachandra, K. Sauer, and M. P. Klein, *Biochemistry* **34,** 10898 (1995).

[162c] T. Ono, T. Noguchi, Y. Inoue, M. Kusunoki, T. Matsushita, and H. Oyanagi, *J. Am. Chem. Soc.* **117,** 6386 (1995).

[162d] P. J. Riggs, R. Mei, C. F. Yocum, and J. E. Penner-Hahn, *J. Am. Chem. Soc.* **114,** 10650 (1992).

[163] P. G. Debrunner, *in* "Biophysical Techniques in Photosynthesis" (J. Amesz and A. J. Hoff, eds.), p. 355. Kluwer Academic Publishers, Dordrecht, 1996.

[164] B. A. Diner and V. Petrouleas, *Biochim. Biophys. Acta* **895,** 107 (1987).

[165] Y. Sanakis, V. Petrouleas, and B. A. Diner, *Biochemistry* **33,** 9922 (1994).

[166] S. G. Johnson, I.-J. Lee, and G. J. Small in "Chlorophylls" (H. Scheer, ed.), p. 739. CRC Press, Boca Raton, 1991.

[167] A. J. Hoff, *in* "Biophysical Techniques in Photosynthesis" (J. Amesz and A. J. Hoff, eds.), p. 277. Kluwer Academic Publishers, Dordrecht, 1996.

[168] S. Malkin, *in* "Biophysical Techniques in Photosynthesis" (J. Amesz and A. J. Hoff, eds.), p. 191. Kluwer Academic Publishers, Dordrecht, 1996.

[169] H.-C. Chang, R. Jankowiak, N. R. S. Reddy, C. F. Yocum, R. Picorel, M. Seibert, and G. J. Small, *J. Phys. Chem.* **98,** 7725 (1994).

[170] H. J. den Blanken, A. J. Hoff, A. P. J. M. Jongenelis, and B. A. Diner, *FEBS Lett.* **157,** 21 (1983).

separation in PSII reaction centers[171] as well as the kinetics of oxygen release.[172]

New developments in time-resolved technology as applied to such methods as UV/vis optical, ENDOR, ESEEM, FTIR, and resonance Raman spectroscopies will undoubtedly have an impact in the near future, allowing a combination of structural and kinetic data to be obtained in both wild-type and mutated PSII reaction centers and core complexes. While there has thus far been little success for PSII in the crystallographic arena, a number of integral membrane protein complexes have recently been solved to atomic or near atomic resolution (e.g., cytochrome oxidase, cytochrome $bc_1$ complex, ATP synthase, PSI). These remarkable successes provide hope that resolution of the x-ray crystallographic structure of PSII and the incomparable structural information that it provides is not far off.

[171] R. Delosme, D. Béal, and P. Joliot, *Biochim. Biophys. Acta* **1185,** 56 (1994).
[172] D. Mauzerall, *Plant Physiol.* **94,** 278 (1990).

# Author Index

Numbers in parentheses are footnote reference numbers and indicate that an author's work is referred to although the name is not cited in the text.

## S

Sabaty, M., 90, 154, 155, 157(12), 165
Sacchi, N., 231
Saeki, K., 95
Saenger, W., 19, 60, 124, 130, 130(4)
Saftic, D., 215
Sahlin, M., 357
Saito, T., 252, 258
Sakurai, H., 108
Salih, G., 166, 167, 170, 172, 176, 178, 179, 180(8), 181, 182, 302
Sallai, A., 347
Salowe, S. P., 357
Saluz, H. P., 159, 163(37)
Sambrook, J., 45, 48(22), 141, 186, 200, 203(15), 283, 286(20), 288(20), 313, 317(13)
Samuelsson, G., 30
Sanakis, Y., 359
Sandlin, D. E., 91
Sandmann, G., 154, 246, 247, 248, 249, 251, 252, 252(16, 24, 49, 50), 261(16), 262, 263, 273
Sandusky, P. O., 351, 352(109)
Sanford, J. C., 311, 312(1), 338
Santel, H. J., 237
Sasakawa, C., 156
Sasamoto, S., 18, 124, 298, 309(14, 15)
Sato, S., 18, 124, 298, 309(14, 15)
Satoh, K., 355, 357
Satoh, T., 87
Satomi, Y., 154
Sauer, K., 340, 342(44), 358
Savereide, P. B., 29
Saygin, O., 347
Sayre, R. T., 321, 345, 346, 347, 350(69), 353(69), 356(67, 69), 358(69)
Scappino, L., 8, 14(37, 38)
Schaefer, M. R., 280, 282, 282(9), 287(19)
Schägger, H., 84
Schatz, G. H., 348
Scheibe, R., 215, 217(31), 319(31)
Scheibe, S., 280
Schelvis, J. P. M., 355
Schenborn, E. T., 187
Schenck, C. C., 76, 324, 350
Schertler, G. F. X., 60
Schiffer, M., 61, 69(33), 76, 337
Schilke, B. A., 87

Schirmer, T., 60
Schirmer-Rahire, M., 39
Schledz, M., 251, 261(44), 263
Schlegel, H. G., 81
Schlodder, E., 313, 347(80), 348
Schloss, J. L., 270(22), 271
Schmetterer, G., 4
Schmidhauser, T., 88
Schmidt-Krey, I., 61
Schnable, P. S., 40
Schnackenberg, J., 271
Schnell, R. A., 27, 28, 28(2), 30(2, 11), 31(2), 32(11), 35(2)
Schneppenheim, R., 46
Schnurr, G., 154
Schrautemeier, B., 4
Schreiber, U., 222, 341, 342(46)
Schreiner, O., 276
Schroder, I., 104
Schröder, W. P., 298
Schubert, W.-D., 19, 60, 124, 130(4)
Schuch, W., 247, 251
Schulten, K., 60
Schulz, A., 252
Schulz, R., 19, 23, 23(6), 271
Schulze, E.-D., 215, 217(31, 32), 219(31)
Schurr, U., 215, 217(31, 32), 219(31)
Schwartz, E., 219
Schwartz, S. H., 257
Schwarz, R., 5
Schwender, J. R., 222, 249
Scolnik, P. A., 87, 247, 251, 251(23), 260, 261, 261(23, 84)
Scott, M. P., 97
Seibert, M., 342, 351, 356(103)
Seidler, A., 321
Selman, B. R., 28, 193
Semenov, A. Y., 106, 115(31)
Sen, P., 165
Senger, H., 255, 256, 271
Servaites, J. C., 220
Servis, R. E., 305
Setif, P., 106, 107, 115(32), 134, 338
Shapiro, R., 305
Shapleigh, J. P., 81, 87
Shark, K. B., 311, 312(1), 338
Shaw, E. K., 222
Shen, G., 294, 297, 301(10), 303(4), 325, 350
Shen, J.-R., 321, 325
Sherman, L. A., 321, 325, 338

# Subject Index

## A

ALA, *see* δ-Aminolevulinic acid
δ-Aminolevulinic acid
  *Arabidopsis* mutants in feedback loop
    controlling synthesis, 243–244
  limitation of precursor for chlorophyll
    biosynthesis, 237, 241–242
Antisense RNA
  biological function in plants, 209
  mechanisms of gene expression downregu-
    lation, 209
  mutant generation
    growth medium for selection, 212
    promoters, 211–212
    Rubisco mutants
      developmental changes in expres-
        sion, 217–219
      feedback regulation analysis,
        219–220
      photosynthesis rate control studies,
        216–217, 219–220
      pulse-labeling and posttranscrip-
        tional control, 213
      transcription analysis, 212–214
      Western blotting, 212–213
    target gene considerations, 210–211
    target sequence selection, 211
Atomic absorption spectroscopy, copper determination, 277–278
ATP synthase, *see* Chloroplast ATP synthase

## B

B800–850, *Rhodobacter sphaeroides* 2.4.1
  components, 151–152
  mutagenesis techniques, 155–156
  regulatory gene isolation
    complementation of regulatory mu-
      tants, 156
    heterologous host gene expression, 157

mapping and cloning, 158
    reverse genetics, 159
    suppressor isolation, 156–157
  regulatory mutant isolation
    *cis*-acting mutations, 152–153
    decreased photosynthesis gene isola-
      tion using transcriptional fusions
      to *sacB*, 154–155
    enrichment using *lacZ* fusions, 153–154
    increased photosynthesis gene isolation
      using transcriptional fusions to
      *aph*, 153
    spontaneous mutant isolation by pig-
      mentation and growth analysis,
      155
    *trans*-acting mutations, 153
  transcriptional regulatory factors
    biochemical characterization, 161–162
    interactions between regulatory factors,
      162
    phenotypic characterization, 159–160
    protein–DNA interactions, 162–163
    quantitation of photosynthetic expres-
      sion, 160–161
    sequence analysis, 159
    structure–function analysis, 161
    types and gene targets, 163–165
B875, *Rhodobacter sphaeroides* 2.4.1
  components, 151–152
  mutagenesis techniques, 155–156
  regulatory gene isolation
    complementation of regulatory mu-
      tants, 156
    heterologous host gene expression, 157
    mapping and cloning, 158
    reverse genetics, 159
    suppressor isolation, 156–157
  regulatory mutant isolation
    *cis*-acting mutations, 152–153
    decreased photosynthesis gene isola-
      tion using transcriptional fusions
      to *sacB*, 154–155